Smart Technologies for Water Sustainability and Policy

This edited volume presents a comprehensive exploration of smart and sustainable water management, combining urban hydrology, digital innovation, cybersecurity, and climate resilience. It covers rainwater harvesting, stormwater control, wastewater management, wetlands, aquifer recharge, and field-based environmental investigations alongside advanced themes such as IoT-enabled monitoring, AI-driven predictive modelling, blockchain governance, homomorphic encryption, and SCADA security. Blending practical case studies with emerging technologies, the book offers a multidisciplinary resource that connects engineering, data science, policy, and environmental management within a single reference framework.

The book integrates traditional water management practices with cutting-edge tools, including smart sensors, machine learning, MNDWI-based impact assessment, bioimpedance water quality systems, and blockchain-enabled governance. Dedicated chapters address cybersecurity risks in water infrastructure and privacy-preserving IoT architectures. Real-world field studies from India provide applied insights, while predictive analytics and climate-resilient strategies support forward planning. Readers gain technical depth, implementation perspectives, and policy alignment guidance in one cohesive volume.

This book is designed for environmental engineers, urban planners, water utility professionals, cybersecurity specialists, policymakers, researchers, and postgraduate students in civil engineering, data science, environmental studies, and smart infrastructure domains. It also supports smart city teams and sustainability practitioners seeking integrated digital water solutions.

Smart Technologies for Water Engineering & Sustainability

With the latest advancements in AI and associated smart technologies, there have been conscious efforts towards a sustainable environment. Water Engineering is one of these areas influenced by recent technologies including smart materials, efficient water management, advanced water systems like dams, irrigation, and so on. This series is a compilation of such books which work at the intersection of latest technology, sustainability, and water engineering, making them extremely relevant in today's world.

Integrated Smart Technologies for Sustainable Water Systems using Computer Applications
Edited by Srikanta Kumar Mohapatra, Monika Sethi, Shivani Sood, Abhineet Anand, and Jayashree Mohanty

Smart Technologies for Water Sustainability and Policy
Edited by Neha Sharma, Abhineet Anand, Abhishek Kumar, Rajeev Tiwari, Anupam Tiwari, and Prateek Thapar

Smart Technologies for Water Sustainability and Policy

Edited by
Neha Sharma, Abhineet Anand, Abhishek Kumar,
Rajeev Tiwari, Anupam Tiwari, and Prateek Thapar

CRC Press
Taylor & Francis Group
Boca Raton London New York

CRC Press is an imprint of the
Taylor & Francis Group, an **informa** business

Designed cover image: Shutterstock

First edition published 2027
by CRC Press
4 Park Square, Milton Park, Abingdon, Oxon, OX14 4RN

and by CRC Press
2385 NW Executive Center Drive, Suite 320, Boca Raton FL 33431

British Library Cataloguing-in-Publication Data
A catalogue record for this book is available from the British Library

ISBN: 9781041245100 (hbk)
ISBN: 9781041245117 (pbk)
ISBN: 9781003742227 (ebk)

DOI: 10.1201/9781003742227

Typeset in Times
by Deanta Global Publishing Services, Chennai, India

Contents

Preface

Water plays a vital role in human survival and development, as well as ecological stability and growth. Nevertheless, urbanization rates, climate variability, industrialization growth, and population increase have placed unprecedented demands on water resources across the world. Although conventional water management practices are vital and foundational in water management and governance, they are no longer sufficient in dealing with new issues such as urban flooding, groundwater depletion, inefficient infrastructure, and cybersecurity threats facing water management and governance systems across the world. In addition, emerging advances in digital technologies such as IoT, AI, and blockchain have created unprecedented opportunities for water management and governance through increased efficiency and sustainability. This book was motivated by the need to bridge conventional water management practices and modern digital innovation in water management and governance.

This book provides a comprehensive and multidisciplinary perspective on water management and governance by integrating hydrology practices, smart sensing technologies, smart forecasting and optimization technologies, secure data architectures, and governance practices. In addition, it covers conventional water management practices such as rainwater harvesting, stormwater management, wastewater treatment, wetlands, and groundwater recharge in the early chapters of the book. Later chapters of the book introduce new and emerging water management and governance technologies such as IoT-based monitoring technologies, machine learning-based forecasting and optimization technologies, blockchain-based governance technologies, and secure data architecture. Moreover, it covers critical and often ignored issues such as cybersecurity threats facing water management and governance systems, cybersecurity of water infrastructure and SCADA systems, and impact studies of water management and governance systems. Overall, the book brings together various contributions from different experts and researchers on how technological innovation can improve the resilience and sustainability of water management and governance systems. A chapterwise summary is provided as follows:

Chapter 1 discusses rainwater harvesting systems as a decentralized and sustainable option that becomes more significant in coping with water stress arising from urbanization, increased water demand, and climate change in conventional water resources. Traditional rainwater harvesting systems have been shown to recharge groundwater and conserve water but are often hampered by inadequate monitoring and manual operation of RWH systems and inadequate use of adaptive controls. It reports the systematic evolution of RWH from conventional frameworks to the smart rainwater harvesting systems based on digital, information, and communication technologies.

Chapter 2 determines the extent to which stormwater management strategies are a viable solution to urban flooding problems and the benefits that blending traditional infrastructure with futuristic (green roofs and permeable pavements) techniques have in controlling runoff, ensuring clean water, and promoting sustainable living. It also deals with various best management practices (BMPs) that include controlling the water supply from its source, how the water flows in the system, and effective end-of-pipe solutions to restore a natural balance in the urban water cycle.

Chapter 3 focuses on wastewater management as a key to the enhancement of sanitation and realization of sustainable development targets. Traditionally, sewage was considered waste, but there is an increase in the understanding of wastewater and excreta as a resource for human well-being and ecological health. As a matter of concern, the waste type and amount determine the effective treatment. The solution is the DWTS, which is especially useful in cities with high levels of population density that have insufficient centralized systems in place and offer quick sanitation benefits.

Chapter 4 highlights the ecological, hydrological, and social aspects of urban wetlands and green infrastructures while introducing their functions, benefits, and design principles. It also discusses the difficulties in implementation, the policy framework, and the worldwide examples that are advocating success. Seeing nature as a vital part of the urban fabric, this chapter stresses the point that wetlands and green infrastructures can make cities not only resilient and sustainable but also livable ecosystems of the future.

Chapter 5 synthesizes recent research developments, highlighting system architectures, sensing technologies, communication protocols, data analytics techniques, energy considerations, and scalability challenges for rainwater harvesting and its potential treatment. Despite significant progress, gaps remain in multi-leak detection, large-scale deployment, sensor optimization, and privacy-preserving analytics.

Chapter 6 focuses on preserving privacy for smart water systems using homomorphic encryption and proposes a framework for ensuring the privacy of smart water management by applying the concept of homomorphic encryption. The proposed framework would allow computations to be performed directly on the water consumption information that is collected through sensors. The feasibility of the proposed framework is demonstrated by writing a Python script for a small-scale smart water management system. The results demonstrate that the proposed approach for ensuring the privacy of smart water management by applying homomorphic encryption has a huge promise for effective water resource management.

Chapter 7 examines the IoT-based water sensor technology, the design, the key functional elements, implementation procedures, and the effects of operations. High-tech sensors will be able to measure features like water flow, water quality, pressure, and chemicals accurately. The chapter also outlines IoT network systems that comprise data collection, transmission, cloud processing, and decision-making. Real-life examples of enhanced efficiency in water, fast leakage detection, accurate irrigation, and better control of the resources have been demonstrated. Technical issues, such as sensor calibration, network stability, data protection, and integration with other legacy systems, are discussed.

Chapter 8 explores blockchain technology's potential to revolutionize water governance systems. Traditional water management faces challenges, including data fragmentation, limited transparency, inefficient resource allocation, and stakeholder distrust. The chapter explains blockchain fundamentals, decentralization, immutability, and consensus mechanisms and examines applications in water rights management, supply chain monitoring, quality control, and equitable distribution systems. Real-world case studies from multiple jurisdictions demonstrate practical implementation outcomes. It emphasizes that successful implementation requires collaborative frameworks to overcome obstacles while enhancing transparency, efficiency, and accountability in water resource management. The chapter underscores blockchain's transformative potential for creating sustainable, trustworthy water governance systems when properly integrated with complementary technologies and supportive regulatory environments.

Chapter 9 explores predictive modeling for water resource management and water shortage forecasting. It applies various forecasting models to obtain the approximate requirement of water in the future depending on the past data and various environmental variables. The objective is to assist government, decision-makers, authorities, and the public in successfully planning water management.

Chapter 10 examines ML-based water demand forecasting, irrigation optimization, water quality control, leakage detection, and climate impact evaluation in water management. It uses numerous examples of cases and international applications to prove that ML-based methods are superior to traditional ones in terms of flexibility, scalability, and predictive accuracy. Also, issues like data quality, computational complexity, interpretability of models, and generalization are critically analyzed, as well as ethical issues. The chapter has also identified the emerging trends, such as IoT integration with ML, energy-saving algorithms, and hyper-local environmental forecasts. It confirms machine learning as an innovative and essential service to the future of weather forecasting and sustainable water management, which can be useful to the research community, policymakers, and environmental science practitioners.

Chapter 11 provides a comprehensive review of the strategies that could be designed to improve the climate resilience of water management systems for urban and rural development. The key factors responsible for affecting water resources are examined, and the foundational principles for sustainable and resilient water governance are outlined. The chapter highlights the importance of nature-based solutions to support engineering interventions. In addition, modern technological innovations are also discussed to support efficient and adaptive water usage. Further, the importance of community participation and policymakers in developing robust and inclusive water management frameworks is discussed with emphasis on the development of solutions that integrate scientific and technological advancements for achieving sustainable and climate-resilient water systems.

Chapter 12 highlights the use of digital technologies, Internet of Things (IoT) sensors, and Artificial Intelligence (AI) based analytics, which are revolutionizing the different levels of Smart Water Management (SWM) in the water cycle. It also analyzes how the future demand for urban water, which is expected to grow by 40–55% by 2050, can be met and how the sustainability and resilience of water resources can be improved through the integration of smart infrastructure, governance, and emerging technologies such as digital twins and nanofiltration.

Chapter 13 presents a remote sensing–based impact assessment of MAR interventions using the Modified Normalized Difference Water Index (MNDWI) across pre-monsoon and post-monsoon periods from 2019 to 2024. By analyzing multi-temporal satellite imagery, the study quantifies changes in surface water extent and evaluates the hydrological response of recharge structures over time. The MNDWI technique enables precise delineation of water bodies, offering a cost-effective and scalable approach to monitor seasonal water dynamics. Results reveal a consistent increase in water spread area post-monsoon, indicating improved recharge efficiency and enhanced water retention due to MAR implementation. The chapter also discusses spatial variability, methodological limitations, and the integration of MNDWI with ground-based observations for robust validation.

Chapter 14 presents a report on Field-Based Investigation of Sinkhole Formation Caused by Excessive Irrigation in the Rice Farm of the Bemetara Region, Chhattisgarh, India. The field-based observations' analysis—offering solid proof of subsurface collapse conditions—supported these findings. Additionally, past groundwater data showed consecutive declines that happened at different stages of sinkhole formation, indicating that variations in groundwater levels serve as necessary preconditions before any obvious surface failure. Overall, the findings show that rapid and ongoing groundwater decreases, primarily brought on by over-irrigation, may be utilized to predict sinkhole development. In order to safeguard agricultural land, groundwater resources, and rural communities in karst-prone areas where sinkhole impacts can be severe and potentially catastrophic, the chapter strongly recommends establishing proactive groundwater monitoring networks, either by installing new observation wells or by using existing ones.

Chapter 15 investigates the integration of the Firefly Algorithm (FA) for the hyperparameter optimization of a Multi-Layer Perceptron (MLP) classifier dedicated to water potability prediction. The work utilized a balanced dataset of 10,000 samples with the objective of eliminating training bias and ensuring robust classification. From a public health perspective, the FA-MLP model demonstrated superior performance in identifying non-potable water samples, as evidenced by confusion matrix and ROC curve analyses. The optimization process required approximately 4.6 hours of computational time, consistent with the iterative nature of metaheuristics. In order to facilitate reproducibility and future research, a comprehensive pseudocode of the FA-MLP workflow is provided. This chapter provides a reliable framework for enhancing decision boundaries in critical environmental monitoring tasks.

Chapter 16 introduced Aqua Sense, a hybrid bioimpedance and multi-parameter IoT system to provide fast, intelligent, and uninterrupted water quality assessments. In order to detect ionic changes, the system also uses Bioimpedance Spectroscopy (BIS) along with other standard sensors (pH, turbidity, conductivity, and temperature) to capture ionic changes, detect the very first microbial activity, and evaluate complex contamination patterns. The work designed and built a bioimpedance water probe that performs frequency sweeps within the range of 1 kHz and 300 kHz to capture the dielectric response of water. An ESP32-based edge processor performs on-board FFT, retrieves impedance spectra, and forwards it to a cloud dashboard for ML-based classification. In contrast to traditional sensor systems, the Aqua Sense hybrid system can identify ionic contamination with a high sensitivity of 98.2%, microbial contamination within 8 seconds, and mixed pollutants with an accuracy of 95.8%. This hybrid real-time water monitoring system is ideal for rural and industrial settings and is economically viable and scalable.

Chapter 17 examines cybersecurity challenges facing Smart Water Systems (SWS) IoT-integrated water infrastructure using SCADA for monitoring and control. Cyber criminals increasingly target these critical systems through ransomware, control system attacks, and nation-state threats, exploiting system vulnerabilities. The analysis explores protection strategies aligned with NIST frameworks, emphasizing fault-tolerant technological approaches applicable across different resource levels. Defenders employ advanced technologies, including AI-powered anomaly detection, zero-trust architecture, and blockchain to strengthen security postures. The chapter discusses predictive cybercrime strategies, business continuity planning, and incident response protocols. It examines compliance with frameworks like the National Water Act to optimize resource allocation for SWS defenders. The comprehensive analysis aims to help water infrastructure operators enhance cyber-resilience while maintaining operational efficiency.

Chapter 18 highlights that water and wastewater utilities rely on SCADA/ICS to run treatment and distribution, but modernization has exposed them to credible process risk. Internet-reachable HMIs, weakly governed remote access, flat networks, legacy platforms, and vendor pathways enable adversaries—from "hacktivist" brands to state groups using living-off-the-land (LOTL) tradecraft—to alter setpoints, suppress alarms, or force manual operations. Incidents across two decades and multiple countries (e.g., Maroochy, Oldsmar, Aliquippa; adjacent "fluids" cases such as Colonial and TRITON) show how small configuration mistakes and governance gaps scale into environmental, safety, or service impacts.

Chapter 19 analyzes the nexus of water security and the space economy and how they represent a critical and synergistic mechanism that serves a mutually reinforcing function for global advancement. Relatively limited to a land-based perspective, the space economy shows that space assets and technologies are radically changing the management of water resources as well. Sustainable and efficient water use is stressed as a key enabler for the sustained, scalable human and industrial activities in space. It further helps to attain water security on Earth and confirms a new age with a sustainable and prosperous future for the human race.

About the Editors

Neha Sharma, Senior Member of IEEE, is an Associate Professor in the Computer Science and Engineering Department at Chandigarh University, Punjab, India. With over 14 years of teaching experience, she has published 200+ peer-reviewed papers and has several national and international patents under the Intellectual Property Rights of the Government of India and abroad. Her research interests include image processing, machine learning, deep learning, and cybersecurity. She has been recognized with numerous accolades, including placement among the top 2% of scientists globally in the Stanford University rankings for 2025.

Abhineet Anand is an Academician and Researcher in the field of Computer Science and Engineering with extensive experience in teaching, research, and academic administration. He is currently serving as Dean Research and Dean CSE at Bahra University. He has contributed significantly to scholarly publications and editorial activities and has edited 12 academic books with reputed publishers. His areas of interest include artificial intelligence, emerging computing technologies, and applied research. He has actively participated in expert talks, academic mentoring, and curriculum development initiatives.

Abhishek Kumar, Senior Member of IEEE, is an Assistant Director and Professor in the Computer Science and Engineering Department at Chandigarh University, Punjab, India. With over 14 years of teaching experience, he has published 200+ peer-reviewed papers and successfully supervised 6 PhD scholars, along with 48+ MTech projects. He holds a postdoctoral research position at Universidad de Castilla-La Mancha, Spain. His research interests span artificial intelligence, renewable energy systems, image processing, and data mining. An award-winning researcher, Dr Kumar has received several accolades, including the Sir C.V. Raman National Award (2018), and holds a patent. An accomplished author and editor, he has authored 7 books and edited 97 volumes with reputed publishers. He has also served as Series Editor for a book series.

Rajeev Tiwari is a Senior Member of IEEE and received his PhD from Thapar University, Patiala, India. He is currently a Professor and the Dean of the School of Artificial Intelligence, Bennett University, Greater Noida, India. He has more than 20 years of research and teaching experience. He has published articles and edited four books with reputed publishers. His research interests include fog computing, ICN, and cloud computing. He has chaired many international conferences globally and delivered keynotes in the field of sensor networks and IoT-fog-cloud-based computing.

Anupam Tiwari, PhD (2022), brings 25+ years in cybersecurity. Certified Ethical Hacker and Cyber Security Professional (CDAC & GFSU), he holds a BE (Jamia Millia Islamia), MTech (JNTU Hyderabad), and postgraduate degrees in Information Security, ERP, and Operations. Focusing on blockchain for cyber-physical systems, he has 25+ publications, 50+ conference presentations, and CQ100 recognition. He has conducted workshops on Blockchain/Quantum at premium government institutions and IIITs, and peer-reviewed multiple IEEE journals.

Prateek Thapar is a distinguished Technologist, Academician, and Veteran of the Indian Air Force, with over two decades of leadership in Artificial Intelligence, Cybersecurity, and Defence Technology. Throughout his career, he has held pivotal roles including Chief Information Officer and Head of the Artificial Intelligence Centre of Excellence, where he led mission-critical initiatives in Generative AI,

Natural Language Processing, and Big Data Analytics for national defence. He holds a PhD in IoT Security using Elliptic Curve Cryptography and has mentored several doctoral researchers in AI and Blockchain. Currently, he is charting a new path as co-founder and CEO of ValoTech AI Pvt Ltd, where he is driving innovation in AI-powered solutions. His ventures include the development of Retrieval-Augmented Generation (RAG) systems, intelligent survey platforms, AI-driven cybersecurity frameworks, and advanced analytics tools for government and strategic agencies.

Contributors

Nirav Amin
Electronics and Communication Engineering Department
Government Engineering College
Bhavnagar, India

Abhineet Anand
School of Computer Science and Engineering
Bahra University
Waknaghat, Solan
Himachal Pradesh, India

Jagdeep Singh Arora
Mittal School of Business
Lovely Professional University
Punjab, India

B. Arundhathi
CSIR-Indian Institute of Chemical Technology (IICT)
Hyderabad, India

Indu Bala
School of Electronics and Electrical Engineering
Lovely Professional University
Punjab, India

Bhausaheb Ashok Botre
CSIR – Central Electronics Engineering Research Institute
CEERI, Pilani
Rajasthan, India
CSIR – Academy of Scientific and Innovative Research
AcSIR, Ghaziabad, India

Aravind Chauhan
Department of Civil Engineering
Bahra University
Waknaghat, Solan, Himachal Pradesh, India

Yogendra Chhetri
Newcastle University
Singapore

Pooja Choudhary
School of Computer Science Engineering and Technology
Bennett University
Greater Noida, India

R. Saroja Devi
Karpagam Academy of Higher Education
Coimbatore, India

Uttam Kumar Giri
Chitkara University School of Engineering and Technology
Chitkara University, Himachal Pradesh, India

Shilpi Harnal
Department of Computer Science and Engineering
Chandigarh University
Mohali, Punjab, India

Deepak Joshi
Department of Electronics and Communication Engineering
Jaypee Institute of Information Technology
Noida, Uttar Pradesh, India

Manikandan Kannappan
Department of Agronomy
Faculty of Agriculture
Annamalai University, Chidambaram, Tamil Nadu, India

P. B. Narendra Kiran
Christ University
Bengaluru, India

Ravi Kumar
Department of Civil Engineering
National Institute of Technology
Raipur, Chhattisgarh, India

Suraj Kumar
Department of Civil Engineering
National Institute of Technology
Raipur, Chhattisgarh, India

Chandan Kumar Singh
Department of Civil Engineering
National Institute of Technology
Raipur, Chhattisgarh, India

Abhishek Mishra
Faculty of Vehicle Technology
Electronics and Mechanical Engineering School
Vadodara, Gujarat, India

Özen Özer
Department of Mathematics
Kırklareli University
Kırklareli, Türkiye

Sudarsan Sahu
Central Ground Water Board State Unit Office
New Delhi, India

Ravi Kumar Sandal
Department of Civil Engineering
Bahra University
Waknaghat, Solan, Himachal Pradesh, India

Caroliny M. da Silva Santos
Federal University of Rio Grande do Norte
Natal, Rio Grande do Norte, Brazil

Deelip Kumar Sarangi
Kalinga University
Naya Raipur, Chhattisgarh, India

Pradeepta Kumar Sarangi
Chitkara University School of Engineering and Technology
Chitkara University, Himachal Pradesh, India

Chetan Sharma
PW Institute of Innovation
Physics Wallah Limited
Lucknow, Uttar Pradesh, India

Lokeshwer Sharma
Department of Civil Engineering
Bahra University
Waknaghat, Solan, Himachal Pradesh, India

Neha Sharma
Department of Computer Science and Engineering
Chandigarh University
Mohali, Punjab, India

Amandeep Singh
Mittal School of Business
Lovely Professional University
Punjab, India

Sandeep Singh
CGC University
Mohali, Punjab, India

Saurav Kumar Singh
Computer Science and Engineering Department
Jaypee University of Information Technology
Solan, HP, India

S. Sridhar
CSIR-Indian Institute of Chemical Technology (IICT)
Hyderabad, India
CSIR – Academy of Scientific and Innovative Research
AcSIR, Ghaziabad, India

Nadir Subaşi
Department of Computer Programming
Kırklareli University
Kırklareli, Türkiye

J. Sudarvel
Karpagam Academy of Higher Education
Coimbatore, India

Prerna Thakur
Department of Pharmacy
Bahra University
Waknaghat, Solan, Himachal Pradesh, India

Ravi Thirumalaisamy
Modern College of Business and Science
Oman

Anupam Tiwari
Department of Computer Science and Engineering
The Institution of Electronics and Telecommunication Engineers
New Delhi, India

Shuchi Tiwari
Department of Chemistry, VNR Vignana Jyothi Institute of Engineering and Technology
Hyderabad, Telangana, India

Ramamani Tripathy
Chitkara University School of Engineering and Technology
Chitkara University, Himachal Pradesh, India

Poomalai Vasanth
Department of Agronomy
Faculty of Agriculture
Annamalai University, Chidambaram, Tamil Nadu, India

Sakshi
Department of Computer Science and Engineering
Amity University
Noida, Uttar Pradesh, India

Saurav
Department of Civil Engineering, Jaypee University of Information Technology
Solan, Himachal Pradesh, India

Rainwater Harvesting and Its Potential Treatment

1

Shuchi Tiwari, Bhausaheb Ashok Botre, S. Sridhar, B. Arundhathi, Pooja Choudhary, and Caroliny M. da Silva Santos

ABBREVIATIONS

AI: Artificial Intelligence
AIoT: Artificial Intelligence of Things
AMRUT : Atal Mission for Rejuvenation and Urban Transformation
CPS: Cyber-Physical System
GIS: Geographic Information System
GSM : Global System for Mobile Communication
ICT : Information and Communication Technology
IoT : Internet of Things
JJM : Jal Jeevan Mission
LoRa : Long Range (Low-Power Wide-Area Network)
MAR: Managed Aquifer Recharge
ML : Machine Learning
RWH : Rainwater Harvesting
SCADA: Supervisory Control and Data Acquisition
SDGs: Sustainable Development Goals
SRWH : Smart Rainwater Harvesting
SuDS: Sustainable Urban Drainage Systems
WASH: Water, Sanitation, and Hygiene

DOI: 10.1201/9781003742227-1

1.1 INTRODUCTION

Rainwater Harvesting (RWH) systems are currently considered an essential parameter in modern infrastructure. They are a key aspect of sustainable water conservation and management. They play a crucial role in managing and conserving the challenges associated with urban growth, increasing populations, and changing climatic conditions [1–3]. Moreover, these systems significantly reduce urban flooding by decreasing surface runoff and regenerating groundwater. They ensure water security, efficiency, reliability, and promote environmentally friendly methods and techniques for water conservation and construction [4–7]. Urban areas can strengthen their resilience and reliability to water scarcity and support smart and green infrastructure goals aligned with national and global sustainability initiatives such as India's Jal Shakti Mission and the Sustainable Development Goals (SDGs) by incorporating RWH into digitally and technologically intensive architecture [8–12]. For reliability, efficiency, and consistency, the key technical parameters and abbreviations used throughout the chapter are summarized in Table 1.1. These abbreviations entail commonly adopted terms and their importance in SRWH literature.

The evolution of RWH management systems worldwide is discussed in this chapter, with particular emphasis on the transformation from manual to digitally integrated and automated architectures. In recent years, digital technologies that enhance performance, monitoring accuracy, promote adaptive control, and overall reliability and efficiency have become more prevalent in RWH systems [13–15]. The conservation and management of water storage, collection, and demand-based distribution have been entirely transformed by the incorporation of smart water meters, IoT-based flow sensors, automatic valves, real-time water quality monitoring devices, and adaptive control systems [16–18]. Moreover, by enabling data-driven decision-making, the mentioned SRWH infrastructures facilitate leak detection, localization, predictive maintenance, and water usage optimization based on real-time data analytics [19]. However, the interrelated network of Internet of Things (IoT) and artificial intelligence (AI) systems operating within

TABLE 1.1 Human, Technological, and Policy Gaps in Smart Rainwater Harvesting (SRWH) Systems

DIMENSION	IDENTIFIED GAPS	KEY NEEDS/INTERVENTIONS
Human/Social	• Low digital literacy and limited technical skills among local operators. • Minimal community participation in system maintenance and monitoring. • Lack of awareness regarding SRWH benefits.	• Capacity-building programs for IoT-based water management. • Community training on digital tools and predictive maintenance. • Awareness campaigns linking RWH to livelihood and sustainability.
Technological	• Fragmented integration of IoT sensors, AI, and automation systems. • Poor data interoperability and network connectivity in remote areas. • Limited use of real-time analytics and predictive models.	• Development of cost-effective, modular IoT platforms. • Adoption of cloud-based or edge-computing frameworks. • Incorporation of AI-driven forecasting and decision-support systems.
Policy/Institutional	• Weak institutional linkages between research, policy, and local governance. • Absence of standardized guidelines for digital RWH systems. • Limited financial incentives for adopting smart water technologies.	• Formulation of digital water governance policies. • Integration of SRWH in urban and rural development plans. • Funding mechanisms for community-level technology deployment.

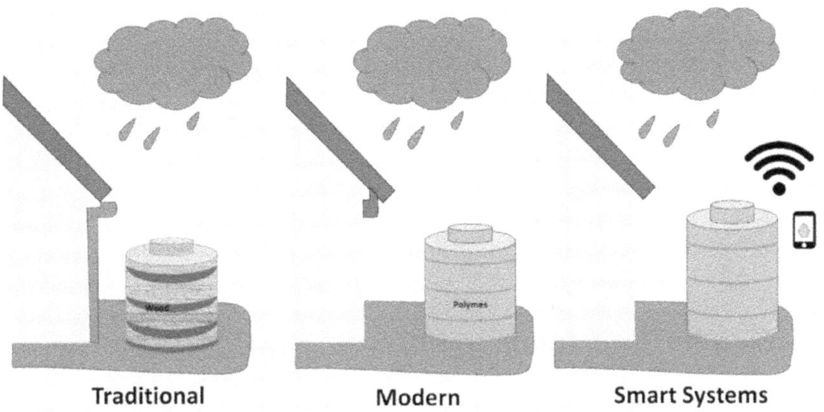

FIGURE 1.1 Development of rainwater harvesting from traditional barrels to smart sensor-based systems.

a CPS architecture is essential to the effectiveness and resilience of these advanced systems [20, 21]. The groundwork for future-generation smart water infrastructures that support reliable, efficient, and sustainable urban water management systems is established by the integration. It ensures smooth communication between digital monitoring and control units and physical water infrastructure [22–24].

RWH management systems is currently experiencing a crucial change with the development of revolutionary digital technologies. It transforms conventional water collecting, storage, and utilization systems into smart, automatic, and data-driven frameworks [13, 25]. Advanced tools such as digital technologies, AI, IoT, CPS, cloud computing, and GIS are being employed to improve the monitoring, adaptive control, and optimization of rainwater systems [26–28]. These technologies enable real-time monitoring and tracking of water quality, levels, flow rates, and quality factors, while predictive and time-series data analytics enable forecasting rainfall patterns, regulating storage capacities, and decreasing water wastage [29, 30].

The discussed digital transition constitutes a substantial difference from earlier manual approaches, resulting in intelligent and adaptive RWH systems. It promotes sustainable urban growth, efficiency, conservation, and resilient water resource management [11, 31]. These innovations not only contribute to operating efficiency and reliability but also help to achieve the more general objectives of climate adaptation, water security, conservation, and environmental sustainability. The progressive transition of rainwater harvesting systems from conventional and rudimentary storage-based practices to smart, AIoT-driven infrastructures entails broader advancements in digitally enabled water management [32–34]. This evolution highlighting the shift from manual monitoring and control to automated is illustrated in Figure 1.1.

1.2 OBJECTIVES

The conventional RWH systems play a significant role in conservation and saving water. However, these systems are usually static, rigid, and labor-intensive. It makes them less effective and unreliable with evolving urban and climatic environments. The gap analysis identifies significant challenges and difficulties, including limited real-time monitoring and adaptive control, and poor integration with decentralized water management systems, particularly in drought, arid, and semi-arid regions. Emerging digital technologies, including the ICT, AIoT, and CPS, hold the potential to evolve conventional RWH into intelligent, smart, and data-centric systems. SRWH systems report current technical and operational barriers and enable continuous monitoring, adaptive control, and systematic decision-making to fulfill sustainable, reliable, and efficient water management.

Objectives

- To study and understand the evolution of the conventional RWH systems to the SRWH systems considering the digitally and ICT-enabled frameworks, technologies, and functional gaps.
- To detail and understand how digital technologies, AIoT, and CPS improve monitoring, adaptive control, and decision-making in RWH systems.
- To evaluate RWH systems for potential treatment based on sustainability, reliability, and policy-aligned water systems focusing on national and global initiatives, including India's Jal Jeevan Mission and SDG-6.

1.3 RESEARCH QUESTIONS

The growing complexity of water shortages, climate variability, and urban development requires a review of the RWH systems design and management methodologies. Although traditional RWH methods remain valid, they are at times hindered by manual execution, by the absence of real-time monitoring, and/or by limited flexibility to environmental and demand changes. Therefore, to address shortcomings in conventional water infrastructure, it is necessary to understand transitions unfolding in RWH systems. As part of this investigation and for the development of RWH, it would also be relevant to determine why digital technologies have become so important in enhancing RWH performance and the extent to which new tools such as digital technologies AioT and CPS may be incorporated within RWH models. These topics are to deepen the analysis of technologies' evolution, the extent to which smart systems may contribute to effectiveness and efficiency, and resilience enhancement and long-term water management options alignment.

1. What constraints of traditional RWH systems highlight the demand for digitally enabled, SRWH frameworks?
2. How could IoT, AI, and Cyber-Physical Systems improve monitoring, control, and operational efficiency in RWH systems?
3. How do SRWH systems offer sustainable, resilient, and policy-aligned water management outcomes?

1.4 ARCHITECTURE OF RWH

RWH offers a harmonizing, decentralized response to this stress. Basically, rainwater harvesting refers to the collection, temporary storage, and beneficial use of rainwater at or near the point where it falls, either for direct use, such as in households or on farms, or to enhance groundwater recharge and local water balances. While the subject and technology may sound modern, the idea is ancient; communities have long built tanks, cisterns, terraces, and earthen bunds to make the most of seasonal rainfall. Modern RWH systems largely refine and engineer these traditional practices, linking them to modern materials, hydrological understanding, and water quality standards. This section provides a concise overview of how rainwater harvesting works, introduces techniques, and discusses their benefits, limitations, and prospects for sustainable water management.

1.4.1 Conceptual Overview

RWH serves two core functions. First, it enables the direct use of captured rainwater for domestic needs, small-scale industry, livestock, and supplemental irrigation. Second, it enhances groundwater recharge by promoting infiltration in areas where runoff has increased due to land-use changes. Often, a single RWH structure delivers both local water supply and downstream recharge benefits.

Most RWH systems share standard components: a catchment surface e.g., roof, courtyard, road, field, or natural slope; a conveyance system such as gutters, channels, or shallow drains; a pre-treatment step with a first-flush device or filter to remove debris; and then a storage or infiltration structure that either holds the water for later use or directs it into the ground. For collecting the water, many parameters have to be considered, such as appropriately sizing and placing these elements to capture meaningful rainfall volumes without causing flooding, seepage, or water-quality issues.

This overview categorizes five main RWH techniques: (i) rooftop harvesting with storage tanks, (ii) ground catchments and small farm ponds, (iii) percolation pits and infiltration trenches, (iv) check dams and retention weirs in drainage lines, and (v) urban stormwater harvesting linked to managed aquifer recharge (MAR). Though varied in form, all are based on shared principles and are applicable across household, agricultural, and urban settings.

1.4.2 Techniques—Main Methods and Their Characteristics

1.4.2.1 Rooftop Rainwater Harvesting with Storage Tanks

Rooftop rainwater harvesting is the collection of rainwater runoff from building houses (roof surface), which is then processed and stored in tanks or cisterns for subsequent use. A system is typical with guttering along the roof line, downpipes, a first-flush system to divert the more contaminated initial runoff, some form of filtering to remove debris and suspended solids and storage tanks, either above or below ground. The system performance is influenced by factors such as roof size, rainfall distribution, storage size, and, to a certain extent, the water demand of the household or institution. The tank size should be determined based on local rainfall data, roof area, demand pattern, and the desired reliability of the system, taking into consideration overflow-related water losses [35, 36]. Quality of water is primarily affected by the roofing materials, the atmospheric pollutants, and maintenance of the system. Reviews in the past emphasize the importance of first-flush diversion, basic filtration, and sporadic disinfection when water is to be used for drinking [37, 38]. More recent work has focused on inexpensive filtration media and system approaches to enhance the removal of microbes and trace metals in stored rainwater [39]. Such systems have been demonstrated to consistently satisfy a significant share of non-potable household water demands, including toilet flushing, laundry, gardening, as well as potable or emergency uses with due treatment [35, 37]. Their decentralized character means that such gains as less reliance on tanker water or on intermittent piped supply accrue directly to users, even if only modestly, while also contributing to stormwater abatement [38]. However, there are certain limitations too, such as the volume of water that can be harvested, which is inherently limited by roof area, and in dense urban areas or during dry periods, per capita yields may be low. Storage tanks involve high upfront costs and require regular maintenance, including cleaning, inspection, and disinfection when water is intended for potable use. Poor overflow design can worsen drainage issues, and neglected systems may become breeding grounds for mosquitoes or sources of microbial contamination [37, 39]. A typical household-level rooftop rainwater harvesting system consists of a catchment roof, conveyance network, first-flush unit, filtration components, and storage tank. The functional layout and interconnection of these elements are schematically represented in Figure 1.2.

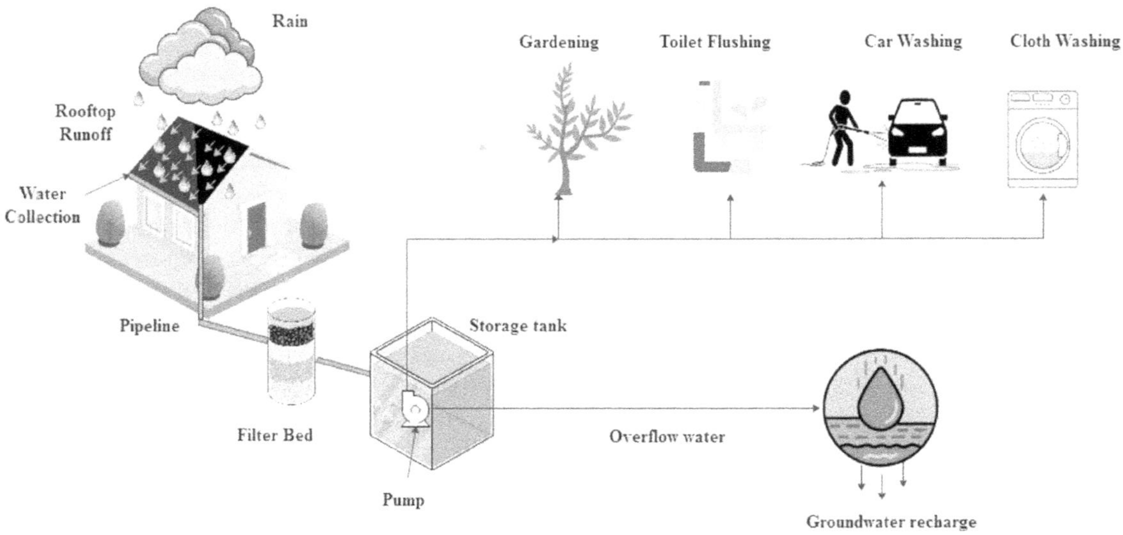

FIGURE 1.2 A conventional schematic of a household rooftop collecting rainwater system.

1.4.2.2 Ground Catchment Systems and Small Farm Ponds

Ground catchment systems are designed to intercept overland flow from small catchments and direct it into purpose-built reservoirs, commonly called farm ponds or on-farm reservoirs. In rainfed agricultural regions, these systems collect runoff from fields or nearby slopes during the rainy season, storing it for supplemental irrigation during critical crop growth periods. The catchment area is typically shaped using contour bunds or graded channels to guide runoff into an excavated pond, which is lined with compacted soil or, where feasible, a geotextile to reduce losses.

Recent studies in semi-arid lands indicate that well-constructed on-farm ponds can substantially increase crop yields and can buffer production from dry spells (making it more stable), if seepage and evaporation losses are controlled through choice of site, linings, and management [40]. Previous hydro-economic modeling suggests that such systems can increase crop yields in dry years by a factor of three, compared to dryland agriculture [36]. Reviews also show the feasibility of community-scale water provision and sustainable urban drainage service delivery by large pond-based RWH systems, especially at peri-urban scales [38]. Other than for irrigation, water stored in ponds may be used for watering livestock, small-scale aquaculture, and in shaping microclimatic conditions in cultivated landscapes. At the basin level, they play a role in diminishing peak runoff and controlling soil erosion and may facilitate recharge of local groundwater supplies under conditions of permeable soil [36, 40]. Nevertheless, small reservoirs are associated with some risks, such as loss of water by seepage that may reduce the availability of water, and in case the pond bottom is intersected by the high water table, waterlogging or salinization may take place in the downslope. Badly maintained ponds can also become habitats for mosquitoes, and competition for stored water may lead to tensions between users. Sedimentation can be prevented from reducing storage capacity by pretreating inflows or by removing sediment. Therefore, system success depends on long-term monitoring and adaptive management [36, 40].

1.4.2.3 Percolation Pits and Infiltration Trenches

Percolation pits and infiltration trenches are small decentralized STPs, which promote groundwater recharge by enabling rainwater and runoff to percolate into the subsurface. Typically, a percolation pit is a vertical or nearly vertical hole filled with graded gravel or coarse sand that captures and retains runoff from adjacent surfaces, such as rooftops, courtyards, or paved surfaces. Infiltration trenches, on the other hand, use the same fundamental concept but horizontally; these are linear trenches (which may be lined with perforated pipes), filled with gravel, acting as temporary stormwater conveyance, allowing storm water onto the soil at a controlled rate. In the city, infiltration trenches are now considered part of low-impact development or urban sustainable drainage systems as they help reduce the amount of runoff and improve water quality via physical filtration and chemical sorption [38]. Hydraulic and economic modeling studies indicate that such systems, if properly sized and placed, can routinely capture a substantial portion of small- and medium-sized storm events, thereby alleviating pressure on downstream drainage infrastructure [41]. Nevertheless, simulations of extended periods of time as well as field observations demonstrate universally encountered problems such as the clogging of the gravel packing and the surrounding soil by fine sediments, which often leads to a significant degradation of infiltration rates after a short time [42].

The feature of these systems is their simplicity and modularity, and the fact that they can be easily retrofitted in urban environments, such as along road medians, in parking lots, or around buildings. As described in recent reviews on rainwater harvesting techniques, percolation tanks are an ideal intermediate solution between simple water storage systems and extensive managed aquifer recharge, especially in places with limited space [43]. But their success hinges on the characteristics of the local soil (slow infiltration rates are characteristic of clayey, low-permeability soils; conversely, highly permeable soils over shallow groundwater tables are a hazard due to potential mobilization of contaminants or their migration into surrounding basements). A strong pre-treatment, e.g. by silt traps or vegetated swales, and adequate provisions for inspection and maintenance (to be performed on a regular basis) are required in order to keep performance [41, 42].

1.4.2.4 Check Dams and Small Retention Weirs in Drainage Lines

Check dams are small structures with low cost built across the course of a seasonal stream or gully that slow down the runoff locally and induce some water retention, thus increasing the potential for groundwater recharge. These structures, which are usually made of masonry, concrete, gabions, or compacted earth and have an integrated spillway, create a series of shallow pools upstream during and after rain events. Water held in these pools slowly permeates through the streambed and banks, elevating local groundwater levels and feeding baseflows to wells and springs located downstream. Field observations in semi-arid areas of India suggest that, within favorable hydrogeological settings, single check dams can recharge groundwater to an extent that is several times its surface storage capacity annually [44, 45]. This also supports operational strategies (such as sluicing to clear accumulated sediments) [46]. More comprehensive accounts of managed aquifer recharge (MAR) report the role of check dams in enhancing not only groundwater quantity but also, under certain conditions, water quality improvement and rural livelihood enhancement [47]. In addition to replenishing groundwater, check dams provide a number of other benefits, including reducing channel erosion, enhancing vegetation cover, and, in certain areas, mitigating some forms of groundwater contamination. When communities are involved in design and maintenance, these types of structures can also empower local institutions and build collective water stewardship [44, 47].

In addition to recharge, check dams provide other co-benefits, such as minimizing channel erosion, enhancing vegetation cover, and in some instances contributing to the dilution of certain groundwater pollutants. If communities are involved in construction and maintenance, these structures can also build local organization and collective water stewardship [44, 47]. Nevertheless, their good effectiveness depends on well-thought-out design and siting. Inadequately designed dams may present flood hazards, hold contaminated urban runoff, or be ineffective in contributing to recharge in certain geologic settings. In addition, multiple check dams within a basin can collectively impact downstream hydrology, leading to reductions in monsoon flows to ecosystems and reservoirs if not incorporated in basin-scale water balance planning [45, 46].

1.4.2.5 Urban Stormwater Harvesting and Managed Aquifer Recharge

In highly urbanized basins, impervious surfaces produce large amounts of stormwater, which is usually considered a source of pollution rather than a resource. Urban stormwater capture in conjunction with MAR represents a paradigm shift in management, treating stormwater to acceptable levels and using it for replenishing aquifers through infiltration ponds, recharge galleries, or via wells for injection. The stored water may be recovered for non-potable uses such as park and sports field irrigation, industrial cooling, and environmental flow, or, with further treatment, for potable use or for non-potable use with advanced treatment [48]. MAR systems have been demonstrated in case studies across Australia, Europe, and North America to be a cost-effective means for water storage, providing low evaporative losses and passive removal of nutrients and pathogens during subsurface residence [48]. While stormwater MAR compares conceptually to more generalized rainwater harvesting systems, it is different in that engineered treatment systems are used and there is an element of risk-based water quality management and (typically) higher tech monitoring [43, 47]. In urban water management, MAR adds to detention basins, constructed wetlands, and other green infrastructure as a means to increase the resilience and flexibility of the system. Conventional MC systems temporarily store stormwater in ponds or wetlands, conduct pre-treatment to remove sediments and pathogens, and then infiltrate or inject the stormwater into an appropriate aquifer. The recovery is either through existing or new drilled wells for managed reuse.

Nevertheless, the success of urban MAR relies on an appropriate institutional and regulatory framework. The following are important considerations: assessment of the suitability of the aquifer, prevention and management of clogging, monitoring of water quality, and long-term operation and maintenance, all require technical know-how and good governance. Public acceptance is overall just as important, especially when potable reuse is planned; open communication on risk management and system performance is key to establish trust [43, 48].

1.4.2.6 Advantages of RWH

From the point of view of hydrology, RWH contributes to the redistribution of water in time and space, attenuating the local water cycle variability. RWH systems mitigate dependence on traditional water sources such as rivers, lakes, and groundwater by harvesting storm runoff, and they also help reduce flash flooding and erosion.

1.4.2.7 Infiltration-Based Structure

Check dams promote groundwater recharge, stabilize water tables, and increase well resilience to drought when they are appropriately installed and maintained [44, 48]. With respect to water quality, rooftop and stormwater harvesting systems can provide relatively clean water for non-potable use after performing adequate filtration and disinfection and can relieve pressure on centralized municipal water supply systems [35]. Infiltration-based urban green infrastructure practices, including infiltration trenches and bioretention cells (rain gardens), also minimize pollutant loads by sedimentation, sorption, and biodegradation prior to recharge into the aquifers or discharge to surface water [41]. Socio-economically decentralized RWH enhances local water security at the household and community scale. Structures such as farm ponds and check dams provide more dependable crop irrigation and water for stock, thus contributing toward reduced rural vulnerability to rainfall variability and enhanced rural livelihoods [40, 44]. In cities, decentralized systems at the building scale can reduce dependency on tanker deliveries, and MAR projects can delay or negate the construction of costly new reservoirs or long-distance pipeline systems [48]. On the whole, RWH presents a promising instrument for climate resilience: distributed, energy-conscious, and stepwise scaling, it complements massive infrastructure while facilitating local adaptation.

1.4.2.8 *Challenges, Limitations, and Future Directions for Rainwater Harvesting*

Design, maintenance, and local participation are essential to the performance of rainwater harvesting systems. Technical problems common to these are undersized channels, poorly constructed spillways, and infiltration structures clogged by sediment. Farm ponds can also experience seepage or unstable banks, and badly situated check dams can have minimal recharge effect. Although water quality poses challenges, dust, bird droppings, and roofing leachates can be found in rooftop runoff, and stagnant water can host algae or serve as breeding sites for disease vectors. Managed aquifer recharge systems can clog and contaminate groundwater unless they are pretreated. Gaps in institutional arrangements also hinder sustainability. Many are ambiguous about maintenance, and shared infrastructure tends to collapse when there's no clear ownership. Regulations may subsidize RWH or encourage it through public education and product standards, but such enforcement and technical support are often feeble. In some cases, upfront costs are a deterrent for low-income users, with wealthier homeowners and institutions demonstrating greater take-up of RWH. Unsupported by specific subsidies or assistance, these systems run the risk of exacerbating disparities in water access. Also, the necessity to store harvested rain for long periods requires large volumes, which makes RWH a civil engineering-demanding technology, particularly in built-up areas. Yet RWH remains a good bet in defiance of these problems. As for the future, progress will entail coupling RWH with nature-based solutions—green roofs, pervious pavements, or even planning at the scale of watersheds. Digital technologies including sensors, data loggers, and GIS can also facilitate adaptive management and siting. Institutionalization of RWH in policy frameworks, training of local users, and strengthening of institutions will be required for scaling it from isolated projects to a mainstream option for climate-resilient water management.

1.4.3 Architecture of SRWH

Traditional rooftop rainwater harvesting systems operate manually, without real-time monitoring and with a high occurrence of water quality problems and reservoir underutilization. SRWH systems tackle this problem by using IoT sensors, automation, and data-driven control to enable around-the-clock monitoring, smart storage, and enhanced water quality management. The evolution from traditional RWH systems to smart, automated systems demands a unified framework encompassing sensing, communication, analytics, and control. Figure 1.3 shows the overall architecture of an SRWH system, depicting the interaction of the physical and digital elements of the infrastructure.

1.4.3.1 *Key Components*

Integrated SRWH systems are designed to consist of various interrelated components that could effectively, reliably, and sustainably manage water in semi-arid and arid regions where water security is under threat of collapse, contrary to those of several existing RWH systems. At their core, these systems consist of a number of sensors and IoT devices that are distributed over catchment, filtration, and storage modules. These devices harvest real-time information such as rainfall intensity, flow rate, tank levels, the pH, turbidity, and conductivity, and enable continual monitoring of the quantity and quality of the harvested water. The gathered information is sent via wireless communication networks, including Wi-Fi, GSM, LoRa, etc., to a centralized platform or cloud-based dashboard [49]. AI- and ML-based systems process the information to forecast patterns of rainfall, manage the storage space, and operate pumps, valves, and filter systems automatically.

1.4.3.1.1 Demand-Driven Water Supply

The automation within the system reduces human interventions, decreases water wastage, and ensures demand-driven water supply. In addition, intelligent RWH systems often integrate with weather prediction

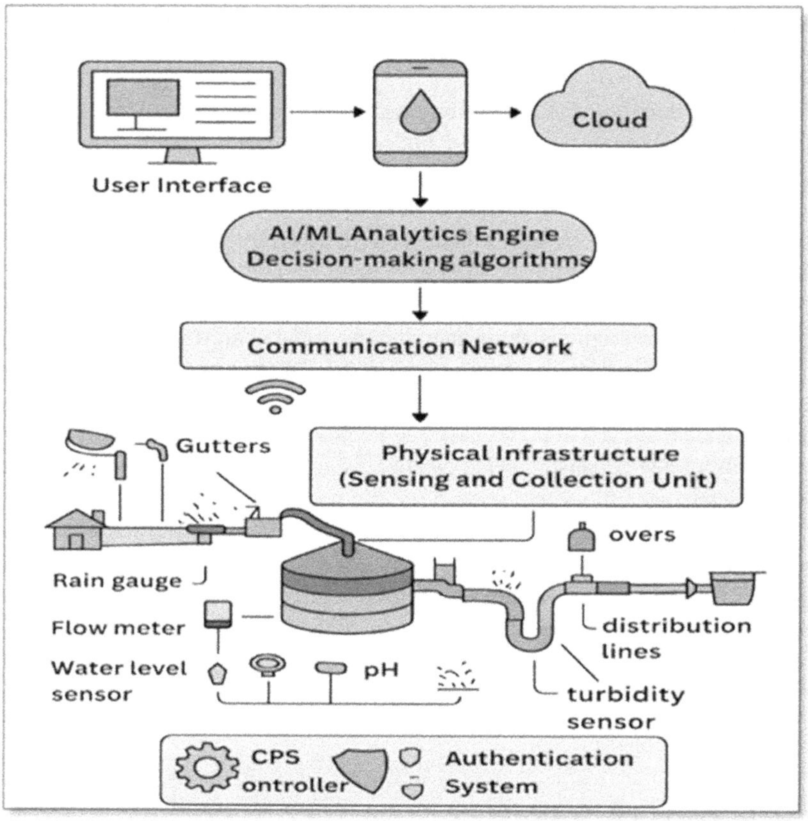

FIGURE 1.3 Integrated SRWH architecture enabling real-time sensing, intelligent control, and automated water management.

services and GIS to estimate rainfall events, which facilitates proactive management activities such as diversions to groundwater recharge or irrigation in the face of excess water. The hardware and software components of the system are intelligently coordinated so that the resilience of the system is enhanced, leading to long-term water provision in regions with high climatic variability. The SRWH system production process operation rainfall sensing to auto storage and supply delivery is facilitated by integrated sensing, analytics, and actuation. This is shown in a flowchart in Figure 1.4, illustrating the joint opportunity of IoT sensors, AI-based control, and cloud analytics to improve the system performance.

1.4.3.2 The RWH Topology

The operational design of SRWH systems is layered, with the physical layer at the bottom serving as the base, followed by the communication, data processing, and user interface layers from bottom to top. All layers have distinct and yet complementary functions in the achievement of an effective and smart water management [50]. The outer layer of the system is made of catchment surfaces, filter modules, storage tanks, and pipes that are equipped with IoT-based sensors to track rainfall, water levels, flow rates, and metrics of water quality including pH and turbidity. The communication layer enables this information to be successfully transferred via wireless networking such as Wi-Fi, GSM, ZigBee, and LoRa from field sensors to centralized or cloud platforms. The aforementioned automation of pump operation, filtration, and spill management is performed in real-time by active automation in the data processing and control layer based on patterns identified by AI/ML for improving system performance, predicting rainfall patterns, and so on. And finally, the user interface layer provides interactive UI means such as

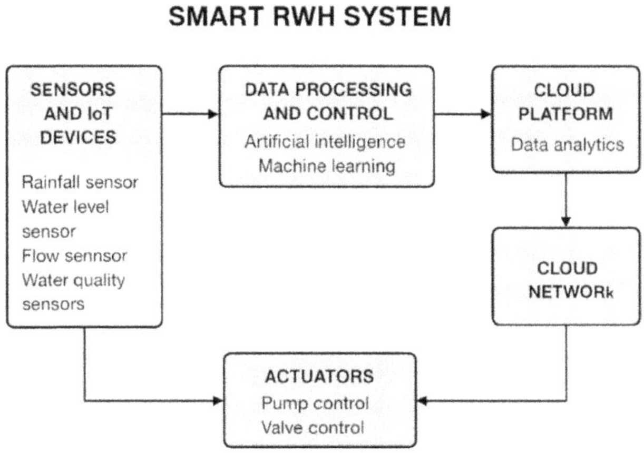

FIGURE 1.4 Demonstration of the architecture of an SRWH system which incorporates IoT sensors, AI-based control, cloud analytics, and automated actuators.

dashboards, mobile apps, and control panels for users and administrators to analyze system status, get notified, and conduct workflow remotely. The hierarchical and interconnected setup enables a data-driven, adaptive, and sustainable management of water, which renders SRWH systems advantageous particularly in urban, desert, and semi-arid regions coping with water scarcity problems [15].

1.4.3.3 Cyber Physical System Convergence

Among the SRWH system functionalities, the most important part is CPS, and it is defined as the integration of physical water infrastructure with digital intelligence, enabling real-time monitoring, interaction, and adaptive decision-making [51, 52]. It involves bringing together physical parts (pipes and pumps) and digital parts (software and data) so that monitoring can happen 24/7 and responses can be automated based on data. In a CPS-empowered SRWH architecture, physical components such as storage tanks, pipelines, valves, pumps, and sensors are tightly integrated with cyber components such as data analytics platforms, AI algorithms, cloud computing systems, and control software. These two domains execute in ongoing feedback loops along with data sensed or sampled from the physical environment transferred to the cyber layer for analysis and the output insight immediately translated into automated physical actions, e.g., opening a valve, activating a pump, or redirecting overflow [53, 54]. This bidirectional connectivity allows real-time modification of water collection, storage, and distribution infrastructure using rainfall data, consumption patterns, and weather prediction models. This kind of integration enhances not just the efficiency of operation but also enables detection of defects, self-correction, and energy-efficient operation, avoiding water loss and increasing reliability in receiving erratic rainfall [15, 55]. The integration of physical water infrastructure with cyber intelligence employs a layered architectural style that facilitates the integration of data and automated response mechanisms. The architecture layer CPS solution for SRWH system is shown in Figure 1.5.

Furthermore, the convergence of CPS allows for scalability and compatibility, such that the SRWH systems can cooperate with wider urban water networks, IoT-based smart city platforms, and environmental monitoring systems [56, 57]. The development of cloud computing, edge analytics, and secure data protocols enhances the system performance and protects data integrity and system security. In a nutshell, Cyber-Physical System convergence leads to the transformation of traditional RWH into a smart, autonomous, and resilient infrastructure that is in line with the principles of Industry 4.0 and the sustainable development of urban areas [58]. It paves the way for next-generation intelligent water management systems that can address the issues of water scarcity and climate variability [11, 59]. The loop of continuous feedback disrupts static monitoring and control and permits real-time monitoring, predictive control, and adaptive decision-making for sound and sustainable water management. SRWH systems depend

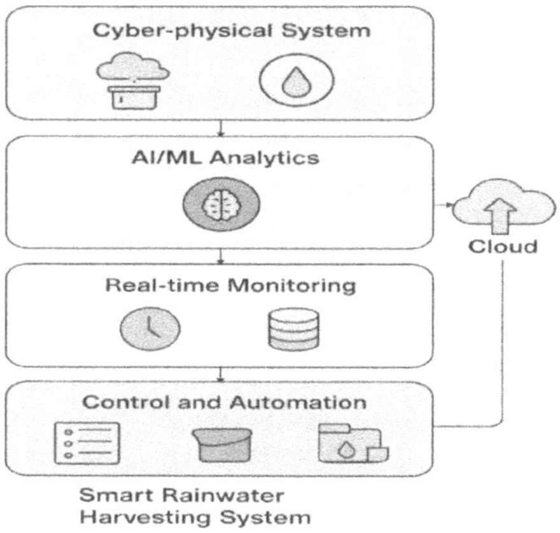

FIGURE 1.5 An architecture of an SRWH system incorporating CPS, real-time monitoring, with automated control.

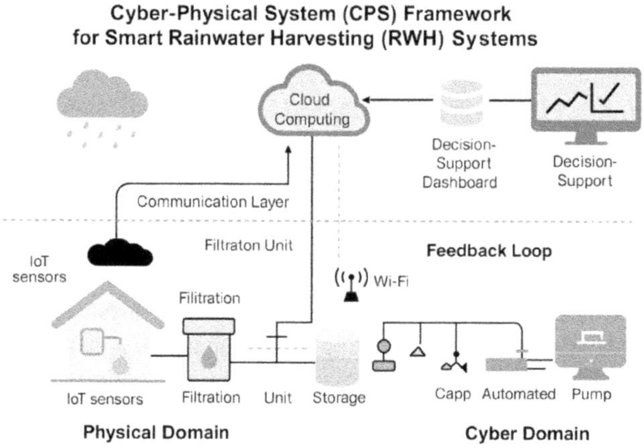

FIGURE 1.6 CPS-based SRWH system which combines IoT sensing, cloud analytics, communication, and automated control.

on closed-loop feedback among sensing, analyses, and actuation for evolving and enduring. Figure 1.6 illustrates the structure of an SRWH system based on CPS, showing the interaction between IoT-enabled physical entities and cloud-based analytics and automated control approaches for real-time decisions.

1.5 TECHNOLOGICAL GAP ANALYSIS AND NEED IN ARID REGIONS

The social and technical aspects of water infrastructure are always interrelated, yet many RWH systems are still insufficient. Sensors, actuators, and automation technologies are often deployed with little or no consideration of local human capabilities, environmental factors, or community involvement. This divergence causes underperformance, inefficient data collection, and sporadic maintenance. In addition, decision-makers in dry and semi-dry lands often struggle with limited access to real-time data, a lack of interoperability between instruments, and an absence of localized forecasting methodologies

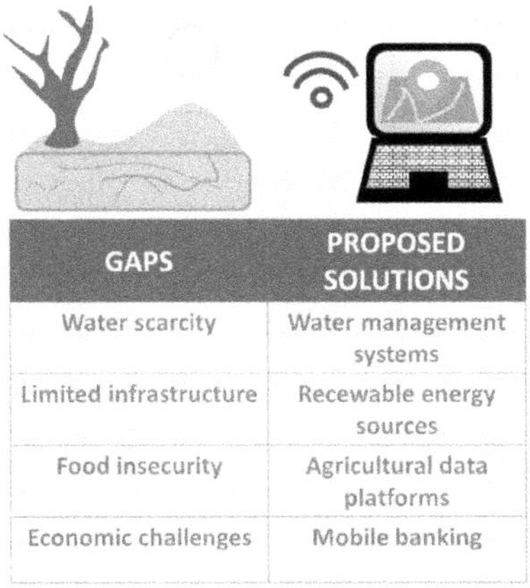

FIGURE 1.7 Mapping gaps in conventional RWH to SRWH solutions.

capturing intra-annual rainfall variability. The disorganized digital framework and poor coordination among institutions hinder RWH from moving from an ineffective collection practice to a forward-looking, self-regulating water management mechanism. On the interpersonal dimension, there are still needs in skills development, democratic governance, and behavioral adaptation to smart technologies. Although the high-end technologies exist, lack of skilled manpower, low awareness of IoT-based solutions, and limited subsidies are hindering the adoption and scaling up. Bridging these divides requires not just technological innovation but also a socio-technical transformation—integration of education, digital literacy, and community capacity building in SRWH systems design. Effective gap analysis, then, must go beyond the physical constraints of hardware to include vulnerabilities in the soft infrastructure such as the distribution of knowledge, levels of institutional trust, and adaptive policy processes that foster long-term sustainability and inclusion. The constraints of traditional rainwater harvesting systems are technological, institutional, and human. Figure 1.7 conceptually illustrates these gaps and how SRWH innovations overcome each obstacle by way of digital and governance-based solutions.

Table 1.1 provides a summary review of the human, technological, and institutional barriers to effective rainwater harvesting. The study restates the challenge that intelligent and integrated solutions are needed to mitigate such constraints. On the one hand, the application of information and communication technologies (ICT) to water system infrastructure has led to dramatic improvements in urban water management, but the diffusion of this application to dry and arid regions is sporadic and scanty [15]. In many RWH systems in such locations, the systems are still manual, decentralized, and with no real-time monitoring. The absence of fully integrated digital technologies, including IoT sensors and data loggers, and automated control systems restricts their ability to predict changes in precipitation, to optimize storage management, and to recognize leakages and contamination events. Also, poor network coverage, lack of technical expertise, and inadequate data infrastructure pose challenges in digitally transforming these systems. Consequently, local operators often use rules of thumb rather than data-based decisions, which constrains system optimization and long-term sustainability. Closing these technological gaps is essential to ensuring water security and climate resilience for the water-stressed conditions. There is a strong need for developing economically viable, locally customizable SRWH designs integrating low-power IoT devices, renewable energy sources (e.g., solar-powered controllers), and high-end cloud-based data analytics. Such systems can also provide real-time visualization of rainfall events, dynamic control of storage capacity, and predictive maintenance using AI-based algorithms. In addition, training programs and digital literacy must be provided to enable communities to adopt and maintain these technologies. By addressing the gap between traditional approaches and intelligent digital interventions, drought-affected regions could move past resilient, self-sustaining water management systems that are aligned with the broader aims of sustainable development and climate transformation. Smart Rainwater Harvesting (SRWH) systems function within a multi-stakeholder environment that includes policymakers, engineers,

TABLE 1.2 Identified Technological Gaps and Proposed Smart Solutions for RWH in Arid Regions

IDENTIFIED GAP	IMPACT ON RWH PERFORMANCE	PROPOSED SMART SOLUTION
Lack of real-time monitoring	Inefficient storage management and undetected leakages	Integration of IoT-based water level and flow sensors for continuous monitoring
Manual operation of valves and pumps	Delayed response during rainfall events and water wastage	Use of automated valves and smart controllers with remote access
Limited data on rainfall and water quality	Poor prediction and quality assurance	Deployment of weather-linked sensors and AI-based water quality analytics
Absence of centralized data system	Fragmented information and poor coordination	Establishment of cloud-based data management platforms for system integration
Low digital literacy among local operators	Poor maintenance and underutilization of systems	Training and capacity-building programs for community-level users
Unreliable power supply	Disruption in sensor and communication functions	Adoption of solar-powered IoT devices and low-energy communication networks (LoRa, ZigBee)
Lack of predictive and preventive maintenance	Increased downtime and repair costs	AI/ML-based predictive maintenance models for system reliability
Weak connectivity in remote areas	Data loss and delayed control actions	Deployment of edge computing and offline data caching for robust operation

technology providers, utilities, and consumers. The functions and interaction between these actors are described in Table 1.2.

1.6 NEED OF SRWH

The increasing demand for SRWH systems is driven by multiple environmental, technological, and socio-economic factors. These key factors from climate variability to digital readiness are summarized in Figure 1.8.

The above-ground vulnerabilities of human populations are well illustrated in historical events when traditional water systems were unresponsive to climate change and urban growth. Water scarcity has been further aggravated in many arid and semi-arid regions by irregular precipitation regimes, prolonged droughts, and overexploitation of groundwater. Traditional RWHs, although they are time-tested and driven by community, often lack sufficient dynamic ability to adjust to changes in rainfall intensity and frequency. Such fixed installations may overflow or underperform in severe weather, leading to questionable efficiency and wastage of water. These challenges show why we need smarter, more flexible systems—ones that use data to sense, analyze, and then act on their own to get the most out of water collection, treatment, and delivery. Intelligent RWH systems take old-school, passive methods and push them further, turning them into active, adaptable tools for managing water in built environments. They use IoT sensors, AI, and real-time monitoring to predict what's coming, adjusting to rainfall forecasts, how fast people are using water, and even tracking water quality as it changes. It's a much more responsive and effective way to handle water. This digital transformation is both technological and strategic, contributing to providing resilience to climate extremes, sustainable urban water reuse, and equitable access to water for rural communities. Hence, the justification for SRWH is derived from its potential to form a viable, self-sustaining, and transparent water ecosystem that is climate-adaptive, Sustainable Development Goal 6 (Clean Water and Sanitation), and national water missions such as Jal Jeevan and Smart Cities Initiative.

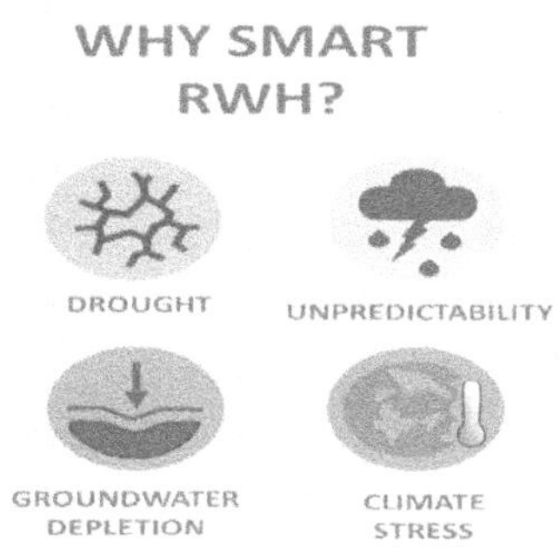

FIGURE 1.8 Key drivers for adopting SRWH systems.

1.6.1 RWH Actors

Although water systems are composed of physical elements such as reservoirs, pipes, sensors, and controllers, their capacity to operate is reliant on the people and organizations who design, manage, and sustain them. RWH is multi-stakeholder at different echelons ranging from policymakers, engineers, and local administrators at one end and community members, maintenance staff, and end-users at another. What makes each of these actors unique is the role they play in keeping the system functioning and societally acceptable. The policymakers establish the SRWH production process legislative and financial framework, and the engineers and the technology providers design, implement, and operate IoT-based components that maximize efficiency. Community volunteers, at the local level, at least, are essential to track local water needs, to maintain structures, and to assure equitable distribution. These jobs have become all the more intertwined in the SRWH era with the rise of digital governance and CPS. Sensors, automated valves, and AI-enabled decision systems necessitate not only technical proficiency but also institutional flexibility and multi-sector coordination. The effective functioning is increasingly dependent on data analysts, ICT professionals, and system integrators that process real-time data and feed this information into decision-making. At the same time, local people need to be given information on technology so that they can use dashes and monitoring screens. Today, RWH practitioners are not just implementers but co-creators of smart water systems, integrating traditional knowledge with modern science and technology to deliver resilience, inclusivity, and sustainability in water resources management. International case studies demonstrate the effectiveness of SRWH systems across diverse climatic and institutional contexts. Selected global examples, along with their technological approaches and outcomes, are presented in Table 1.3.

1.6.2 RWH Frameworks and Best Practices

At the city and regional scales, water infrastructures have evolved from isolated compartments (e.g., storage tanks) to integrated management systems that consider hydrological, technological, and social dimensions. Rainwater harvesting system (RWH) frameworks describe generalized models for the design, execution, and management of strategies for the collection of water, storing it, and reusing it. A sound rainwater harvesting framework consists of four basic elements, which are catchment assessment, system design, treatment, and utilization.

TABLE 1.3 RWH Actors and Multi-Level Governance Framework for Smart Rainwater Harvesting Systems

LEVEL	KEY ACTORS/STAKEHOLDERS	PRIMARY ROLES AND RESPONSIBILITIES	INTERACTION AND LINKAGES
Policy/Governance Level	• National and State Water Authorities • Urban Development Departments • Municipal Corporations • Policy Makers and Funding Agencies	• Formulate regulatory frameworks for RWH integration. • Allocate budgets and subsidies for SRWH deployment. • Set performance benchmarks and sustainability standards.	• Collaborate with research institutions and technology providers to design evidence-based policies. • Ensure policy alignment with national missions (e.g., Jal Jeevan Mission, Smart Cities Mission).
Technology/Institutional Level	• Engineers and Designers • IoT and AI Solution Providers • ICT Specialists and Data Analysts • Academic and Research Institutions	• Design and implement SRWH systems. • Develop and manage digital infrastructures (sensors, control systems, data platforms). • Analyze system performance and provide predictive analytics for decision-making.	• Coordinate with policymakers for technical feasibility and standards. • Share real-time data and reports with governance and community levels. • Collaborate with local agencies for field deployment and maintenance.
Community/Grassroots Level	• Local Households and Farmers • NGOs and Self-Help Groups (SHGs) • Village Water Committees and User Associations • Local Maintenance Workers	• Operate and maintain RWH units. • Participate in water monitoring, feedback, and awareness programs. • Adopt sustainable water-use behaviors and report faults or irregularities.	• Provide ground-level data and experiential feedback to institutional and policy levels. • Receive training and capacity-building from higher levels. • Ensure inclusivity and local ownership for long-term success.

1.6.2.1 IoT-Enabled SRWH Systems

A modern-day RWH system also caters to a wide variety of sophisticated features, including IoT-based sensors, automated valves, and data-driven control systems to enable continuous monitoring and dynamic management. These devices enhance the efficiency of harvesting and reduce loss in evaporation/leakage and predict maintenance. In addition, geospatial mapping, hydrological modeling, and real-time analytics lead to the discovery of appropriate RWH locations, prediction of rain patterns, and evaluation for long-term strategies by the policymakers and engineers for this task. From around the world, a number of good practices exist, highlighting the need for community participation, digital governance, and eco-design to effectively implement RWH. For example, Singapore's "Active, Beautiful, Clean Waters" program integrates RWH in urban planning and design, while India's "Jal Shakti Mission" is centered on decentralized rainwater harvesting and recharge models powered by local data networks. The application of green infrastructure, such as bioswales, permeable pavements, and living roofs, enhances water filtration and ecosystem restoration. The successful practices demonstrate that the effectiveness of RWH is not solely determined by technology but also by institutional cooperation, societal involvement, and sustaining policies. Consequently, futuristic frameworks need to seamlessly combine traditional wisdom with digital innovation such that SRWH systems developed through future frameworks are sustainable, inclusive, and are capable of adapting to diverse climatic and socioeconomic conditions.

1.6.2.2 Standard Frameworks for RWH Infrastructure

People look at water management from a lot of different angles, and honestly, each approach brings something unique to the table. When it comes to managing RWH in big, complex water systems, there's no single answer. You need a mix of integrated hydrodynamic modeling, digital monitoring, sustainability checks, and solid water governance. All of this together paves the way for water management that's not just efficient, but also fair. Take ISO 24512:2007 and ISO 24518:2015, for example. These standards push for RWH solutions that are both tough and adaptable, fitting them into the bigger picture of water supply and emergency management. Then you've got frameworks from UNESCO's International Hydrological Programme and the Global Water Partnership. They make sure local rainwater harvesting projects don't just sit in isolation—they tie them into larger national policies through integrated water resources management. What's really important here is how these global standards work together. They make it easier for different systems to talk to each other, keep things sustainable, and ensure everything's transparent—from the first design sketches to daily operations and monitoring. All these efforts help move rainwater harvesting out of small, scattered setups and turn it into smart, city-wide infrastructure. Smart water systems are the future, and these approaches are laying the groundwork. Smart water management systems such as the European Water Framework Directive (WFD) and ISO 37101:2016 (Sustainable Development in Communities) encourage the utilization of cyber-physical systems, IoT-based data analytics, and digital twins as enablers to enhance efficiency, traceability, and participation of communities. These systems support a system-level approach to physical infrastructure that incorporates digital information to provide continuous feedback and performance optimization. In India, some national initiatives like the Jal Jeevan Mission, AMRUT 2.0, and Smart Cities Mission are beginning to conform to international norms by bringing together IoT monitoring, GIS-based mapping, and community-driven data sharing into RWH design. Common RWH infrastructure frameworks therefore integrate engineering practice and regulatory compliance and enable innovation and a globally harmonized, yet locally adaptable, approach for sustainable water security in the context of digital transformation.

1.6.2.3 Key Components and Controls

An SRWH system relies on integrating physical infrastructure with digital intelligence and automated control systems. The system consists of basic physical elements such as the catchment surfaces, conveyance pipes, filtering units, and storage reservoirs, which are enhanced by the application of IoT-enabled sensors, automated valves, and smart pumps. These sensors continuously keep track of rainfall, water level, quality indicators (pH, turbidity, conductivity), and flow rate, and the gathered information is sent as a stream in real-time to microcontrollers or cloud-based platforms for further processing. This information is processed by embedded AI and ML algorithms which, in turn, optimize the water harvest, predict maintenance, and automatically control flow. Integrated control systems (e.g., SCADA dashboards, mobile applications) provide an additional layer of remote monitoring and allow for energy-efficient operation with a high level of security. Collectively, when combined together, all these components and controls constitute a cyber-physical system where data-driven automation, human supervision, and environmental feedback ensure sustainable, efficient, and adaptive water management.

1.6.2.4 RWH for Utilities

For a long time, water utilities relied on big, centralized systems—think groundwater wells, surface reservoirs, and sprawling networks of pipes. But cities keep growing, the climate keeps shifting, and water is getting harder to come by. The old ways just aren't cutting it anymore. That's where RWH steps in. It's a simple, nature-based fix: collect rain where you need it, right at the source. This cuts down on how much we depend on outside water supplies and eases the load on city sewers, especially during heavy storms. Utilities can use harvested rainwater in all sorts of ways: for watering lawns and gardens, cooling machinery in factories, and even topping up underground aquifers. Some homes use it for non-drinking purposes, like flushing toilets or washing cars. When utilities build RWH into their systems, they help create a circular water economy. That means water gets used, reused, and cycled back, which makes cities tougher and more resilient when the unexpected hits.

Tech is making this even smarter. With IoT sensors, AI-powered maintenance, and GIS tools, utilities can keep an eye on rainwater tanks, track water quality, and make sure distribution runs smoothly—all in real time. They can even link up these systems with the main city supply, so rainwater steps in when treated water runs low. But for RWH to really take off, utilities need solid policies, some funding help, and buy-in from the community. When rainwater harvesting becomes part of how cities manage water, it's not just about sustainability. It's a chance to get people involved, build awareness, and move everyone closer to real water security.

1.6.2.5 Awareness and Training

At the end of the day, it's really people's knowledge, skills, and willingness to get involved that make rainwater harvesting work—not just better technology or new infrastructure. When folks actually understand how RWH affects the environment, economy, and their own lives, they start changing how they use water. That's where awareness campaigns and training come in. These programs lay the groundwork for real, lasting change. Training gives technical staff, local governments, and everyday users the tools they need to design, run, and manage efficient systems. They learn to think about details like local water quality, how to build storage, and how the seasons shape water needs. Public education matters, too. When schools, universities, and people living in rural areas get involved, it sparks a mindset that values saving water and caring for resources. To push these efforts further, governments can add digital courses, hands-on demo projects, and certification programs. These extra steps help drive the point home and make the benefits of RWH even more real for communities.

1.6.2.6 Key Focus Areas for Awareness and Training

- Community Participation: Get people involved through campaigns, workshops, and hands-on local projects. The idea is simple—when folks take part, they're more likely to care about water conservation and stick with it for the long haul. That's how you build real social and environmental resilience.
- Capacity Building and Technical Skills: Offer practical, hands-on training for engineers, technicians, and community workers. Focus on everything from designing to installing and maintaining RWH systems. Throw in some ICT, IoT, and GIS-based learning tools so people can manage and monitor these systems the smart way.
- Policy and Governance Awareness: Make sure everyone knows about the government programs, legal rules, and incentives that support RWH. When stakeholders understand what's out there, they're more likely to get on board.

1.7 CASE STUDIES FOR RWH AND WATER SYSTEMS

RWH technology really stands out as a game changer for water sustainability, especially in places facing water scarcity, unpredictable rainfall, or excessive groundwater depletion. You can find examples all over the world where clever RWH setups—think IoT sensors, GIS mapping, even AI-powered predictions—are shaking up how local communities manage water. Take Israel and Australia, for example. They've combined RWH with graywater recycling and digital sensors, and it's paid off. These systems make water use way more efficient, even in those tough, dry climates. These cases illustrate how RWH, when integrated with smart technologies and policy measures, shifts from a water conservation measure to become an instrument of resilient urban water management. Participatory RWH methodologies have been applied with similar success even in rural and semi-urban areas. In Table 1.4, include Singapore, Japan, United States, Sri Lanka, and South Africa.

1.7.1 RWH Projects in India

RWH has turned into a citywide movement in Bengaluru, propelled by policy as well as need, considering that urban growth had drained groundwater supplies. In 2009, the Bangalore Water Supply and Sewerage Board (BWSSB) made it compulsory to have RWH systems in all buildings measuring 60 × 40 feet or above. This initiative made Bengaluru a major center for Indian urban water conservation, with over 200,000 RWH installations in the residential, institutional, and industrial areas. For instance, the Association for People with Disability (APD) Campus at Kothanur, where a rooftop catchment and groundwater recharge system integrated with a rainwater harvesting system, has significantly reduced the dependence on tankers and enhanced the local aquifer levels. The Indian Institute of Science (IISc) and the Infosys campus, both in Electronic City, have also incorporated SRWH systems that utilize IoT-based sensors to gauge water levels, flow rates, and quality to optimize storage and usage. These case studies demonstrate how Bengaluru's proactive policies, citizen engagement, and technological integration can serve as a model for other Indian cities to develop sustainable and intelligent urban water management. This section provides a few more examples. SRWH projects have been successfully implemented in the Indian scenario through policy interventions and techno-innovation. Some representative Indian case studies and their outcomes are outlined in Table 1.5.

1.7.2 RWH Projects in Arid and Semi-Arid Regions

RWH in dry semi-dry zones like Rajasthan and Gujarat have become lifesavers for the city dwellers and rural populations witnessing acute water shortages and erratic monsoons. These areas—once dependent on traditional arrangements such as the Johads, Kunds, Baoris, and Tankas—are transforming through the resurgence of indigenous techniques, aided by scientific mapping, GIS-based hydrological planning, and low-cost engineering solutions. The Tarun Bharat Sangh in Rajasthan, under Rajendra Singh—the "Waterman of India"—revived over 8,000 Johads, in Alwar district alone, and thereby rejuvenated five dried-up rivers and recharged groundwater in over 1,000 villages is nothing short of miraculous. Correspondingly, rooftop RWH systems and underground "Kund" tanks supported by local NGOs have ensured water availability for drinking all year round in Jaisalmer and Barmer, largely eradicating the pressing need to carry water over long distances. In Gujarat, RWH measures have been promoted in agricultural and urban water supply systems to address salinization and water table depletion. The Bhuj City Water Resilience Project is a case in point, with decentralized RWH structures check dams constructed under the Kutch Jal Sankalp programme to recharge aquifers and stabilize the city's water supply. Mission KhushalJivan performed percolation tanks and farm ponds in the semi-arid Saurashtra region, leveraging

TABLE 1.4 International Case Studies of Smart Rainwater Harvesting Systems and Their Outcomes

COUNTRY	PROJECT/LOCATION	TECHNOLOGY/APPROACH	KEY FEATURES	OUTCOMES/IMPACT
Australia	Sydney Water Harvesting Initiative	Urban rooftop collection with IoT monitoring	Integrated stormwater and rainwater system for public facilities	Reduced urban flooding; enhanced water reuse efficiency by 35%
Israel	Negev Desert RWH System	Catchment-based micro-reservoirs with digital flow sensors	Smart irrigation and predictive modeling for arid conditions	Increased agricultural yield by 25%; improved water availability year-round
Singapore	Marina Barrage & ABC Waters Programme	SRWH integrated with urban landscape design	Combines stormwater collection, filtration, and IoT-based water quality monitoring	Enhanced city's freshwater supply by 10%; improved flood control
Japan	Fukuoka City Rainwater Utilization Project	Building-integrated storage with automated control valves	Recycled rainwater used for toilets and cooling systems	Reduced municipal water demand by 15%; promotes eco-friendly urban planning
United States	Tucson, Arizona RWH Model	Rooftop and landscape-based harvesting using smart controllers	AI-based control for irrigation scheduling in arid regions	Saved ~30% irrigation water annually; community-scale resilience improved
Sri Lanka	Dry Zone Household Tank System	Rooftop RWH tanks with IoT leak detectors	Designed for drinking and domestic water storage	Ensured potable water security in drought seasons; low-cost model replicated regionally
South Africa	eThekwini Municipality Smart RWH	Public school installations with cloud-based monitoring	Community-level awareness programs linked with real-time dashboards	Improved school water access; enhanced public participation in water conservation
India	Urban and rural SRWH deployments under Jal Jeevan Mission, AMRUT, and Smart Cities Mission	IoT-enabled RWH integrated with AI analytics, smart sensors, and cloud-based control systems	Real-time water level/quality monitoring, automated valve control, rainfall-driven storage optimization, leak and contamination detection	Enhanced groundwater recharge, reduced urban flooding, improved system reliability, lower dependence on centralized supply, and strengthened climate-resilient water governance

geospatial information to maximize rainwater collection for irrigation. These projects demonstrate that combining traditional knowledge, modern hydrological science, and community involvement could lead to sustainable water solutions even in the most unforgiving environments. The success of both Rajasthan

TABLE 1.5 Smart Rainwater Harvesting Projects in India and Their Outcomes

PROJECT/LOCATION	TECHNOLOGY/APPROACH	KEY FEATURES	OUTCOMES/IMPACT
Chennai SRWH Initiative, Tamil Nadu	Rooftop RWH with IoT-based level and quality sensors	Automated monitoring of storage tanks; integration with municipal recharge wells	Improved groundwater levels; reduced urban flooding during monsoon seasons
Indore Smart Water Management Program, Madhya Pradesh	Building-scale RWH integrated with smart meters	Real-time data on rainwater capture and usage, linked with smart city dashboard	Reduced dependence on municipal water supply; enhanced urban water efficiency
Auroville Rainwater Harvesting Systems, Tamil Nadu	Decentralized catchment-based RWH with digital flow monitoring	Community-managed storage and recharge structures; data-supported water budgeting	Year-round water availability; improved resilience during drought periods
Jaipur Urban RWH Projects, Rajasthan	Rooftop and surface runoff harvesting with automated control valves	Designed for arid climate; optimized storage and recharge using sensor feedback	Reduced groundwater depletion; improved water availability in dry seasons
IIT Madras Smart Campus RWH, Tamil Nadu	Campus-wide RWH integrated with IoT sensors and analytics	Monitoring of rainfall, storage, and recharge; decision-support for water reuse	Achieved near self-sufficiency in non-potable water demand
Jal Jeevan Mission RWH Pilots, Multiple States	Household and community-level RWH with digital monitoring	Linked with rural water supply schemes; focus on sustainability and resilience	Improved drinking water security; reduced seasonal water stress in rural areas
Telangana Mission Kakatiya (Digitally Assisted Phase)	Tank restoration combined with GIS and monitoring tools	Data-driven restoration and performance assessment of traditional tanks	Enhanced surface water storage; revival of traditional water systems

and Gujarat serves as a precedent for how RWH supported by digital monitoring technologies and policy interventions can convert arid regions into water-secure ecosystems.

1.7.3 RWH in Jal Jeevan Mission in India

Rainwater harvesting is quickly becoming a core part of India's Jal Jeevan Mission, the government's push since 2019 to get working tap water into every rural home. It's not just about laying pipes—people running the program know that piped water won't last unless local sources get a boost. That's why the mission puts a big focus on things like rainwater harvesting, recharging groundwater, and managing greywater. To keep water flowing during dry spells, states like Maharashtra, Telangana, and Rajasthan have rolled out their own village-level rainwater harvesting plans, following the mission's guidelines. But it's not all smooth sailing. Plenty of places still struggle with getting these water systems running well and keeping them fixed. Tech is starting to make a real difference. With IoT sensors, automated data loggers, and dashboards that let people keep an eye on things remotely, these harvesting systems are better maintained and tracked. Officials can now check water quality and quantity in real time. One standout innovation is the use of blockchain. Under the Jal Jeevan Mission, rural water projects in places like Maharashtra and Telangana use blockchain-based systems to track how money gets spent, confirm installations, and monitor performance—all live. It's a way to limit corruption, hold people accountable, and make sure subsidies actually go to the folks who need them. Local communities aren't left out, either. Water committees—Pani Samitis—are getting trained to run these harvesting setups, keep recharge pits in shape, and connect older catchment systems to people's homes. Bringing together traditional rainwater harvesting and modern digital tools is changing the game in rural water management. This new approach helps build water systems that are more transparent, better monitored, and ready to handle a changing climate—all while relying on local know-how.

1.8 EMERGING TECHNOLOGIES FOR SRWH INFRASTRUCTURE

Emerging technologies are revolutionizing RWH systems by offering sophisticated, high-performance, and security-centric solutions that have the potential to strengthen aged water infrastructures and allow for smart, automated functioning.

1.8.1 Blockchain for Security and Integrity

With decentralization, immutability, and cryptographic verification, blockchain acts as a rock-solid, tamper-proof data hub for SRWH. In traditional water management, you always run into fraud, mistakes, or even cyberattacks—whether you're tracking water levels, flow rates, subsidy payments, or maintenance schedules [60]. Blockchain wipes out those worries. Every sensor reading, every payment, every schedule update—everything lands on a distributed ledger that no one can fudge after the fact [61]. That means everyone involved—governments, utility companies, engineers, even the local community—gets information they can actually trust. It's not just about transparency; it's about proof that things happened, written into the blockchain [62].

Then there's the power of smart contracts. These aren't just lines of code; they handle the routine stuff automatically: releasing funds only when a new RWH unit is installed and verified, sending out alerts as soon as tank levels get too low, updating maintenance schedules right when the sensors say it's time. No more long waits, no more middlemen, way less room for fake reports—exactly the headaches that used to drag down big, spread-out water projects. This kind of automation and accountability just wasn't possible before, and it's changing the game for RWH systems. Subsidy monitoring through blockchain in Maharashtra and Telangana under the Jal Jeevan Mission are examples of how blockchain can bring transparency to the allocation of resources and the manner of working. When IoT-enabled sensors are combined with blockchains, an end-to-end secure digital ecosystem emerges that can provide for the sustainable long term, enhancing governance and providing fair access to water in both urban and rural areas.

1.8.2 IoT-Enabled Leak Detection in Rainwater Harvesting Systems

SRWH technology plays an important role in sustainable water management; yet, poor efficiency of RWH systems due to unnoticed leakages in tanks, pipes, and fittings is common. IoT-based solutions for leak detection offer a smart and automated mechanism that employs sensors such as flow meters, water level sensors, rain sensors, and soil moisture sensors. These sensors perform continuous monitoring of the system parameters and send real-time data to a cloud or edge platform. Unnatural non-rain flow, unexpected

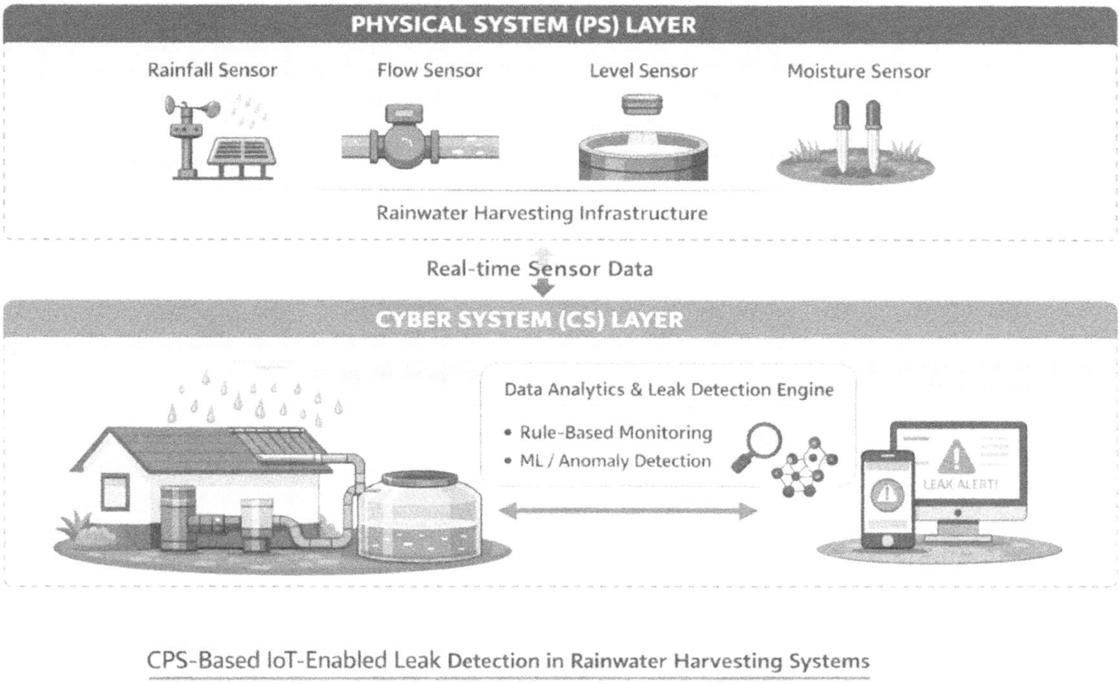

FIGURE 1.9 Cyber–physical architecture of an SRWH system with leak detection.

tank level reductions, or moisture increases around pipes could be signs of leakage. Data analytics and ML-based anomaly detection techniques contribute to improved detection accuracy by separating leak events from normal operating variations. IoT-based system monitoring can facilitate fault detection at an early stage, reduce water loss, and lower maintenance costs. Although dealing with limitations associated with sensor installation and power consumption, the integration of IoT technology for leak detection within RWH systems offers a promising approach to enhance the reliability, efficiency, and sustainability of rainwater harvesting systems. Smart Rainwater Harvesting systems also need the right type of implementation framework encompassing technology, policy, capacity building, and governance for them to be scaled up effectively. Figure 1.9 illustrates the cyber–physical structure of a smart rainwater harvesting system, where real-time sensors in the physical layer are processed in the cyber layer for analysis for the purpose of monitoring and leak detection monitoring and leak detection [63–66].

1.8.3 Digital Twins

Digital twin computing is revolutionizing the design, operation, and performance analysis of water infrastructures in general, and SRWH systems in particular. A digital twin is a real-time, virtual water system constructed by integrating data collected from SCADA systems, IoT-based sensors, telemetry units, and historical data [67]. The virtual simulation allows engineers and decision makers to simulate, control strategy tests, optimize flow management, and predict system performance in rainfall, demand, or climate change scenarios without risk or interruption to the actual RWH system. Since these simulations can effectively mimic the real world, digital twins have also become powerful tools for pre-deployment testing, operational tuning, and strategic management of water storage, recharge, distribution, and contingency planning systems [68]. In addition, digital twins are significant in cybersecurity, risk assessment, and resilience planning for smart water systems. Operators may conduct penetration tests, discover vulnerabilities, assess sensor failures, or model cyber-attacks on the prototypical infrastructure (simulated) to their heart's content without the slightest risk of perturbing the real one. Enabling policy and regulatory

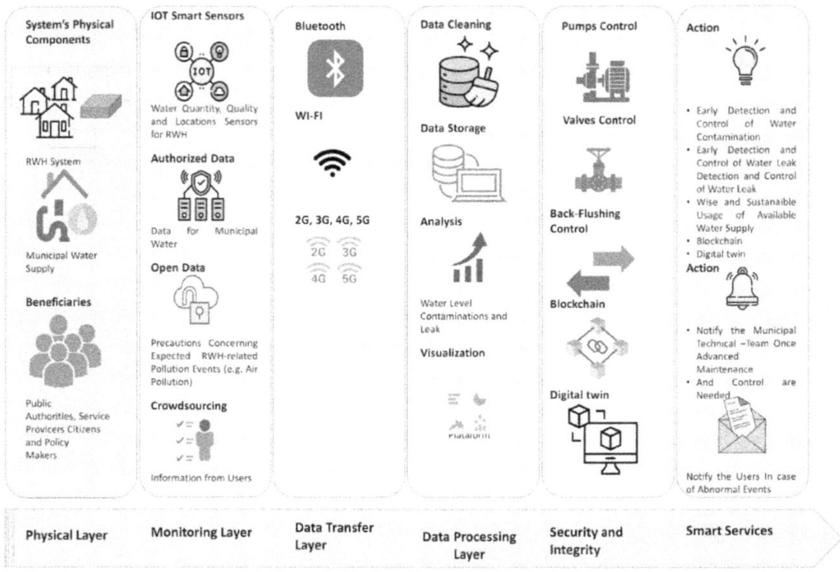

FIGURE 1.10 Layered design of emerging technologies for SRWH infrastructure.

frameworks are critical to the successful adoption of Smart Rainwater Harvesting technologies. Figure 1.10 shows an IoT-based smart rainwater harvesting and city water management system in a layered format, where sensing, data transmission, analytics, secure control, and smart decision-support systems are seamlessly integrated.

1.9 THEMATIC RELEVANCE OF RWH AND THEIR APPLICATIONS

Amid depleting freshwater supplies in the world, the RWH has emerged as a critical approach for enhancing climate resilience, sustainable water management, and water governance at a decentralized level. There has been a rise in water stress on a global scale due to the increase in urbanization, over-exploitation of groundwater, and erratic climate, which has brought RWH into the spotlight as a cost-effective, environmentally sound, and region-specific technology. RWH is directly aligned to SDG 6 (Clean Water and Sanitation) and contributes to achieving long-term water security in urban and rural contexts by enabling the harvesting and utilization of rainwater at source. Integrated with IoT sensors, AI-based analytics, and digital monitoring platforms, RWH solutions are now more than just storage. These are enabling automated distribution, predictive storage optimization, real-time water quality monitoring, and efficient system management. RWH also assists in achieving green building ratings, increases sustainability adherence, and reduces water footprints in commercial and industrial applications. Resilience in RWH infrastructure relies on effective preparedness, detection, response, and recovery processes. Preparations (hydrologic evaluation, catchment studies, training of stakeholders, maintenance of the system, contingency planning) during the preparation stage are known beforehand. Continuous monitoring with smart construction of the system enables the invigilation of leakage, contamination, overflow hazards, or system paralysis. Damage and water loss are minimized by automated containment through valve isolation, flow diversion, and alert systems. Recovery after the incident involves assessing damage, conducting root-cause analysis, recalibrating systems, and incorporating lessons learned to drive future improvements. In general, RWH is regarded as a pan climate-resilient, adaptable, socially inclusive water solution applicable for urban planning, agriculture, industry, and disaster management. Its integration with digital technologies makes RWH the foundation for next-generation and resilient water infrastructure.

1.10 POLICY, REGULATION, AND FUTURE OUTLOOK

The investments, operational management, and accountability of public and private rainwater harvesting (RWH) infrastructure are influenced by complex and dynamic regulatory systems at national, local, and sectoral levels shaped by national water policies, local regulations, environmental directives, and multilateral commitments to sustainability. Cities, utilities, developers, and community groups are coming together in new ways, thanks to a mix of policy and technology. We see mandatory rainwater harvesting, strict recharge rules, building codes, and subsidies all lining up with digital tools—think blockchain audits, IoT sensors keeping tabs on compliance, and real-time dashboards tracking performance. Sure, these systems bring transparency and make standards clearer. But the real shift happens locally, where all these actors connect and coordinate more smoothly than before. Looking ahead, the regulatory landscape doesn't stand still. New instruments and digital approaches keep shaping how we govern water. Predictive analytics, digital twins, and AI-powered systems are moving into the mainstream, giving planners sharper insights for water management and climate resilience. We'll see smarter policies—ones that demand smart meters, catch leaks automatically, and roll out integrated water-energy-climate reporting. There's also a push for nature-based solutions and circular water models, not just as buzzwords but as practical strategies. Put together, these trends point to a future where technology, sustainability, and active community involvement aren't just parallel tracks—they converge. This is how rainwater harvesting can become a cornerstone for safe, productive, and climate-resilient water systems.

1.10.1 Global Policy Frameworks

The development of advanced RWH technologies has led to a series of policy interventions at national and international levels to promote water security, operational resilience, and climate adaptation in critical water infrastructure. These policies highlight the convergence of digital (IoT sensing, real-time monitoring platform, data-driven decision system) with traditional RWH to enhance efficiency, transparency, and sustainability of RWH practices. Global frameworks such as the UN Sustainable Development Goals (SDG-6: Clean water and sanitation), Sendai Framework for Disaster Risk Reduction, and Paris Agreement on Climate Change require countries to promote nature-based and technology-enabled water management solutions. Several national standards and policies for RWH are in place, such as Australia's Water Efficiency Labelling Standards (WELS), Singapore's PUB Water Sustainability Guidelines, India's Jal Shakti smart water directives, and Japan's urban water resilience regulations, among others that allow for RWH system integration, smart metering, and leakage management. Together, they are key enablers for the transition towards resilient and digitally managed water systems that protect communities from the impacts of climate variability, groundwater over-extraction, and infrastructure failure. Table 1.6 provides an overview of both national and international rainwater harvesting and smart water governance promoting policies, guidelines, and institutional mechanisms.

1.10.2 Indian Context

The Indian RWH sector is also experiencing a digital transformation with increasing penetration of digital technologies and smart water management solutions. The combination of IoT-enabled flow meters, ultrasonic level sensors, SCADA-based water distribution systems, and RTUs is changing how urban

TABLE 1.6 Comparative Analysis of Traditional Rainwater Harvesting and AIoT-Driven SRWH Systems

PARAMETER	TRADITIONAL RWH	AIOT-DRIVEN RWH
Monitoring Method	Manual inspection of tanks, gutters, and recharge pits; irregular monitoring schedules	Real-time monitoring via IoT sensors, ultrasonic level indicators, smart flow meters, and cloud dashboards
Water Quality Assessment	Periodic manual sampling; results obtained after lab analysis	Continuous sensing of pH, turbidity, conductivity, microbial load with instant alerts
Operational Control	Manual valve operations; human-dependent flow adjustments	Automated valves, SCADA-based pump control, AI-optimized flow regulation
Leak Detection	Detected only after visible water loss or reduced harvesting efficiency	AI-driven leak analytics, pressure sensors, ultrasonic leak detection, predictive fault identification
Data Availability	Minimal or no recorded data; difficult to analyze trends	High-resolution datasets stored on cloud; time-series analytics and digital twin integration
Forecasting Capability	No mechanism for predicting rainfall or demand	AI-based rainfall forecasting, consumption prediction, and storage optimization
Maintenance Approach	Reactive maintenance after clogging or malfunction	Predictive maintenance triggered by AI anomaly detection and sensor alerts
Scalability	Limited to individual buildings or small communities	Scalable to campuses, industries, cities with centralized monitoring
User Interaction	Community participation; manual cleaning and oversight	Smartphones, automated notifications, dashboards for utilities and households
Energy Consumption	Generally low, but inefficiencies arise due to manual pump control	Energy-optimized operations using AI to reduce pump runtime and wastage
Resilience to Drought	Low adaptability; storage often insufficient	AI-enhanced planning improves drought resilience and water rationing strategies
Cybersecurity Needs	Minimal; mostly offline systems	Requires cybersecurity protocols (firewalls, encryption, intrusion detection)
Integration with Other Systems	Usually standalone	Integrates with GIS, weather stations, smart water grids, and digital twins
Cost (CAPEX/OPEX)	Low initial cost but higher long-term inefficiencies	Higher initial investment but reduced life-cycle cost due to automation and optimization
Environmental Impact	Beneficial but limited by operational inefficiencies	Maximizes recharge, minimizes wastage, supports sustainable water management

municipalities, rural panchayats, and private homes utilize harvested rainwater. Influenced by government schemes such as the Jal Jeevan Mission, Atal Mission for Rejuvenation and Urban Transformation (AMRUT), and several Smart City projects, digital RWH elements are finding a place in pan water security blueprints. These digital systems allow monitoring of tank levels in real time, rain patterns, predictive analytics, leak notifications, and automatic re-routing to recharge pits, as well as guarantees of transparency and operational efficiency. Simultaneously, state-level regulations in Karnataka, Tamil Nadu, Rajasthan, and Maharashtra have started requiring digital compliance submissions for high-rise buildings with RWH systems. Together, these trends represent a transformative shift for India from conventional rainwater harvesting (RWH) infrastructures to smart, accountable, and climate-resilient RWH networks that can bear the brunt of urbanization, groundwater depletion, and climate variability. SRWH systems have an impact on at least six Sustainable Development Goals (SDGs) and are in accordance with the concepts of India's National Education Policy (NEP) 2020 through education, innovation, and community participation focused on sustainability. This pairing is systematically outlined in Table 1.7.

TABLE 1.7 Global Policy and Regulatory Frameworks Supporting Rainwater Harvesting and Smart Water Management

REGION	KEY REGULATION/POLICY FRAMEWORK	FOCUS AREAS	REMARKS
EU (European Union)	EU Water Framework Directive (2000/60/EC); Urban Water Management Guidelines; Sustainable Urban Drainage Systems (SUDS) standards.	Urban stormwater reuse, flood management, green infrastructure, sustainable drainage, climate adaptation.	Strong emphasis on environmental compliance; RWH integrated with green roofs and SUDS; high policy maturity across Western Europe.
Asia (General)	National water security acts (varies by country), Smart City Water Missions, Urban rainwater mandates in Japan, Singapore, and China.	Urban water reuse, disaster resilience, groundwater recharge, digital water technologies.	Rapid adoption of smart sensors, AI-based water grids; policies often linked to climate resilience and scarcity management.
India	National Water Policy (2012), Jal Shakti Abhiyan, Building by-laws requiring RWH (varies by state), Jal Jeevan Mission.	Mandatory RWH for buildings, groundwater recharge, rural water sustainability, source sustainability, community participation.	Strong regulatory push; implementation varies by state; integration of digital tools (IoT, blockchain) in select regions.
Sri Lanka/ Bangladesh	Sri Lanka National Rainwater Policy (2005), Bangladesh National Water Policy (1999), Coastal Zone Policies.	Household RWH tanks, flood control, drinking-water security in cyclone-prone/ coastal zones, rural water supply.	Widely used in rural and coastal regions; driven by vulnerability to climate extremes; policies emphasize low-cost, community-based systems.

1.11 CONCLUSION

Rainwater harvesting, as a purely traditional and static conservation practice, has progressively evolved into a dynamic and strategic component of modern water infrastructure. Conventional harvesting systems remain environmentally sustainable and cost-effective over long periods; however, their effectiveness is increasingly constrained by manual operation requirements, limited monitoring capabilities, and low adaptability to climate variability, rapid urban expansion, and rising water demand. These challenges are especially critical in arid and semi-arid regions, where uncertain rainfall patterns and intensive groundwater extraction demand more adaptive and resilient water management approaches.

In response, SRWH have developed through the integration of digital technologies such as the Internet of Things (IoT), artificial intelligence, machine learning, cloud-based platforms, and cyber-physical systems. This technological framework enables automated control of storage and distribution processes, predictive maintenance strategies, data-driven optimization of harvesting and recharge practices, and continuous monitoring of both water quantity and quality. Advanced tools, including digital twins, further enhance system performance by allowing virtual scenario testing and risk-informed operational planning, while IoT-enabled leakage detection substantially reduces distribution losses. At the same time, blockchain-based solutions contribute to data reliability by ensuring transparency, traceability, and secure information exchange, thereby strengthening governance and accountability mechanisms. Beyond technological advancement, the successful implementation of smart rainwater harvesting also depends on socio-institutional conditions. Digital capacity building, active community involvement, and coordinated governance frameworks are essential to guarantee long-term functionality, operation, and maintenance. Experiences from international and Indian case studies indicate that the most effective outcomes are achieved when technological innovation is combined with inclusive policy frameworks, stakeholder participation, and sustained institutional support. Within this context, national and global initiatives, including India's Jal Jeevan Mission, AMRUT, the Smart Cities Mission, and the United Nations Sustainable Development Goals, have played a pivotal role in promoting smart rainwater harvesting as a decentralized, climate-resilient strategy that contributes to urban sustainability, rural development, and long-term water security.

1.12 FUTURE SCOPE

Future research should prioritize the integration of AI–based forecasting, digital twin frameworks, and blockchain-enabled governance within rainwater harvesting systems to enhance operational autonomy, transparency, and system resilience. The convergence of these technologies can support predictive decision-making processes, improve data traceability, and strengthen information management, thereby increasing the reliability and overall performance of smart harvesting infrastructures. In addition, forthcoming studies should emphasize the development of cost-effective and energy-efficient solutions tailored to arid and semi-arid regions, ensuring alignment with climate adaptation strategies and reinforcing the contribution of rainwater harvesting to long-term water security and the achievement of Sustainable Development Goal 6. Equally important is the expansion of community-centered initiatives, the integration of harvesting systems with smart urban water networks, and the deployment of real-time water quality monitoring technologies, which together can enable inclusive, scalable, and policy-oriented applications.

REFERENCES

1. Szołdrowska, D.; Włóka, D.; Smol, M. Circular Economy of Water: Rainwater Harvesting (RWH) and Reuse Solutions. *Desalin. Water Treat.*, **2025**, *323* (March). https://doi.org/10.1016/j.dwt.2025.101257.
2. Rahman, S.; Khan, M. T. R.; Akib, S.; Din, N. B. C.; Biswas, S. K.; Shirazi, S. M. Sustainability of Rainwater Harvesting System in Terms of Water Quality. *Sci. World J.*, **2014**, *2014* (1), 721357. https://doi.org/10.1155/2014/721357.
3. Kakoulas, D. A.; Golfinopoulos, S. K.; Koumparou, D.; Alexakis, D. E.; Kakoulas, D. A.; Golfinopoulos, S. K.; Koumparou, D.; Alexakis, D. E. The Effectiveness of Rainwater Harvesting Infrastructure in a Mediterranean Island. *Water,* **2022**, *14* (5). https://doi.org/10.3390/W14050716.
4. Pachouri, V.; Kothari, P.; Kathuria, S.; Gehlot, A.; Singh, R.; Thakur, A. K.; Gupta, L. R.; Dogra, S.; Priyadarshi, N.; Mohamed, H. G. Revolutionizing Urban Water Resilience: Innovative Strategies and Advancements in Sustainable Urban Drainage Systems (SuDS). *Desalin. Water Treat.*, **2025**, *323*, 101407. https://doi.org/10.1016/J.DWT.2025.101407.
5. Fathy, I.; Abd-Elaty, I.; Eltarabily, M. G.; Bali, K. M.; Abd-Elhamid, H. F.; Elmeghany, F. A.; Nosair, A. M.; Nassar, A.; Elbagory, I. Integrated Flood Mitigation and Groundwater Recharge Strategies for Sustainable Water Resource Management in Arid Regions. *Urban Water J.*, **2025**. https://doi.org/10.1080/1573062X.2025.2589067;SUBPAGE:STRING:ACCESS.
6. Muhammed, S. P.; Holla, B. R. K.; Balakrishna, K.; Warrier, A. K. Developing Rejuvenation Strategies and Artificial Flood Mitigation Plans for Indrani River: A Case Study. *Sci. Reports*, **2025**, *15* (1), 30344. https://doi.org/10.1038/s41598-025-12439-z.

7. Alao, J. O.; Eze, S. U.; Onyenweife, G. I.; Ibe, A. A.; Otorkpa, O. J.; Ayejoto, D. A.; Abubakar, F.; Abdulsalami, M.; Abdulmalik, D. O. Enhancing Water Security through Integrated Storage Mechanisms and Rainwater Harvesting for Sustainable Development. *Discov. Sustain.*, **2025**, *6* (1), 941. https://doi.org/10.1007/S43621-025-01914-2.
8. Raimondi, A.; Quinn, R.; Abhijith, G. R.; Becciu, G.; Ostfeld, A.; Raimondi, A.; Quinn, R.; Abhijith, G. R.; Becciu, G.; Ostfeld, A. Rainwater Harvesting and Treatment: State of the Art and Perspectives. *Water,* **2023**, *15* (8). https://doi.org/10.3390/W15081518.
9. Sharma, S.; Das, A.; Bhattacharya, S. P. Strengthening WASH Resilience in Flood-Affected Urban Poor Communities: Insights from Patna. *Cities*, **2025**, *159*, 105758. https://doi.org/10.1016/J.CITIES.2025.105758.
10. Kumar, R.; Mishra, A.; Goyal, M. K. Water Neutrality: Concept, Challenges, Policies, and Recommendations. *Groundw. Sustain. Dev.*, **2024**, *26*, 101306. https://doi.org/10.1016/J.GSD.2024.101306.
11. Das, A.; Mishra, S. Smart and Sustainable Water Management: Policy, Technology, and Innovation Pathways for Resource Resilience. *Green Technol. Resilience, Sustain.*, **2025**, *5* (1), 5. https://doi.org/10.1007/S44173-025-00022-8.
12. Home | Department of Water Resources, River Development and Ganga Rejuvenation. https://www.jalshakti-dowr.gov.in/ (accessed December 24, 2025).
13. Bansal, V. Integrated Rainwater Harvesting and Analytics Model for Sustainable Urban and Industrial Water Management in India. *Univers. Res. Reports*, **2025**, *12* (3), 31–49. https://doi.org/10.36676/URR.V12.I3.1567.
14. Lepcha, R.; Kumar Patra, S.; Ray, R.; Thapa, S.; Baral, D.; Saha, S. Rooftop Rainwater Harvesting a Solution to Water Scarcity: A Review. *Groundw. Sustain. Dev.*, **2024**, *26*, 101305. https://doi.org/10.1016/J.GSD.2024.101305.
15. Judeh, T.; Shahrour, I.; Comair, F. Smart Rainwater Harvesting for Sustainable Potable Water Supply in Arid and Semi-Arid Areas. *Sustain.,* **2022**, *14* (15). https://doi.org/10.3390/SU14159271.
16. Okoli, N. J.; Kabaso, B.; Okoli, N. J.; Kabaso, B. Building a Smart Water City: IoT Smart Water Technologies, Applications, and Future Directions. *Water,* **2024**, *16* (4). https://doi.org/10.3390/W16040557.
17. Abebe, A. Internet of Things (IoT) Enabled Water Distribution System for Smart Water Management. *Int. J. Wirel. Commun. Mob. Comput.*, **2024**, *11* (1), 1–10. https://doi.org/10.11648/J.WCMC.20241101.11.
18. Jamadarkhani, M.; Raphael, R.; Ramprasad, S. H. P.; Babu, H.; Narasimhan, S. IoT Enabled Smart Water Metering Using Multi Sensor Data and Machine Learning Techniques. *Front. Water*, **2025**, *7*, 1586916. https://doi.org/10.3389/FRWA.2025.1586916/BIBTEX.
19. Oberascher, M.; Rauch, W.; Sitzenfrei, R. Towards a Smart Water City: A Comprehensive Review of Applications, Data Requirements, and Communication Technologies for Integrated Management. *Sustain. Cities Soc.*, **2022**, *76*, 103442. https://doi.org/10.1016/J.SCS.2021.103442.
20. Bibri, S. E.; Huang, J. Artificial Intelligence of Things for Sustainable Smart City Brain and Digital Twin Systems: Pioneering Environmental Synergies between Real-Time Management and Predictive Planning. *Environ. Sci. Ecotechnology*, **2025**, *26*, 100591. https://doi.org/10.1016/J.ESE.2025.100591.
21. Ahmed, S. F.; Alam, M. S. Bin; Hoque, M.; Lameesa, A.; Afrin, S.; Farah, T.; Kabir, M.; Shafiullah, G. M.; Muyeen, S. M. Industrial Internet of Things Enabled Technologies, Challenges, and Future Directions. *Comput. Electr. Eng.*, **2023**, *110*, 108847. https://doi.org/10.1016/J.COMPELECENG.2023.108847.
22. Aldehim, G.; khan, S.; Shahzad, T.; Khan, M. A.; Ghadi, Y. Y.; Jiang, W.; Mazhar, T.; Hamam, H. Balancing Sustainability and Security: A Review of 5G and IoT in Smart Cities. *Digit. Commun. Networks*, **2025**. https://doi.org/10.1016/J.DCAN.2025.06.007.
23. Dada, M. A.; Tega Majemite, M.; Obaigbena, A.; Daraojimba, O. H.; Oliha, J. S.; Queen, Z.; Nwokediegwu, S. Review of Smart Water Management: IoT and AI in Water and Wastewater Treatment. *https://wjarr.com/sites/default/files/WJARR-2024-0171.pdf*, **2024**, *21* (1), 1373–1382. https://doi.org/10.30574/WJARR.2024.21.1.0171.
24. Tiwari, S.; Botre, B. A.; Sridhar, S.; Santos, C. M. Smart Water Grids in India: A Systematic Review of Purification, Conservation, and Emerging Digital Trends. *Proc. Indian Natl. Sci. Acad.*, **2025**, 1–21. https://doi.org/10.1007/S43538-025-00621-W.
25. Lasya, B.; Bhavana, Y.; Deekshitha, B.; Priya, B. K. An Innovative and Effective Electronic Based Automatic Rainwater Harvesting System. Proceedings of the 3rd International Conference on Smart Systems and Inventive Technology (ICSSIT), **2020**, 114–121. https://doi.org/10.1109/ICSSIT48917.2020.9214118.
26. Zaigham Abbas Naqvi, S. M.; Hussain, S.; Awais, M.; Tahir, M. N.; Saleem, S. R.; Al-Yarimi, F. A. M.; Ashurov, M.; Saidani, O.; Khan, M. I.; Wu, J.; et al. Climate-Resilient Water Management: Leveraging IoT and AI for Sustainable Agriculture. *Egypt. Informatics J.*, **2025**, *30*, 100691. https://doi.org/10.1016/J.EIJ.2025.100691.
27. Chatrabhuj; Meshram, K.; Mishra, U.; Omar, P. J. Integration of Remote Sensing Data and GIS Technologies in River Management System. *Discov. Geosci.*, **2024**, *2* (1), 67. https://doi.org/10.1007/S44288-024-00080-8.

28. Ansari, S. A.; Vidyarthi, V. K. Use of Internet of Things in Water Resources Applications: Challenges and Future Directions: A Critical Review. *Discov. Internet Things*, **2025**, *5* (1), 96. https://doi.org/10.1007/S43926-025-00193-7.
29. Expósito, A.; Díez Cebollero, E. How the Digital Revolution Is Reshaping Water Management and Policy: A Focus on Spain. *Util. Policy*, **2025**, *96*, 102020. https://doi.org/10.1016/J.JUP.2025.102020.
30. Forhad, H. M.; Uddin, M. R.; Chakrovorty, R. S.; Ruhul, A. M.; Faruk, H. M.; Kamruzzaman, S.; Sharmin, N.; Jamal, A. S. I. M.; Haque, M. M. U.; Morshed, A. M. IoT Based Real-Time Water Quality Monitoring System in Water Treatment Plants (WTPs). *Heliyon*, **2024**, *10* (23). https://doi.org/10.1016/j.heliyon.2024.e40746.
31. UFM. UFM WATER AGENDA: Digital Transformation for Water and Sustainable Development Final Report. **2024**, January, 1–29. https://ufmsecretariat.org/wp-content/uploads/2024/05/Digital-Transformation-for-Water-Sustainable-Development-in-the-Med_Jan2024.pdf
32. Karimi, A.; Khataee, A.; Vatanpour, V.; Safarpour, M. High-Flux PVDF Mixed Matrix Membranes Embedded with Size-Controlled ZIF-8 Nanoparticles. *Sep. Purif. Technol.*, **2019**, *229* (November 2018), 115838. https://doi.org/10.1016/j.seppur.2019.115838.
33. Dai, Y.; Huang, Z.; Khan, N.; Labbo, M. S.; Dai, Y.; Huang, Z.; Khan, N.; Labbo, M. S. Smart Water Management: Governance Innovation, Technological Integration, and Policy Pathways Toward Economic and Ecological Sustainability. *Water,* **2025**, *17* (13). https://doi.org/10.3390/W17131932.
34. Bhadauria, S. S.; Singh, A. Sustainable Water Management Practices in the Face of Climate Change. *Environ. Reports*, **2023**, *5* (2), 23–28. https://doi.org/10.51470/ER.2023.5.2.24.
35. Bañas, K.; Robles, M. E.; Maniquiz-Redillas, M.; Bañas, K.; Robles, M. E.; Maniquiz-Redillas, M. Stormwater Harvesting from Roof Catchments: A Review of Design, Efficiency, and Sustainability. *Water,* **2023**, *15* (9). https://doi.org/10.3390/W15091774.
36. Pandey, P. K.; Van Der Zaag, P.; Soupir, M. L.; Singh, V. P.; Pandey, P. K.; Soupir, M. L.; Van Der Zaag, P.; Singh, V. P. A New Model for Simulating Supplemental Irrigation and the Hydro-Economic Potential of a Rainwater Harvesting System in Humid Subtropical Climates. *Water Resour. Manag.*, **2013**, *27* (8), 3145–3164. https://doi.org/10.1007/S11269-013-0340-1.
37. Meera, V.; Ahammed, M. M. Water Quality of Rooftop Rainwater Harvesting Systems: A Review. *J. Water Supply Res. Technol.*, **2006**, *55* (4), 257–268. https://doi.org/10.2166/AQUA.2006.0010.
38. Zabidi, H. A.; Goh, H. W.; Chang, C. K.; Chan, N. W.; Zakaria, N. A. A Review of Roof and Pond Rainwater Harvesting Systems for Water Security: The Design, Performance and Way Forward. *Water,* **2020**, *12* (11), 1–22. https://doi.org/10.3390/W12113163.
39. Gupta, S.; Mishra, P. K.; Khare, D. Advancements and Challenges in Roof-Top Harvested Rainwater Filtration: A Review. *Environ. Process.*, **2025**, *12* (1), 7. https://doi.org/10.1007/S40710-025-00748-4.
40. Manikandan, M.; Saliha, B. B.; Narsimlu, B.; Prasad, J. V. N. S.; Baskar, K.; Sanjivkumar, V.; Manoharan, S.; Guru, G.; Chary, G. R.; Rao, K. V.; et al. Design of Rainwater Harvesting Pond for Runoff Storage and Utilization in Semi-Arid Vertisols. *Water,* **2025**, *17* (21). https://doi.org/10.3390/W17213034.
41. Wang, J.; Diao, Y.; Cao, S.; Wang, J.; Jia, J.; Guo, Y. Towards the Cost-Effective Design of Stormwater Infiltration Trenches: A Hybrid Model Integrating Cost–Benefit Analysis and an Analytical Stochastic Approach. *Environ. Sci. Water Res. Technol.*, **2024**, *10* (5), 1108–1121. https://doi.org/10.1039/D3EW00963G.
42. Freni, G.; Mannina, G. Long Term Efficiency Analysis of Infiltration Trenches Subjected to Clogging. *Green Energy Technol.*, **2019**, 181–187. https://doi.org/10.1007/978-3-319-99867-1_32.
43. Davis, E. Hydrology: Current Research Rainwater Harvesting Techniques for Sustainable Water Use. *Hydrol. Current Res.*, **2024**, *15*, 550. https://doi.org/10.37421/2157-7587.2024.15.550.
44. Mozzi, G.; Pavelic, P.; Alam, M. F.; Stefan, C.; Villholth, K. G. Hydrologic Assessment of Check Dam Performances in Semi-Arid Areas: A Case Study From Gujarat, India. *Front. Water*, **2021**, *3*, 628955. https://doi.org/10.3389/FRWA.2021.628955/BIBTEX.
45. Djuma, H.; Bruggeman, A.; Camera, C.; Eliades, M.; Kostarelos, K. The Impact of a Check Dam on Groundwater Recharge and Sedimentation in an Ephemeral Stream. *Water,* **2017**, *9* (10). https://doi.org/10.3390/W9100813.
46. Parimalarenganayaki, S.; Elango, L. Quantification of Groundwater Recharge and River Bed Clogging by Daily Water Level Measurements in a Check Dam. *Arab. J. Geosci.*, **2018**, *11* (8), 174. https://doi.org/10.1007/S12517-018-3511-9.
47. . S. P. R. A Review on Managed Aquifer Recharge By Check Dams: A Case Study Near Chennai, India. *Int. J. Res. Eng. Technol.*, **2013**, *2* (4), 416–423. https://doi.org/10.15623/ijret.2013.0204002.
48. Page, D.; Bekele, E.; Vanderzalm, J.; Sidhu, J.; Page, D.; Bekele, E.; Vanderzalm, J.; Sidhu, J. Managed Aquifer Recharge (MAR) in Sustainable Urban Water Management. *Water,* **2018**, *10* (3). https://doi.org/10.3390/W10030239.

49. Sunar, N. M.; Kosat, A. T.; Harun, H.; Hamid, N. H. A.; Khalid, A.; Zaidi, E.; Azhar, A. T. S. Development of Smart Kit Rainwater Harvesting System for Potable Water Purpose. *Int. J. Eng. Technol.*, **2018**, *7* (4.30), 338–341. https://doi.org/10.14419/IJET.V7I4.30.22304.
50. Biantoro, A. W.; Majid, R. B. A.; Vidayanti, D.; Hidayat, A.; Hidayat, S.; Citra, Z. Design and Development of Smart Rainwater Harvesting Systems in Green Buildings Using Hybrid Energy and IoT. *BIO Web Conf.*, **2025**, *189*. https://doi.org/10.1051/bioconf/202518903003.
51. Yang, T.-H.; Yang, S.-C.; Kao, H.-M.; Wu, M.-C.; Hsu, H.-M. Cyber-Physical-System-Based Smart Water System to Prevent Flood Hazards. *Smart Water*, **2018**, *3* (1), 1. https://doi.org/10.1186/S40713-018-0008-3.
52. Adedeji, K. B.; Hamam, Y. Cyber-Physical Systems for Water Supply Network Management: Basics, Challenges, and Roadmap. *Sustain*, **2020**, *12* (22), 1–30. https://doi.org/10.3390/SU12229555.
53. Oberascher, M.; Dastgir, A.; Li, J.; Hesarkazzazi, S.; Hajibabaei, M.; Rauch, W.; Sitzenfrei, R.; Oberascher, M.; Dastgir, A.; Li, J.; et al. Revealing the Challenges of Smart Rainwater Harvesting for Integrated and Digital Resilience of Urban Water Infrastructure. *Water*, **2021**, *13* (14). https://doi.org/10.3390/W13141902.
54. Infant, S. S.; Vickram, S.; Saravanan, A.; Mathan Muthu, C. M.; Yuarajan, D. Explainable Artificial Intelligence for Sustainable Urban Water Systems Engineering. *Results Eng.*, **2025**, *25*, 104349. https://doi.org/10.1016/J.RINENG.2025.104349.
55. Pierre, N.; Ishimwe Viviane, I. V.; Lambert, U.; Viviane, I.; Shadrack, I.; Erneste, B.; Schadrack, N.; Alexis, N.; Francois, K.; Theogene, H. AI Based Real-Time Weather Condition Prediction with Optimized Agricultural Resources. *Eur. J. Technol.*, **2023**, *7* (2), 36–49. https://doi.org/10.47672/EJT.1496.
56. na, P. Cyber-Physical System Design for Smart Water Distribution Networks. *Int. J. Res. Mod. Eng. Emerg. Technol.*, **2022**, *10* (11), 10. https://doi.org/10.63345/IJRMEET.ORG.V10.I11.2.
57. Tricomi, G.; Giacobbe, M.; Ficili, I.; Peditto, N.; Puliafito, A.; Tricomi, G.; Giacobbe, M.; Ficili, I.; Peditto, N.; Puliafito, A. Smart City as Cooperating Smart Areas: On the Way of Symbiotic Cyber–Physical Systems Environment. *Sensors*, **2024**, *24* (10). https://doi.org/10.3390/S24103108.
58. Awang, N.; Johari, A.; Safie, S. I.; Januddi, M. A. F. M. S.; Radzi, A. R. M.; Ismail, M. A.; Hassan, C. H. C.; Jalil, A. F. A.; Zulkifli, S. N.; Nordin, N. N. M. Introduction to Computational Automation in Water Quality Management. *Comput. Autom. Water Secur. Enhancing Water Qual. Manag.*, **2025**, 277–323. https://doi.org/10.1016/B978-0-443-33321-7.00001-9.
59. Parra-López, C.; Ben Abdallah, S.; Garcia-Garcia, G.; Hassoun, A.; Trollman, H.; Jagtap, S.; Gupta, S.; Aït-Kaddour, A.; Makmuang, S.; Carmona-Torres, C. Digital Technologies for Water Use and Management in Agriculture: Recent Applications and Future Outlook. *Agric. Water Manag.*, **2025**, *309*, 109347. https://doi.org/10.1016/J.AGWAT.2025.109347.
60. Banerjee, B.; Sen, U.; Pramanik, S.; Banerjee, A. Smart Water Governance with Blockchain Technology: Blockchain Water Management. *Converg. Blockchain, Internet Everything, Fed. Learn. Secur.*, **2025**, 311–344. https://doi.org/10.4018/979-8-3373-1424-2.CH011.
61. Xia, W.; Chen, X.; Song, C. A Framework of Blockchain Technology in Intelligent Water Management. *Front. Environ. Sci.*, **2022**, *10*, 909606. https://doi.org/10.3389/FENVS.2022.909606/BIBTEX.
62. Le Thuy, T.; Nguyen, M.-K.; Bui, T. D.; Phan, H.; Yen, H.; Hoai, N. T.; Vo, N.; Ngan, C.; Pradiprao Khedulkar, A.; Pham Van, D.; et al. Blockchain-Enabled Water Quality Monitoring: A Comprehensive Review of Digital Innovations and Challenges. **2025**. https://doi.org/10.20944/preprints202507.1849.v1.
63. Choudhary, P.; Botre, B. A.; Akbar, S. A. 1-D Convolution Neural Network Based Leak Detection, Location and Size Estimation in Smart Water Grid. *Urban Water J.*, **2023**, *20* (3), 341–351. https://doi.org/10.1080/1573062X.2022.2164732;WGROUP:STRING:PUBLICATION.
64. Choudhary, P.; Modi, A.; Botre, B. A.; Akbar, S. A. Leak Detection in Smart Water Distribution Network. *AIP Conf. Proc.*, **2021**, *2335* (1). https://doi.org/10.1063/5.0044005/1000709.
65. Nagaraj, A.; Kotamreddy, G. R.; Choudhary, P.; Katiyar, R.; Botre, B. A. Leak Detection in Smart Water Grids Using EPANET and Machine Learning Techniques. *IETE J. Educ.*, **2021**, *62* (2), 71–79. https://doi.org/10.1080/09747338.2021.1984317.
66. Porwal, S.; Vijay, M.; Jain, S. C.; Botre, B. A. A Classification Approach for Monitoring and Locating Leakages in a Smart Water Distribution Framework. *Commun. Comput. Inf. Sci.*, **2018**, *827*, 418–429. https://doi.org/10.1007/978-981-10-8657-1_32.
67. Ramos, H. M.; Kuriqi, A.; Besharat, M.; Creaco, E.; Tasca, E.; Coronado-Hernández, O. E.; Pienika, R.; Iglesias-Rey, P.; Ramos, H. M.; Kuriqi, A.; et al. Smart Water Grids and Digital Twin for the Management of System Efficiency in Water Distribution Networks. *Water*, **2023**, *15* (6). https://doi.org/10.3390/W15061129.
68. Bonilla, C. A.; Zanfei, A.; Brentan, B.; Montalvo, I.; Izquierdo, J. A Digital Twin of a Water Distribution System by Using Graph Convolutional Networks for Pump Speed-Based State Estimation. *Water*, **2022**, 14, 514. https://doi.org/10.3390/w14040514

Stormwater Management in Reducing Urban Flooding

2

Ravi Kumar Sandal, Saurav, Lokeshwer Sharma, and Saurav Kumar Singh

2.1 INTRODUCTION

Urban flooding is one of the most significant environmental problems in cities today. As cities grow, they replace natural landscapes with impervious surfaces (asphalt, concrete, and roof tiles) that prevent rainwater from reaching the soil and create excessive runoff that causes the city to flood [1]. Stormwater management refers to the planning, design, and implementation of systems to collect, convey, store, and treat stormwater. A well-managed stormwater management system not only avoids flooding but also improves water quality, recharges groundwater, and improves urban ecology. Urban flooding is disruptive to transportation systems, irreparable for property and property damage, and a source of waterborne diseases and economic losses. Therefore, integrated stormwater management is essential to building resilient and sustainable urban cities. The rate of urban flooding has been increasing due to rapid suburbanization and shifts in natural drainage systems. As vegetation and open land are replaced by hard surfaces such as concrete, the ground infiltration capacity dramatically decreases. This results in higher runoff quantities and shorter times to peak flows [2]. Unplanned development in metropolitan areas also results in the development of natural water resources and floodplains that limit the area available for storage and flow of floodwater. Thus, even moderate rainfall events can cause severe flooding in metropolitan areas. From a hydrological point of view, the balance between rainfall, infiltration, evapotranspiration, and runoff is disturbed in urban landscapes. The runoff coefficient (which is a measure of the proportion of rainwater that can become surface runoff) increases significantly in urban areas compared to natural ones. Such a hydrological change not only elevates flood risk but also reduces groundwater recharge, thus causing water shortages in the future. Global warming, however, exacerbates the issue by increasing the strength and frequency of rainfall events and also the length of time during which they can happen. Stormwater management therefore requires a shift in approach from traditional drainage-focused designs to integrated and sustainable systems focusing on infiltration, retention, and reuse.

DOI: 10.1201/9781003742227-2

2.2 URBAN HYDROLOGY AND STORMWATER GENERATION

The hydrological cycle is the movement of water between the atmosphere, land, and oceans. Rainwater falls in natural surroundings (rainwater that falls on the ground reaches local aquifers and recharges them) and then slowly flows into rivers and lakes. Urbanization pushes against this flow, so [3] rainwater in cities cannot pass through land or pavement, leading to rapid surface runoff, which increases the volume and velocity of stormwater reaching drainage systems. Flooding becomes more frequent and possibly worse in even moderate rainfall events.

In addition, the change in the vegetation cover alters evapotranspiration, further decreasing rainfall water absorption (the role of vegetation in drought simulations). In addition to limiting the natural rainfall water absorption to sites within a certain size range (e.g., an urban area at a small city, an urban area closer to the urban center), buildings, roads, and parking lots cluster local [4] stormwater into a "heat island effect" that changes local weather conditions and leads to more frequent and severe recurrence of rainfall. As well, many pollutants such as oils, heavy metals, sediment, and organic matter accumulate in urban stormwater runoff. When discharged into a river, these contaminants can enter the water table and damage both flood protection and natural and environmental quality.

Figure 2.1 compares the natural and urban hydrological cycles, illustrating how urbanization alters infiltration, runoff generation, and stormwater movement. The most vital factors that impact the production of stormwater are:

- **Impenetrable Area**: The higher number of paved areas lead to less infiltration.
- **Landform**: The steeper the slope, the higher the velocity of the runoff.
- **Precipitation Intensity**: Short, but heavy rain episodes [5] create large volumes of runoff.
- **Soil Properties**: Soils made of clay have low permeability.

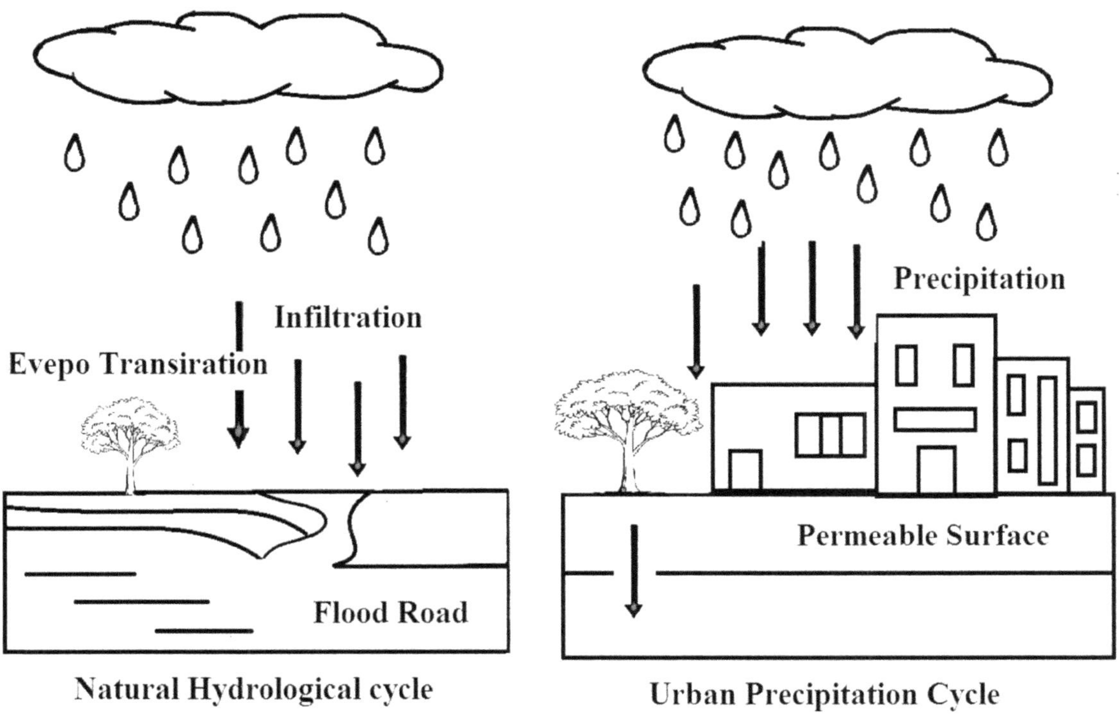

FIGURE 2.1 Natural vs. urban hydrological cycle.

Successful stormwater control through various means attempts to regain the lost equilibrium by imitating natural water cycle processes, making infiltration, storage, and vaporization of rainwater more efficient.

2.3 CAUSES AND IMPACTS OF URBAN FLOODING

Urban flooding happens when there is too much rainfall to contain the stormwater drainage system. There are several related factors that cause this problem, and these often happen at [6] the same time to make flooding more extreme and often more frequent.

2.3.1 Causes of Urban Flooding

Corrupt Drainage Systems: Aged or undersized drains do not have the capacity to control the amounts of rainfall that are occurring today. In many older cities, drainage systems were built decades ago to handle smaller populations and lower runoff rates. As cities continue to expand, these systems become overloaded and cause surface water to accumulate even after a few hours of heavy rain.

Flooding: Constructing on floodplains, wetlands, and riverbanks impedes floodplain water flow and stores water. Farmers building on floodplains or creating artificial storage lakes cause water to trickle off floodplain waterways and flood homes and businesses.

Climate Change: Warming global temperatures have intensified the hydrological cycle, leading to heavy downpours over shorter duration.

Poor Maintenance: If drains and culverts are clogged with solid waste, silt, and vegetation, they seriously reduce the flow of the drains and culverts. Uncontrolled solid waste disposal and inadequate periodic cleaning can make these drainage networks choke points during storms.

2.3.2 Impacts of Urban Flooding

Urban flooding is a complicated issue that affects a city economically, socially, and environmentally. The level of the influence relies on the duration, the depth, and the volume of the flooding, as well as the city's readiness and the amount of its infrastructure. Floodwater stops [7] the performance of a city at its most basic level. These can end traffic and power supplies, as well as communication and public health systems. Short-duration floods, for instance, in highly urbanized areas, may cause a lot of damage to the local economy, namely the closure of businesses and the destruction of the most important assets.

Besides the direct physical harm to cities, urban flooding leads to a big socio-economic gap because low-income residents usually live in areas that are vulnerable to [8] floods and have a low level of infrastructure development. The floodwaters carry sediments and pollutants (together with untreated sewage) into rivers and groundwater systems, causing air, water, and soil pollution. In addition, the flooded areas become breeding grounds for mosquitoes and other pathogens that can spread to the rest of the area.

Table 2.1 highlights the relationship between common urban flood drivers and their consequences, emphasizing the need for effective stormwater management strategies. Flooding harms the facilities that form the foundation of a city, dislocates traffic, and brings about losses for the businesses and communities of the affected region. In addition, it pollutes water sources, encourages the spread of diseases carried by vectors, and harms ecosystems.

TABLE 2.1 Causes vs. Effects of Urban Flooding

CAUSE	EFFECT
Impervious surfaces	Increased surface runoff and reduced groundwater recharge
Inadequate drainage	Localized flooding and property damage
Encroachment on floodplains	Reduced storage capacity of natural systems
Climate change	Increased frequency and intensity of floods
Poor maintenance	Blocked water flow and stagnant water causing health risks

2.4 STORMWATER MANAGEMENT APPROACHES

Stormwater management includes both traditional (gray) and modern (green/blue) infrastructure systems, aiming at controlling the quantity and quality of the runoff, as well as incorporating sustainability into urban planning. Advanced stormwater management is not necessarily about drainage alone, but the attempt to replicate the natural water cycle—which industrialization has ruined in many ways—to slow down, store, filter, and reuse rainwater near the source.

In recent years, there has been a shift from end-of-pipe drainage to decentralized and nature-based systems. Decentralized and nature-based systems are focused on regulating stormwater at multiple scales (from individual buildings to entire catchments) to reduce peak discharges and improve water quality. Combining stormwater management with parks, green roofs, and urban landscapes not only increases flood risk but also enhances aesthetics and biodiversity. In addition, stormwater is increasingly seen as a valuable resource to supplement the urban water supply through collection and reuse.

2.4.1 Traditional (Gray) Infrastructure

Traditional stormwater management involves rapidly collecting and conveying stormwater away from built-up communities through the use of engineered storm drains, pipes, culverts, and detention basins. such systems are usually designed on the basis of hydrological and hydraulic calculations to account for a specified design storm—usually a rainfall event that has a specific probability of occurring in a particular region, such as a 10-year or 25-year storm. The system is normally designed for speed and efficiency: it is hoped that stormwater will be removed from streets and built environments as quickly as possible so that surface-water flooding does not occur. However, while such schemes have proven effective in managing stormwater quantity, the emphasis in these schemes has been essentially on stormwater quality and ecological features. Gray infrastructure occurs in a linear and centralized form. This means that water, collected from various sources and fed into a single outlet—typically a river, lake, or sea—without any natural filtering or treatment, leads to downstream flooding and air pollution of the receiving body of water.

Besides, it is also necessary to invest continuously in the traditional systems that have been established in order to maintain and repair them. In fact, the periodic replacements or maintenance of the underground pipelines, pumping stations, or concrete channels that have been constructed can be quite expensive. Their performance may be significantly deteriorated even though it is less anticipated due to sedimentation, clogging, or corrosion over time. Moreover, in extreme rainfall events that go beyond the system's design requirements, they usually fail prematurely and thus flash floods occur in urban areas. If left unrestrained, gray infrastructure still plays a vital role in stormwater management, being the last resort in densely built-up areas, where natural systems are hardly accessible. The drainage arrangement depicted in Figure 2.2 demonstrates how conventional stormwater systems rapidly remove runoff from urban areas to reduce localized flooding. Thus, contemporary planning is about the interaction of gray

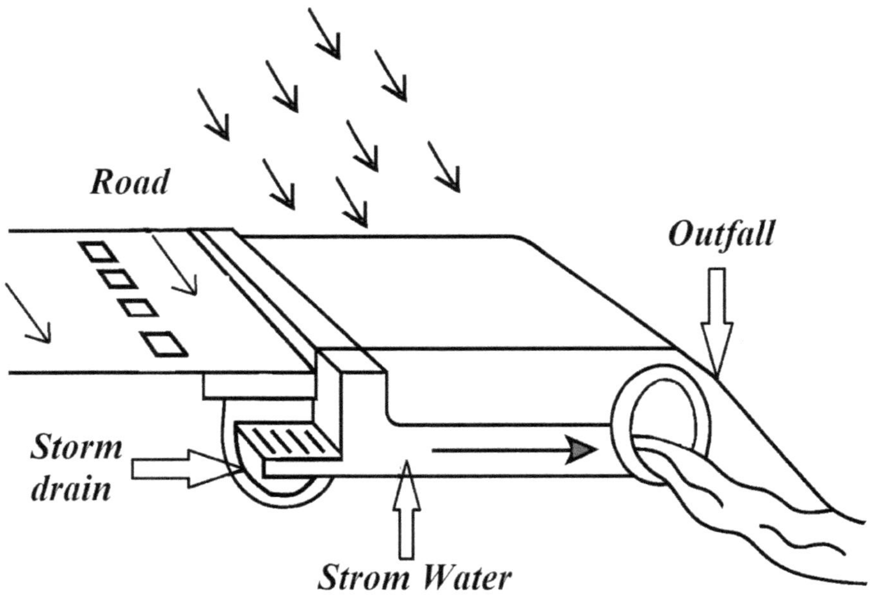

FIGURE 2.2 Diagram of a traditional storm drain system.

and green systems: the use of traditional conveyance for large flows together with the use of natural methods for runoff control at the closest point.

2.4.2 Green Infrastructure

Green infrastructure follows the examples of nature in managing water from storms by allowing water to infiltrate, evaporate, and be reused within the urban environment. It is a decentralized system that manages stormwater near where it falls rather than directly into drains or downpipes. The use of methods such as vegetated swales, green roofs, rain gardens, and permeable [9] pavements slows runoff, helps with infiltration, and works toward naturally filtering pollutants. The use of green infrastructure to restore nature's way of working reduces flooding and peak volume of stormwater. In addition, it improves soil health, which supports biodiversity and increases the quality of life in cities. It makes urban areas more attractive and provides some recreational and microclimatic benefits. Yet, green infrastructure has to be implemented with enough space, good soil conditions, and regular maintenance to ensure its performance and functionality in the long run.

2.4.3 Blue-Green Infrastructure Integration

Blue-green infrastructure is a water (blue) and vegetation (green) infrastructure system that results in multifunctional landscapes, parks, wetlands, retention ponds, and recreational uses. It attempts to balance engineered water management systems with natural ecosystems to manage stormwater and improve the quality of life of local residents. By connecting the bodies of water with green corridors, planners can build resilient drainage networks that hold excess rainwater over time and release it slowly to mitigate floods, improve biodiversity, enhance air quality, and provide aesthetic and recreational value. Blue-green infrastructure can also help adapt to climate change by responding to urban temperatures and increasing groundwater recharge. Table 2.2 compares the characteristics of gray and green infrastructure approaches used in urban stormwater management, highlighting their functional, economic, and environmental differences.

TABLE 2.2 Comparison of Gray vs. Green Infrastructure

ASPECT	GRAY INFRASTRUCTURE	GREEN INFRASTRUCTURE
Primary Function	Conveyance and drainage	Infiltration and retention
Design Philosophy	Engineered control	Nature-based approach
Cost	High initial cost	Moderate, long-term savings
Environmental Impact	Often negative	Positive (biodiversity, aesthetics)
Maintenance	Technical	Community-based possible

2.5 BEST MANAGEMENT PRACTICES (BMPS)

Runoff Estimation (Pre-BMP)

Peak runoff is computed using the Rational Method:

$$Q = 0.278 C I A$$

where

Q = peak runoff (m³/s),
C = runoff coefficient,
I = rainfall intensity (mm/hr),
A = catchment area (ha).
Assume:

$$A = 2.0 \text{ ha}, I = 75 \text{ mm/hr}, C = 0.80 Q_{\text{pre}} = 0.278 \times 0.80 \times 75 \times 2 = 33.36 \text{ m}^3/\text{s}$$

Runoff Reduction (Post-BMP)

BMP implementation reduces imperviousness and increases infiltration:

$$C_{\text{BMP}} = 0.45 Q_{\text{post}} = 0.278 \times 0.45 \times 75 \times 2 = 18.77 \text{ m}^3/\text{s} \; Peak\ flow\ reduction = \frac{33.36 - 18.77}{33.36} \times 100 = 43.7\%$$

Infiltration Performance (Source Control BMPs)

$$f = k_s \times A_i$$

Assume:

$$k_s = 0.015 \; m/hr, A_i = 800 \; m^2 f = 12 \; m^3/hr$$

For storm duration t = 3hr:

$$V_i = f \times t = 36 \; m^3$$

Flow Velocity Control (Conveyance BMPs)

Using Manning's equation:

$$V = \frac{1}{n} R^{2/3} S^{1/2}$$

Assume:

$$n = 0.035, R = 0.25\ m, S = 0.003V = 0.45\ m/s$$

Velocity < 0.6 m/s; hence, erosion is controlled.

Detention Storage Requirement (End-of-Pipe BMPs)

$$V_d = (Q_{in} - Q_{out}) \times t$$

Assume:

$$Q_{in} = 18.77\ m^3/s, Q_{out} = 8.0\ m^3/s, t = 900\ sV_d = 9{,}693\ m^3$$

Water Quality Improvement

Pollutant removal efficiency:

$$\eta = \frac{C_i - C_o}{C_i} \times 100$$

Assume:

$$C_i = 180\ mg/L, C_o = 60\ mg/L\eta = 66.7\%$$

Groundwater Recharge

$$V_r = (P - R) \times A$$

Assume:

$$P = 40\ mm, R = 22\ mm, A = 20{,}000\ m^2 V_r = 360\ m^3$$

2.5.1 Source Control BMPs

These management systems remove large amounts of stormwater at the source as opposed to using large, centralized disposal and storage systems. Source control measures (BMPs) capture, infiltrate, or reuse rainwater near the point of generation, which reduces the total quantity of stormwater runoff and delays peak discharge so that in some areas floodplains do not develop. As shown in Figure 2.3, green roofs consist of multiple functional layers designed to support vegetation while facilitating stormwater retention and drainage. Green roofs act like living tanks that absorb rainfall, promote evapotranspiration, and allow for improved building insulation, which lowers energy use and mitigates the urban heat island effect. Rain gardens are shallow vegetation depressions built around or on rooftops to contain and filter runoff from rooftops and other paved areas. The water is allowed to soak into the ground directly while sediments and pollutants are broken up by plant and soil processes. Permeable pavements (made from porous concrete, pavers, or gravel) allow rainwater to percolate through surface voids into sub-layers, reducing surface runoff and recharging groundwater.

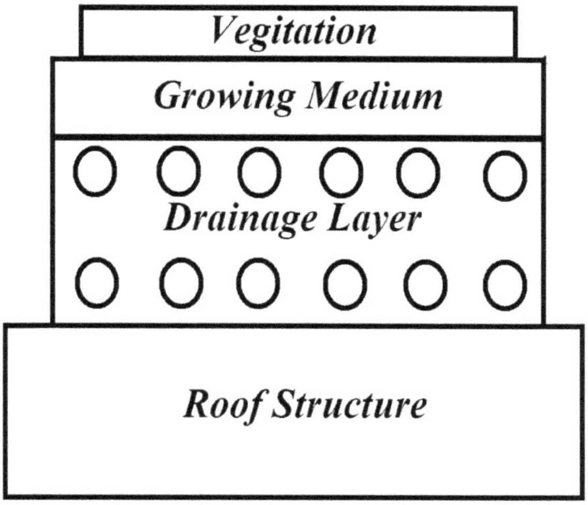

FIGURE 2.3 Cross-section of a green roof system.

2.5.2 Conveyance Control BMPs

The BMPs listed in Table 2.3 demonstrate how decentralized stormwater management practices can effectively reduce runoff volumes and improve water quality. These measures bring stormwater handling to a new level by going beyond the traditional methods. Instead of dumping runoff water into underground pipes at high speed, conveyance control BMPs allow water to flow through areas that are planted and have soil and gravel. This process not only slows down the water but also purifies it and provides it with plenty of time to infiltrate. Hence, the benefits are less erosion, cleaner water, and slower flows before stormwater reaches creeks or rivers [11]. Apart from that, these systems are not only great in functionality but also in aesthetics. Install them along streets, parking lots, or even in vacant areas, and you will have an integrated green network that is compatible with other stormwater solutions, thus resulting in a healthier and more efficient drainage plan. Figure 2.4 illustrates the cross-sectional configuration of a bioswale, highlighting its vegetation layer, gravel bed, and perforated pipe system for stormwater conveyance and infiltration.

a) Bioswales are vegetated, depressed areas that collect, retain, and filter stormwater as it percolates through.
b) Vegetated filter strips are normally just rows of grass or other plants that absorb water running off hard surfaces and purify it by spreading out and catching it.
c) Infiltration trenches are underground pits or trenches filled with gravel that temporarily hold stormwater and let it filter into the earth gradually.

TABLE 2.3 Source Control BMPs and Their Benefits

BMP	FUNCTION	BENEFITS
Green roofs	Absorb rainfall	Reduce runoff, improve insulation
Rain gardens	Infiltrate runoff	Reduce localized flooding
Permeable pavements	Allow infiltration	Reduce peak discharge, filter pollutants

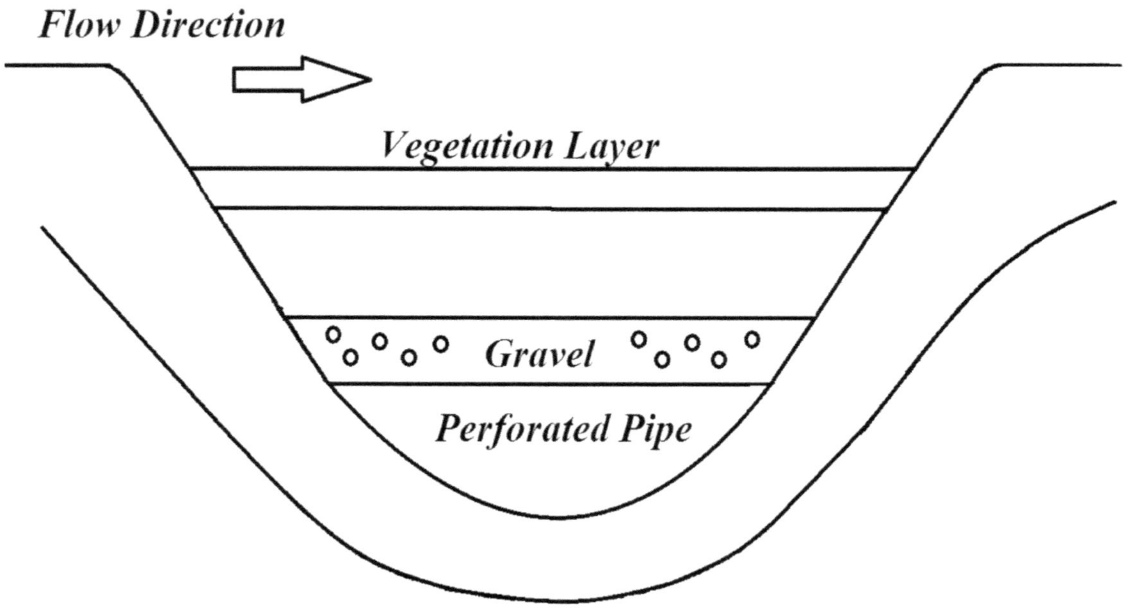

FIGURE 2.4 Diagram showing bioswale cross-section and flow direction.

2.5.3 End-of-Pipe Solutions

After stormwater has been channeled through roads or pipes, end-of-pipe solutions are the next step. These systems deal with runoff that is about to be discharged from the infrastructure, so they are designed to locate pollutants and slow down the flow at the last stage. You may compare them to the last checkpoint before water enters rivers or lakes. They are basically the last stop before the water goes on. In simple terms, they are the final stop before including water into the local environment. Their main jobs are to store, treat, and release the runoff gradually so that there will be no downstream floods, dirty water, or damage to local ecosystems. By retaining water, sometimes just temporarily, sometimes permanently, these systems allow solids to settle and give nature's processes time to break down pollutants. They provide you with the opportunity for water reuse and can convert areas that are susceptible to flooding into parks or wetlands. The entire process of handling stormwater becomes more efficient when you combine them with upstream fixes. The BMP categories summarized in Table 2.4 illustrate a multi-stage approach to stormwater management, ranging from runoff prevention to treatment and reuse. The problem of stormwater is solved in a way that really benefits the environment.

a) Detention and retention basins are similar in operation. Detention basins store stormwater for a short time, thus allowing things to settle before the water is released. Retention basins keep water for a longer period, so that it can infiltrate the earth or simply evaporate.
b) Constructed wetlands have a different method. People create these wetlands to purify stormwater in a natural manner. The plants take in pollutants, while the dirt and [12] debris settle at the bottom.
c) Stormwater harvesting systems collect runoff so that you may use it for watering lawns, for instance, or any other activity that does not require water of drinking quality. Figure 2.5 highlights the key structural elements of a constructed wetland system, illustrating its role as an effective nature-based solution for urban stormwater management.

TABLE 2.4 Summary of BMP Categories

CATEGORY	EXAMPLES	PRIMARY FUNCTION
Source Control	Green roofs, rain gardens	Reduce runoff generation
Conveyance Control	Bioswales, infiltration trenches	Slow and filter flow
End-of-Pipe	Retention ponds, wetlands	Store, treat, and reuse water

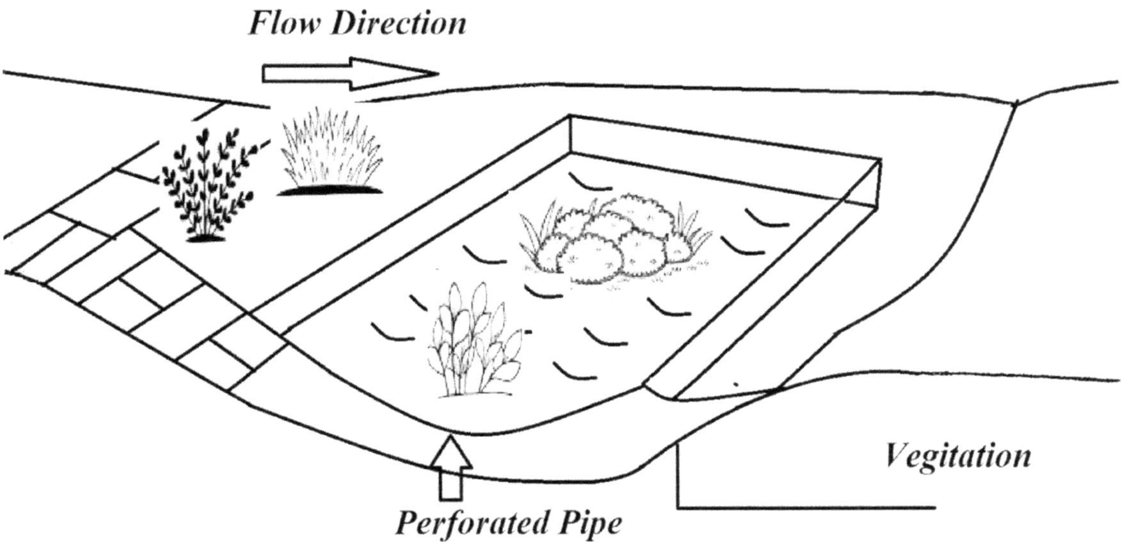

FIGURE 2.5 Example layout of a constructed wetland for stormwater treatment.

2.6 DESIGN AND PLANNING FOR STORMWATER MANAGEMENT

Gone are the days when stormwater management [10] was simply about putting pipes underground. It's a combination of precise engineering, thoughtful landscape design, and, to be honest, a lot of listening to the people who live in these neighborhoods. The cleverest schemes treat stormwater as a resource, not just a problem to be hidden away. Hydrological modeling has become a tool in city planning providing insight into the water movement in various areas across town and hence the most appropriate instruments can be selected for each area. Local climate, soils, and the land slope matter. Disregard them, and the whole thing will collapse. And it is never just the engineers' job. City leaders, developers, and residents all need to be on board if you want a system that is efficient and durable.

2.6.1 Key Design Principles

- Mimic natural systems
- Site-based solutions
- Integrated planning
- Safety and maintenance

TABLE 2.5 CommonSoftware Tools

SOFTWARE	PURPOSE	APPLICATION
SWMM (EPA)	Stormwater Management Model	Urban runoff simulation
HEC-RAS	River Analysis System	Floodplain and channel flow modeling
Info Works ICM	Integrated Catchment Modeling	Combined sewer and drainage analysis
Mike Urban	Water and stormwater management	Real-time simulation and design

2.6.2 Design Tools and Modeling Software

Such models are the mainstay of contemporary stormwater management. They simulate rainfall scenarios, demonstrate the efficiency of drainage, and predict locations of potential floods. By entering data about the land, weather, and soil, planners can figure out the perfect place, the [13] optimal size, and the required strength of the stormwater component. Moreover, they assess the effectiveness of newly introduced cleaning measures, thus enabling decision makers to choose the most efficient solutions without incurring excessive costs. In fact, these models are invaluable in preparing for extreme weather events, as they allow one to ascertain the system's response to a severe storm.

The tools mentioned in Table 2.5 are instrumental in determining peak flow rates, detention storage volume, and overall system capacity, thus providing the basis for data-driven stormwater design and flood risk mitigation.

2.6.3 Policy and Regulation Frameworks

Governments play a major role in shaping stormwater management through the introduction of policies and setting the norms for sustainable urban growth. Rules make [14] sure new projects control runoff and protect water quality. In the United States, Low Impact Development advises ways to reduce site disturbance and keep stormwater on-site. Australia's Water Sensitive Urban Design adds water management to city plans, reusing stormwater and protecting waterways. China's Sponge City Program stores rainwater to help cities handle heavy rain and floods. These structures ensure urban growth aids flood defense and environmental care, instead of worsening runoff and contamination.

2.7 CLIMATE RESILIENCE AND SUSTAINABILITY

2.7.1 Stormwater and Climate Change

Changes in climate have, in a major way, influenced rainfall patterns such that there have been more intense and largely unpredictable storms. This, therefore, adds a big problem to traditional drainage systems that are usually built based on old rainfall data which no longer reflect either the present or the future. Because of the increase in rainfall peaks and elongation of dry periods, it is often the case that the capacity of the existing stormwater infrastructures is exceeded, thus resulting in flash floods, erosion, and damage to the infrastructure. Stormwater management is still very important in the establishment of climate-resilient cities through the introduction of adaptive systems that are capable of handling variable rainfall patterns and, at the same time, promoting water reuse. The use of SuDS, permeable surfaces, and retention features not only helps cities to absorb stormwater but also enables them to store it, thus reducing their flood risk and drought vulnerability. Besides that, the use of climate projections in urban planning gives more room for engineers to come up with flexible and redundant infrastructure designs. By doing so, the management of stormwater becomes a support system rather than merely flood control [15] of a short-term nature. It, therefore, provides a way for the mitigation of climate change effects and also the efficient use of resources on a long-term basis.

2.7.2 Nature-Based Solutions (NbS)

In contrast to traditional engineered systems, NbS interact with natural processes like infiltration, evapotranspiration, and storage in order to regulate the flow of water over the land. Green areas, wetlands, and permeable surfaces not only absorb and retain rainwater but also provide some other beneficial effects such as carbon sequestration, habitat creation, and urban temperature regulation. As these systems reintroduce nature to the water cycle, they thus become both local flood mitigation measures and tools for long-term climate adaptation [26].

Table 2.6 presents common nature-based stormwater management measures and highlights their multifunctional benefits beyond flood mitigation. Some of the examples include urban wetlands functioning as temporary flood storage areas, river restoration projects that reopen floodplains and improve [16] ecological connectivity, and permeable urban parks that also serve as retention areas during storm events. These multifunctional landscapes, thus, turn stormwater from a threat into a resource, which is the way to create more livable and sustainable cities. Also, NbS integration in the urban planning process supports community well-being, as it provides more opportunities for nature interaction and at the same time raises biodiversity and facilitates the implementation of green infrastructure networks.

2.7.3 Community-Based Adaptation

The involvement of local communities makes stormwater systems more sustainable, and their effectiveness can be retained even after a long period of time [17–20]. Residents can take up the responsibility of green infrastructure maintenance, establishing rainwater harvesting, and watching over the local drainage channels to ensure their accessibility. The involvement of the community not only helps city authorities to reduce their costs of maintenance but also creates among the people a feeling of ownership and environmental consciousness. When people are aware of the fact that wrongly disposed waste and blocked drains lead to flooding, they are more willing to adopt those practices that are safe for the urban water system [21–23].

The gap between the policy and the practical side can be closed through public awareness programs, neighborhood clean-up activities, and participatory planning, which in turn will have a positive impact on the stormwater issues [24]. Besides that, local care for the environment also contributes to social resilience because communities will become more capable of responding together during flood events. Incorporating community-based adaptation with formal infrastructure planning allows cities to have more inclusive, fair, and sustainable stormwater management systems [25].

TABLE 2.6 Nature-Based Solutions and Their Benefits

NATURE-BASED SOLUTION	PRIMARY FUNCTION	ADDITIONAL BENEFITS
Urban Wetlands	Retain and filter stormwater	Habitat creation, water purification
Green Corridors	Convey runoff naturally	Improves urban biodiversity
Permeable Parks	Temporary storage	Recreation and cooling effect

2.8 CONCLUSION

Storm water management, if done efficiently, can considerably lessen the problem of urban flooding and increase the capacity of cities to withstand the changes of climate. The use of gray, green, and blue methods of storage and treatment not only guarantees hydraulic performance but also maintains ecological equilibrium, thus allowing stormwater to be dealt with as a resource that has value. Environmentally friendly roofs, rain gardens, bioswales, and constructed wetlands, for example, help in the reduction of flood risks, impure water purification, and also the attraction of animals and plants, of which water is a source. The involvement of local communities in the establishment of such systems, however, increases their personal stake, which is very important for the continuation of these systems through their active participation in maintenance and awareness. Hence, efficient stormwater management is at the core of sustainable urban development, as it not only safeguards public health and safety, but it is also in line with the trend of creating residential areas that are flood-resistant and offer a high quality of life, thus being able to withstand environmental challenges in the future.

REFERENCES

1. M. Qi, A. Lehmann, H. Huang, L. Liu, and X. Chen, "A SWMM-based evaluation of the impacts of LID and detention basin retrofits on urban flooding," *Urban Water J*, vol. 22, no. 1, pp. 51–65, 2025, doi: 10.1080/1573062X.2024.2426590.
2. S. Dobkowitz, A. Bronstert, and M. Heistermann, "Water retention by green infrastructure to mitigate urban flooding: A meta-analysis," *Urban Water J*, vol. 22, no. 5, pp. 477–492, 2025, doi: 10.1080/1573062X.2025.2472325.
3. S. Towsif Khan, T. Wynn-Thompson, and D. Sample, "Stormwater controls for channel stability: Focusing on bed material transport prevents degradation," *J Environ Manage*, vol. 374, 2025, doi: 10.1016/j.jenvman.2024.123651.
4. F. Qin, L. Huang, X. Qi, L. Sun, J. Cui, and Y. Wei, "Optimal LID designs based on SWMM simulations regarding the sustainable efficacy of stormwater management in port areas," *Sustainability (Switzerland)*, vol. 17, no. 6, 2025, doi: 10.3390/su17062544.
5. N. Sharma, C. Chakraborty, and R. Kumar, "Optimized multimedia data through computationally intelligent algorithms," *Multimed Syst*, vol. 29, no. 5, pp. 2961–2977, 2023.
6. F. R. Aderyani, K. Jafarzadegan, and H. Moradkhani, "A surrogate machine learning modeling approach for enhancing the efficiency of urban flood modeling at metropolitan scales," *Sustain Cities Soc*, vol. 123, 2025, doi: 10.1016/j.scs.2025.106277.
7. Y. Pan, F. Zhou, Q. Wang, C. Tong, C. Song, and L. Wang, "Urban rainstorm flood rapid simulation in plain river network area based on cellular automata," *Natural Hazards*, vol. 121, no. 7, pp. 7891–7914, 2025, doi: 10.1007/s11069-025-07116-3.
8. D. Srivastava, N. Sharma, D. Sinwar, J. H. Yousif, and H. P. Gupta, eds., *Intelligent Internet of Things for Smart Healthcare Systems*. CRC Press, 2023
9. S. Ahmad *et al.*, "Quantifying LID impact: A modified metric for enhanced flood mitigation and urban resilience," *Resour Conserv Recycl*, vol. 215, 2025, doi: 10.1016/j.resconrec.2024.108089.
10. N. Sharma and U. Batra, "An enhanced Huffman-PSO based image optimization algorithm for image steganography," *Genet Program Evolvable Mach*, vol. 22, no. 2, pp. 189–205, 2021.
11. L. C. Dickey, C. R. Rehmann, M. A. Perez, and K. Ikuma, "Evaluation of a pretreatment intake system for reducing particulate loading to a bioretention cell," *J Sustain Water Built Environ*, vol. 11, no. 2, 2025, doi: 10.1061/JSWBAY.SWENG-589.
12. D. Pumo, M. Ippolito, F. Alongi, A. Francipane, and L. V Noto, "Modelling the hydrological response of blue-green roofs: ECO-PRO model," *Sci Total Environ*, vol. 976, 2025, doi: 10.1016/j.scitotenv.2025.179299.
13. N. Sharma, M. Soni, S. Kumar, R. Kumar, N. Deb, and A. Shrivastava, "Supervised machine learning method for ontology-based financial decisions in the stock market," *ACM Trans Asian Low-Resour Lang Inf Process,* vol. 22, no. 5, pp. 1–24, 2023.

14. X. Wang, J. Xia, J. Feng, and B. Dong, "Numerical modelling of the impact of drainage system clogging on urban flood processes," *J Environ Manage*, vol. 388, 2025, doi: 10.1016/j.jenvman.2025.125969.
15. WuW., , Jamali, B., Marshall, L., Deletic, A., and Zhang, K. A water sensitive urban design (WSUD) planning framework for catchment-scale urban pluvial flood mitigation targets. *Water Research*, vol. 285, 124095, 2025.
16. N. Sharma and U. Batra, "A review on spatial domain technique based on image steganography," 2017 International Conference on Computing and Communication Technologies for Smart Nation (IC3TSN). IEEE, 2017, October, pp. 24–27.
17. H. Dui, H. Zhang, and S. Wu, "A IoT-based novel methodology to optimize multidimensional flood resilience of drainage systems for sponge city," *Sustain Cities Soc*, vol. 130, 2025, doi: 10.1016/j.scs.2025.106523.
18. N. Sharma and U. Batra, "A study on integrating crypto-stego techniques to minimize the distortion," International Conference on Recent Developments in Science, Engineering and Technology. Singapore: Springer Singapore, 2017, October, pp. 608–615.
19. R. Javidi Sabbaghian, M. Fereshtehpour, and M. R. Goli Hosseinabad, "Integrated hydrologic-economic modeling for urban flood risk mitigation using SWMM, HEC-RAS, and HAZUS: A case study of the Bronx river watershed, NYC," *Sustain Water Resour Manag*, vol. 11, no. 5, 2025, doi: 10.1007/s40899-025-01263-y.
20. S. Gill, N. Sharma, C. Gupta, and A. Samanta, "Attendance management system using facial recognition and image augmentation technique," 2021 International Conference on Intelligent Technology, System and Service for Internet of Everything (ITSS-IoE). IEEE, 2021, November, pp. 1–6.
21. C. Bosco *et al.*, "An excel-based tool to support decisions on the selection of outflow devices for blue roofs using historical maximum rain events," *Water Sci Technol*, vol. 92, no. 8, pp. 1145–1159, 2025, doi: 10.2166/wst.2025.150.
22. M. Beri, B. Kumar, S. Tiwari, N. Sharma, H. Vashishtya, and P. Chaudhary, "IoT based health monitoring system built on ESP32," 2022 2nd International Conference on Advance Computing and Innovative Technologies in Engineering (ICACITE). IEEE, 2022, April, pp. 454–458.
23. T. Lazzarin, P. Costabile, and D. P. Viero, "An efficient physics-based modeling strategy for pluvial floods in urban areas with a subgrid scheme for the stormwater drainage network," *J Hydrol (Amst)*, vol. 661, 2025, doi: 10.1016/j.jhydrol.2025.133617.
24. N. Sharma, D. Srivastava, and D. Sinwar, eds., *Artificial Intelligence Technology in Healthcare: Security and Privacy Issues*. CRC Press, 2024.
25. T. Grodek and G. Benito, "Reevaluating flood protection: disaster risk reduction for urbanized alluvial fans," *Natural Hazards and Earth System Sciences*, vol. 25, no. 11, pp. 4343–4360, 2025, doi: 10.5194/nhess-25-4343-2025.
26. M. Kumari, N. Sharma, R. Chauhan, K. Joshi, V. Madaan, and G. Sunil, "Investigation of OWC system incorporating different filters," 2024 4th International Conference on Innovative Practices in Technology and Management (ICIPTM). IEEE, 2024, February, pp. 1–4.

Sewage and Wastewater Management for Improved Sanitation

3

Manikandan Kannappan and Poomalai Vasanth

ABBREVIATIONS

AI: Artificial Intelligence
AnMBR: Anaerobic Membrane Bioreactor
AOX : AOX Adsorbable Organic Halides
BOD: Biochemical Oxygen Demand
BOD_5: 5-day Biochemical Oxygen Demand
CH_4 : Methane
CHP: Combined Heat and Power
CNT: Carbon Nanotube
CO_2: Carbon Dioxide
COD : Chemical Oxygen Demand
DEWATS: Decentralized Wastewater Treatment Systems
DPR: Direct Potable Reuse
DWTS: Decentralized Wastewater Treatment System
EU: European Union
GHGs: Greenhouse Gases
GO: Graphene Oxide
IoT: Internet of Things
JMP: Joint Monitoring Programme
MRC: Microbial Recycling Cell
PN/A: Partial Nitritation/Anammox
RNG: Renewable Natural Gas
SCADA : Supervisory Control and Data Acquisition
SDG: Sustainable Development Goal

DOI: 10.1201/9781003742227-3

TN: Total Nitrogen
TP: Total Phosphorus
TSS: Total Suspended Solids
UNICEF: United Nations Children's Fund
WASH: Water, Sanitation, and Hygiene
WHO: World Health Organization
WRRF: Water Resource Recovery Facility
WWTP: Wastewater Treatment Plant

3.1 INTRODUCTION AND GLOBAL SANITATION CHALLENGES

Good hygiene, clean water, and proper sanitation are pillars of the global socioeconomic development and the health of the population. Nevertheless, billions of people still do not have access to adequate water, sanitation, and hygiene (WASH) services, which put them at risk of acquiring diseases and increases social exclusion. Sustainable Development Goal 6 aims at providing universal access to safe drinking water, sanitation and hygiene by the year 2030 but has to be accelerated in order to provide safely managed services. Smart technologies play a vital role in promoting water sustainability and the formulation of policies regarding sewage and wastewater management. In this chapter, the author discusses issues related to sanitation in the world, focusing on the lack of access and how it is interdependent with the Sustainable Development Goals [70].

3.2 IMPORTANCE OF WASTEWATER MANAGEMENT IN SANITATION

The effective disposal of wastewater is a central component in the quest to achieve better sanitation and alleviation of the additional health and environmental threats that accompany the issue on the face of the world. The excreta and other liquid effluents contained in wastewater, which are not properly managed, are a source of waterborne diseases and widespread water pollution, and, as such, they violate human dignity and the welfare of the community. The International Organization of the Promotion of Children Welfare [69] has outlined a change in paradigm that is no longer to offer children improved sanitation but to offer sustainable sanitation [68]. In this modern construct, excreta and wastewater are not seen only as waste but have the potential to be used as resources. In this respect, the shift to wastewater reuse instead of disposal is easier to enable the recovery of precious resources such as energy, drinking water, and nutrients [4, 53].

As an example, the treatment of wastewater can contribute to developing 11 out of 17 Sustainable Development Goals (SDGs) by increasing the supply of water, improving human health, providing streams of income, transforming waste into energy, and reducing the environmental impact [53]. Furthermore, wastewater treatment significantly improves water quality and makes it reusable or dischargeable to the ecosystems and, consequently, is very important in reducing environmental pollution, especially in urban areas [40]. The academic sources are becoming more and more infused with the notion that the rheological properties of sewage sludge should be understood to maximize the design and operation of the wastewater treatment facility to cost less and work better [17].

Decentralized systems, waste-stream separation, low- or non-flushing sanitation, and conversion of fecal sludge to energy represent only some of the innovative approaches with proven potential to mitigate a range of development issues and promote a variety of SDGs [4]. However, effective implementation of these solutions requires thorough planning, financial resources, social awareness, and well-developed institutional frameworks, especially in low-income countries where infrastructural insufficiency and a lack of funds are the most severe [3]. The incorporation of wastewater management into extended strategies of public health, climate, and antimicrobial resistance is also becoming accepted, and its multisided nature is emphasized in the context of resilient and sustainable health outcomes [9].

3.3 GLOBAL ACCESS DEFICIT AND SUSTAINABLE DEVELOPMENT LINKAGES

By 2024, 2.1 billion people worldwide still lacked access to a safely managed supply of drinking water, and 106 million continued using surface sources that are not treated. At the same time, 3.4 billion of the world's population does not have access to safely managed sanitation, and 354 million of the population defecate in the open air, based on the SDG 2025 statistics. Furthermore, 1.7 billion individuals lacked access to basic hygiene services at home, of whom 611 million had no facilities at all.

These inequalities are increased in the least developed nations, where the likelihood of not having access to basic drinking water and sanitation is over twice, and the likelihood of not having access to basic hygiene services is over threefold the probability rate in other countries [15]. In weak settings, the reporting of safe drinking water is 38% points lower than in other nations, thus showing gross inequalities. Although the rural areas have experienced a rise in coverage, increasing 50% to 60% in terms of safe drinking water under management in the period between 2015 and 2024 and 52% to 71% in basic hygiene, in the urban sector, it has been stagnant in both areas [76].

The SDG sanitation goal aims at providing access to safely managed sanitation facilities by the year 2030 globally, alongside the absence of open defecation. To reach this ambitious goal, one will require that some 5.6 billion more people have access to safely managed sanitation, and that some 1.3 billion will change to utilize sanitation facilities instead of open defecation by 2030 [48]. The total amount of money that is needed annually to achieve such targets in the cities is estimated to be about US $46 billion, and in the rural regions, it is estimated to be about US $25 billion during 2016–2030.

To achieve universal coverage of basic WASH services and ending the practice of open defecation by 2030 will demand even more urgent measures because the current patterns indicate that access to safely managed services is becoming more and more elusive. A sixfold increment in the advancement rates is necessary to ensure safely managed drinking water, a fivefold improvement in sanitation, and a threefold increment in basic hygiene services [36]. The Joint Monitoring Programme (JMP) was created in 1990, and this program has 35 years of oversight of the global progress, and it has the mandate of monitoring SDG targets 1.4, 6.1, and 6.2, which require universal access to safe water, sanitation, and hygiene and the elimination of open defecation by 2030 [76].

3.4 EVOLUTION, CHARACTERISTICS, AND IMPACTS OF WASTEWATER

The current state of wastewater management is on a massive paradigm shift, where it has ceased to be a mere issue of disposal but rather an opportunity for resources. This metamorphosis is informed by rising global needs for water and nutrients and by an awareness of wastewater as a source of recoverable resources, including nutrients, energy, and bioproducts, which can be used to meet the aim of sustainable development and a circular economy. Wastewater is widely classified on the basis of its origin into municipal, agricultural, and industrial wastewater, each with its own particular features and profile of pollutants [52]. It follows that the overall comprehension of such types and their complex constructions is necessary to facilitate the formulation of effective management interventions to reduce the impact they have on the environment and human health.

3.5 CHANGING PERCEPTION: FROM WASTE TO RESOURCE

The attitude towards wastewater has changed radically, from being seen as a waste that requires disposal to an apparent source of useful resources. This transition fits into the principles of a circular economy and becomes essential in the conditions of increasing world requirements for freshwater and the exhaustion of non-renewable natural resources, such as phosphorus [37]. Urbanization and industrialization have only increased water shortages and pollution, and hence the need to consider wastewater as a possible remedy to reduce these demands. As recent research has shown, wastewater and sewage sludge do harbor high levels of organic and inorganic compounds, both vital nutrients (phosphorus, ammonium) and biomolecules, including proteins, lipids, and other biomolecules [37].

The use of these resources not only solves the environmental issues related to eutrophication caused by uncontrolled discharge but also provides the economic incentive of creating value-added products and decreasing the use of virgin resources [37]. As one such example, phosphorus, an essential and finite resource of the planet estimated to have about 100 years, can be obtained via municipal wastewater and human excreta, which could potentially serve much of the world's phosphorus needs [37].

Furthermore, the valorization of wastewater enables the generation of various by-products, such as biofuels, biofertilizers, biopesticides, biopolymers, vitamins, enzymes, dyes, pigments, and phenolic compounds [6, 24]. Such a shift towards a linear to circular economic model of wastewater management highlights a new way of thinking that redefines environmental issues as a resource to regenerate and develop sustainably. The strategy is also becoming important in establishing sustainable food systems as well as in meeting the Sustainable Development Goals of the United Nations.

3.6 TYPES AND COMPOSITION OF WASTEWATER

There are three main types of wastewaters, namely, municipal, agricultural, and industrial wastewater, that have different characteristics and compositions (Table 3.1). Municipal wastewater, also known as domestic wastewater, is residential and commercial in origin and usually consists of about 99% water and 1% solids, dissolved, suspended, or colloidal organic and inorganic substances. Farm wastewater is a result of agricultural practices and could include pesticides, fertilizers, and animal waste. Industrial wastewater is highly toxic and heterogeneous in its properties, types, and concentration of contaminants, which is produced by various types of industrial processes and is usually more toxic than municipal wastewater. This division can also be broken down into cooling, washing, and processing wastewater.

To illustrate, the latest wastewater of the cement industry is comprised of suspended solids (calcium carbonate), dissolved solids (potassium hydroxide), heavy metals (iron, zinc, and manganese), biochemical oxygen demand (BOD), and chemical oxygen demand (COD) [14]. Pollutants that are found in

TABLE 3.1 Composition of Municipal Wastewater

PARAMETER	TYPICAL CONCENTRATION
BOD_5	200–300 mg L^{-1}
COD	400–600 mg L^{-1}
Total Suspended Solids (TSS)	200–350 mg L^{-1}
Total Nitrogen (TN)	20–50 mg L^{-1}
Total Phosphorus (TP)	4–12 mg L^{-1}
Pathogens	High (bacteria, viruses, protozoa)

Summarizes the typical physicochemical and biological characteristics of municipal wastewater, highlighting its high organic load, nutrient content, and presence of pathogens that necessitate effective treatment, modified from the source: Kato and Kansha (2024).

petrochemical wastewater, which is formed after refining crude oil and chemical manufacturing, include aromatics, hydrocarbons, sulfides, ammonia, and heavy metals, including chromium, iron, nickel, and copper [14]. This is because food industry wastewater contains high concentrations of COD, BOD, total suspended solids (TSS), total nitrogen (TN), and total phosphorus (TP) due to its various processing and non-processing water utilities. Certain industries, including the processing of meat, feed a significant amount of BOD, COD, TN, TP, and TSS due to blood, debris, and protein [14].

The iron and steel industry also generates COD, ammonia-nitrogen, volatile phenol, cyanide, TSS, heavy metals, and petroleum, the release of which occurred significantly through coking processes. Pharmaceutical wastewater is intricate, and it often consists of non-biodegradable organics (antibiotics, anti-inflammatories, steroids, and hormones), high levels of organic contaminants, salinity, and biological toxicity [16, 43]. The products include high COD, TSS, low biodegradability, and refractory organic compounds in the form of adsorbable organic halides (AOX) that result from several processes, including wood preparation, pulp manufacturing, and bleaching, which is a result of the pulp and paper industry [14]. Wastewater used in the textile industry is highly contaminated, with wet processes being the most contaminated, and is characterized by a high BOD, total solids, TSS, and various dyes, chemicals that are used during sizing, desizing, scouring, bleaching, mercerization, dyeing, and printing [14]. The quality of wastewater is determined by the presence of such diverse pollutants as organic matter content, BOD, COD, TSS, nutrients (TN, TP), and micropollutants. Such unique compositions require unique, and usually combined, physical, chemical, and biological treatment processes in order to reach success in the removal of contaminants and fulfill regulatory requirements.

3.7 ENVIRONMENTAL AND PUBLIC HEALTH IMPACTS OF INADEQUATE TREATMENT

Poor handling of wastewater poses major and complex threats to both the environment and human health worldwide. It is estimated that more or less 80% of all the global wastewater is released unsuccessfully into natural systems without adequate treatment, leading to the massive pollution of natural systems and the destruction of crucial natural systems [41]. This raw wastewater is usually made up of a convoluted blend of pollutants, such as pathogens, pharmaceuticals, microplastics, heavy metals, endocrine disruptors, and other innovative chemicals [75]. These materials, when introduced in rivers, lakes, and oceans, have gross ecological impacts including eutrophication, depleted dissolved oxygen, developing hypoxic dead zones, loss of habitat, loss of fish species, and destruction of aquatic biodiversity [66].

The habitats of the coasts, such as mangrove forests and seagrass beds that play a fundamental role in the process of carbon sequestration, face an increasing threat of degradation caused by the pollution of wastewater. In addition, industrial wastewater releases that are poorly treated also lead to the degradation of the receiving water bodies, which, in turn, may increase eutrophication and create conditions that promote the growth of toxin-producing waterborne pathogens [35, 54]. Oil-and-gas wastewaters contain radioactive pollutants such as radium that can be washed to groundwater and surface water and lead to other risks to the environment and human health [64].

In the context of public health, untreated or poorly treated wastewater is known to cause a broad range of diseases when exposed to it. Wastewater is often associated with viral, bacterial, and protozoan infections, such as salmonellosis, shigellosis, cholera, giardiasis, amoebiasis, and hepatitis A, which occur due to the exposure of the host to contaminated crops or direct contact with feces during farming. Farmers, along with their families, particularly in low-income nations where irrigation with untreated wastewater is widespread because of water scarcity, are at increased risk of helminth infections, dermatologic irritations, and chronic medical consequences of the build-up of heavy metals [41]. Children are particularly at risk, as they are more affected by infection and developmental deficits because of their exposure during their play in contaminated zones or by being engaged in working in the fields [23].

Poor wastewater treatment is also a factor that causes the contamination of the groundwater, weakening drinking water systems and enhancing the proliferation of infectious diseases due to the absence of clean water to maintain hygiene. The world health cost is very high, and a substantial percentage of diseases and deaths among children all over the world can be attributed to poor quality water. The financial implications of poor wastewater management include high-cost health care spending and high environmental cleanup costs. These widespread effects require robust regulatory frameworks, improved treatment facilities, and a paradigm shift in viewing wastewater as a resource and not a waste.

3.8 WASTEWATER TREATMENT SYSTEMS: CENTRALIZED AND DECENTRALIZED APPROACHES

The wastewater treatment systems are important in the treatment of anthropogenic and industrial effluents and offer protection of the health of people and maintenance of ecological integrity by means of solids, organic components, and nutrients removal through physical, chemical, and biological procedures. After treatment, the effluent can be discharged safely or reused, depending on the intensity of treatment required and the purpose of application; this would reduce pollution and promote the survival of viable ecosystems. Centralized and decentralized systems are adopted, both possessing their own advantages and disadvantages that reflect the peculiarities of the population and environmental requirements of a certain community. Centralized schemes are efficient and scaled in highly populated city environments, while decentralized systems are flexible and incur lower operating costs in low-density environments, and the choice has a huge impact on the development of infrastructure, cost structure, and environmental performance.

3.9 OVERVIEW OF CONVENTIONAL CENTRALIZED TREATMENT SYSTEMS

The traditional centralized wastewater treatment plants typically consist of a series of sequential steps such as preliminary, primary, secondary, and sometimes tertiary or advanced treatment, often finishing

TABLE 3.2 Comparison of Centralized and Decentralized Wastewater Treatment Systems

PARAMETER	CENTRALIZED	DECENTRALIZED (DEWATS)
Capital cost	Very high	Low–moderate
Energy requirement	High	Low
Sewer network	Extensive	Minimal
Resource recovery	Limited	High
Suitable areas	Large cities	Rural, peri-urban
Climate resilience	Low	High

Compares centralized and decentralized wastewater treatment systems, emphasizing differences in cost, energy demand, infrastructure needs, and suitability under varying climatic and settlement conditions. Modified from the source: Saadatinavaz et al. (2024); Milićević et al. (2024).

with a disinfection step [27]. The first step mainly aims at eliminating unrefined effluent of large size and coarse debris to protect downstream equipment [27]. This phase usually includes crude screening and stripping off grit, in which grit chambers maintain high flow velocities, or they add aeration to prevent settling of the organic particulates. Primary treatment then removes settleable organic and inorganic substances by sedimentation in a process that is accompanied by eliminating surface scum through skimming [25]. The primary clarifiers, which are sedimentation basins, are typically 3 to 5 m deep with a hydraulic retention time between 2 and 3 hours with an achievement of 25–50% biochemical oxygen demand (BOD_5) and 50% total suspended solids (SS) [27].

Primary effluent is further treated in secondary treatment, which removes the remaining organics and suspended solids, and this is mainly done by aerobic biological processes (Table 3.2) wherein the microorganisms oxidize the organic substances in the presence of oxygen. Activated sludge systems and trickling filters are some of the high-rate biological processes that are frequently used at this stage. The activated sludge technique is an aerobic biological treatment based on the use of microorganisms that use organic substances in an aeration basin [8]. In this system, the microorganisms are suspended in the mixed liquor located in the aeration basin; aeration equipment adds oxygen and provides intense mixing. Upon completion of the aeration stage, the microorganisms are separated from the liquid through the process of sedimentation in a secondary clarifier, and a portion of the biological sludge that is formed is pumped back into the aeration basin.

Trickling filters, also known as trickling biofilters or biological filters, are fixed-bed biological reactors (Figure 3.1), which are usually aerobic in nature. Pre-settled influent is sprayed between filter media,

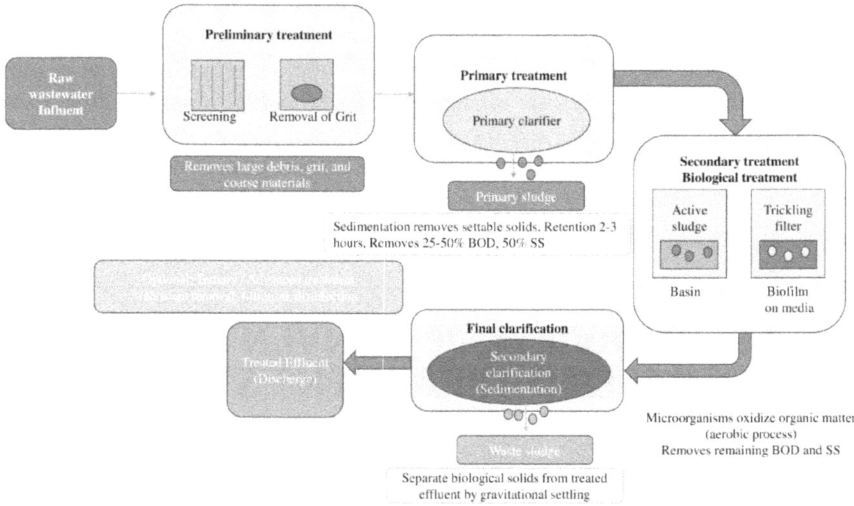

FIGURE 3.1 Conventional municipal wastewater treatment process flow diagram.

where a biofilm of microorganisms oxidizes organic components between water flowing through the filter pores. Natural or forced airflow supplies oxygen into the biofilm. With time, the biofilm becomes thicker, and some of it eventually sloughs off and is afterwards separated from the liquid in a second clarifier [63]. In both activated sludge and trickling filter systems, a clarifier is used to separate solids and liquids after the biological treatment. Clarifiers, which are basically sedimentation basins, aid in the gravitational settling of the particulate matter. Specifically, the secondary clarifiers have the effect of separating the biological solids and the treated effluent such that the clarified water can be discharged or undergo further treatment. The sludge obtained with both primary and secondary clarifiers is then transferred to further treatment.

3.10 LIMITATIONS IN RAPIDLY URBANIZING AND DENSELY POPULATED REGIONS

The centralized wastewater treatment systems, though effective in handling the large amounts of wastewater produced by the urban centers, face significant challenges in densely populated and socio-culturally oriented urbanization. A significant disadvantage is that establishing the infrastructure or creating huge systems of sewer pipes and pumping stations is a very costly undertaking in terms of capital. In regions that lack proper infrastructure or have deteriorated infrastructure, improvement or increase in size of these networks is not only difficult, but also expensive. Besides, the design nature of such systems makes them inherently resistant to changes in the conditions of operation, making them difficult to embrace sustainability or resiliency [59].

These systems are susceptible to natural disasters and sudden changes in climate because their vulnerability to the hierarchical collection network means that a breakdown will affect many people. The amount of wastewater generated by the highly populated urban centers is also an extra challenge, especially when the energy intensity of the highly sophisticated treatment technologies is taken into account in the pursuit of the attainment of high discharge standards [18]. The increased use of energy is a characteristic of centralized systems since it requires pumping wastewater through great distances, which, consequently, leads to a higher carbon footprint [16, 59]. In addition, industrial waste streams are mixed with domestic wastewater in centralized systems, which makes resource recovery and reuse efforts more difficult.

Rapid urbanization often leads to organic development and the lack of control over development, thus having a massive impact on wastewater treatment and making the process of centralized systems implementation and maintenance more complicated. The large centralized facilities that need vast land areas can also make no sense in the metropolitan areas that can be characterized by space constraints, thus leading to the need to explore ways that can fit in limited spaces in a manner that is compatible with the urban aesthetic. Although they are able to produce high-quality effluent with the help of sophisticated treatment procedures, the high cost of operation in terms of energy use and maintenance is still a huge challenge. These restrictions highlight the growing necessity of other or complementary methods of wastewater management in these dynamic settings.

3.11 CONCEPT, PRINCIPLES, AND ADVANTAGES OF DECENTRALIZED WASTEWATER TREATMENT SYSTEMS (DEWATS)

Decentralized Wastewater Treatment Systems (DEWATS) is a shift in attitude of wastewater treatment systems, which focuses on treating the wastewater at its source or at the closest point instead of transporting it long distances to a central plant. The main difference of this approach to the traditional framework of its creation is that the traditional model is to build one large wastewater treatment plant (WWTP) that serves an entire growing territory, most of the time before it is developed. Rather, DEWATS entails building smaller numbers of WWTPs, which can act independently or in combination to cover particular locations or groups of households and companies.

The major concept of DEWATS is the principle of localized treatment, which significantly decreases or removes the need to provide a large-scale and expensive sewer network and pumping facilities, particularly in a rural or remote environment [51]. The process may also be tailored in response to the wastewater specifications, thus making it more efficient and effective by treating wastewater on-site. Resource recovery is another important principle and, in this case, it is possible to discuss the numerous DEWATS that should recover various resources, e.g., water to be used in irrigation, nutrients to be used in agriculture and even energy in the form of biogas, contributing to sustainability and economic efficiency. Localized treatment and recovery of resources enable re-use of water at the community scale and help to implement the principles of a circular economy and minimize the use of freshwater.

The benefits of DEWATS are multi-dimensional, i.e., they are economic, environmental, and social. DEWATS are more economical to install and operate than centralized systems, especially in areas with limited population, since they eliminate the need to have a large, centralized system of pipes and lift stations of large diameter. Initial capital costs are significantly reduced since smaller and modular plants can be built in stages according to the demand, and the cost of financing can be shared over a long-term period [7]. This modularity also gives more flexibility, and the systems can be expanded or reduced to meet the needs of the population or volume of wastewater.

DEWATS can be deemed environmentally friendly and sustainable solutions as they do not use water to treat wastewater; instead, accelerated natural processes are used to treat wastewater and reduce water pollution transmitted downstream [21]. They usually have a less significant impact on the environment because they use less energy to pump the water, and they also promote the growth of green space and recharge of the aquifer [10]. In social matters, DEWATS support community participation and local control of local resources, development of a feeling of ownership and responsibility regarding environmental health [21]. They also allow growth in development areas without leading to tax increases or suburban sprawl, and they also increase property value by providing enhanced local infrastructure. Moreover, DEWATS help ensure that there is better health for the people as they offer sound sanitation, hence minimizing the expenses incurred in medical care and also boosting productivity due to the prevention of waterborne diseases.

3.12 RESOURCE RECOVERY AND CIRCULAR ECONOMY IN WASTEWATER MANAGEMENT

The current state of wastewater management is in the process of fundamental change, with a shift in paradigm from a linear model (disposal) to a circular economy, which views wastewater as a quasi-valuable resource. This development is driven by the increasing demand for water in the world and the need for a sustainable operational mode, thus repositioning the treatment plants as resource recovery centers that can extract valuable materials and energy. The new paradigm focuses on the reduction of waste, the enhancement of resource use, as well as environmental and economic sustainability due to its strategies of water reuse, nutrient recovery, and energy generation. These plans coincide with the goals of sustainable development and the philosophy of a cradle-to-cradle, bio-based economy (Figure 3.2).

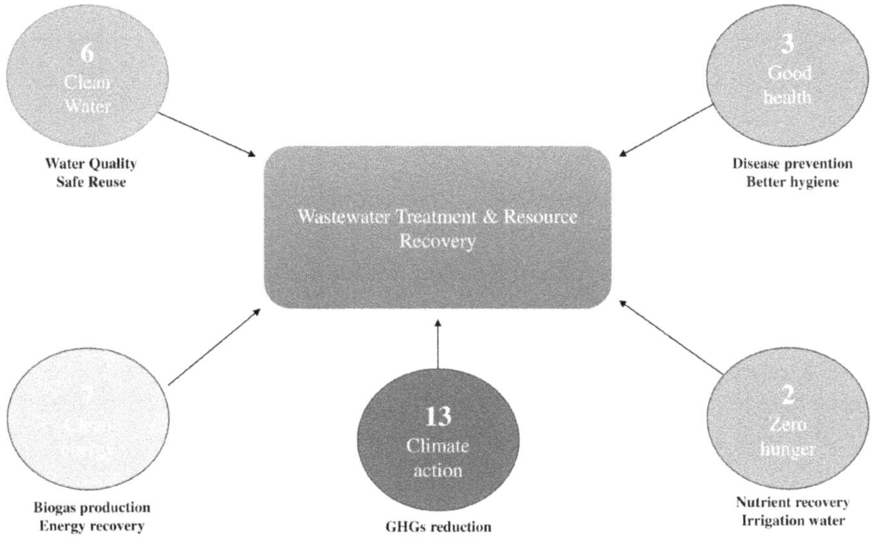

FIGURE 3.2 Linkages between wastewater Treatment and Sustainable Development Goals (SDGs).

3.13 WASTEWATER AS A RESOURCE WITHIN A CIRCULAR ECONOMY FRAMEWORK

In the framework of a circular economy, wastewater is also becoming less of a waste product and more of a resource (Figure 3.3) that can be used to some effect to produce a variety of benefits, instead of a use-and-dispose paradigm. This transformation essentially re-invigorates the purpose of a wastewater treatment plant (WWTP) and rebrands it as a Water Resource Recovery Facility (WRRF), which is an active extractor and valorizer of embedded resources. These resources include treated water for reuse, essential nutrients such as phosphorus and nitrogen for agriculture, and energy in the form of biogas as well as heat.

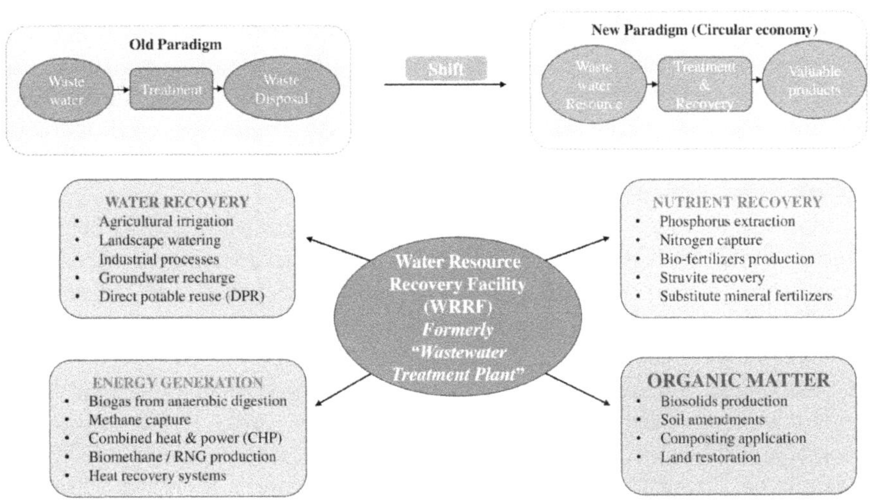

FIGURE 3.3 Transition from conventional wastewater treatment to water resource recovery facility (WRRF) under circular economy framework.

TABLE 3.3 Resource Recovery Potential from Wastewater

RESOURCE	RECOVERY METHOD	END USE
Organic matter	Anaerobic digestion	Biogas
Nitrogen and phosphorus	Sludge processing	Biofertilizer
Treated water	Advanced treatment	Irrigation, potable reuse

The application of circular economy principles in wastewater management delivers strong environmental benefits, including reduced freshwater abstraction, minimized waste generation, lower carbon emissions, and enhanced system resilience. Organic sludge generated during sewage treatment can be used for renewable energy production through biogas generation and as a phosphorus-rich fertilizer for sustainable agriculture [62]. Traditionally landfilled materials such as grit can now be reused in construction, reducing quarrying activities and associated emissions. Heat recovery from treated wastewater for district heating further illustrates a holistic approach to resource recovery.

However, achieving these benefits requires overcoming technological, regulatory, financial, and social barriers. In many regions, conventional wastewater systems still result in limited nutrient recovery, contributing to eutrophication and climate-related impacts [69]. Source-separation systems in low-density areas have demonstrated higher nutrient recovery efficiencies, particularly for phosphorus and nitrogen, while reducing the presence of harmful contaminants compared to conventional centralized systems [37].

For example, blackwater separation allows recovery of organic matter and nutrients suitable for anaerobic digestion and energy production (Table 3.3). Diverted urine, although requiring careful management to prevent struvite precipitation, enables recovery of highly concentrated nutrients. Such systems can significantly reduce reliance on mineral fertilizers and contribute toward nutrient self-sufficiency in agriculture, aligning wastewater treatment facilities with global goals of water security, climate resilience, and resource efficiency.

3.14 ORGANIC MATTER RECOVERY FOR SOIL IMPROVEMENT

Organic matter recovery from wastewater plays a crucial role in circular economy-based soil enrichment strategies. Wastewater by-products such as sewage sludge and digestate contain valuable organic matter and nutrients that can be recycled for agricultural use [62]. Traditionally, these organic wastes were disposed of in landfills; however, thermochemical conversion technologies now offer alternative pathways for resource recovery [13]. Biosolids derived from domestic wastewater can be treated to minimize pathogens and odors, producing safe agricultural fertilizers.

Similarly, anaerobic digestion of organic sludge produces biogas while leaving behind digestate rich in stabilized nutrients, making it suitable for direct application as a biofertilizer and soil conditioner [24, 55]. Composting further stabilizes digestate, enhances nutrient recycling, and reduces dependence on virgin resources. However, land application must comply with strict sanitation standards to avoid soil and groundwater contamination.

Challenges associated with organic matter reuse include product quality control, environmental safety, regulatory compliance, and social acceptance. Nevertheless, recycling composted sewage-based materials into organic fertilizers contributes significantly to a circular bioeconomy [22]. Emerging technologies such as microbial recycling cells (MRCs) also enable recovery of organic carbon and nutrients, converting waste streams into economically valuable fertilizer products.

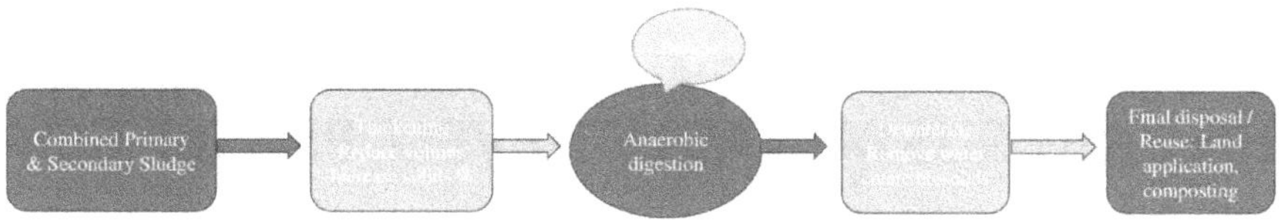

FIGURE 3.4 Sludge treatment and resource recovery pathway in wastewater treatment plants.

3.15 BIOGAS GENERATION AND LOW-CARBON ENERGY PRODUCTION

Anaerobic digestion of wastewater for biogas production represents a cornerstone of low-carbon energy generation (Figure 3.4) in wastewater treatment systems, offering both environmental and economic benefits. This biological process converts organic matter into biogas, primarily composed of methane (CH_4) and carbon dioxide (CO_2), under oxygen-free conditions. Sewage sludge is an ideal feedstock due to its high organic content [72].

Co-digestion of external organic wastes such as food residues, fats, oils, and grease can significantly enhance biogas yields [26]. The resulting biogas, typically containing 50–75% methane, can be utilized for combined heat and power (CHP) generation, enabling wastewater treatment plants to meet their own energy demands and potentially export surplus electricity to the grid [72].

Advanced biogas upgrading removes impurities such as CO_2 and hydrogen sulfide, producing biomethane or renewable natural gas suitable for pipeline injection or vehicle fuel use [29]. Methane recovery further reduces greenhouse gas emissions while improving energy efficiency [19, 57]. Overall, wastewater-based biogas production exemplifies circular economy principles by reducing waste, lowering emissions, and transforming treatment plants into net energy producers [56, 58].

3.16 TREATED WASTEWATER REUSE FOR IRRIGATION AND POTABLE PURPOSES

Treated wastewater reuse is a critical component of circular water management, supporting both irrigation and potable water applications. In agriculture, reclaimed wastewater conserves freshwater resources, enhances drought resilience, and supplies essential nutrients such as nitrogen and phosphorus, reducing dependence on synthetic fertilizers [46]. This practice is particularly valuable in water-scarce regions, where it also mitigates nutrient pollution in natural water bodies [39].

Safe wastewater reuse requires adherence to strict quality standards to protect public health and the environment [12]. Regulatory frameworks such as the EU regulation on water reuse establish minimum quality requirements and risk management protocols for agricultural reuse [29, 78]. While non-potable reuse is widely accepted, direct potable reuse (DPR) is increasingly adopted in regions facing severe water scarcity.

Successful DPR initiatives in countries such as Namibia, South Africa, and parts of the United States demonstrate the feasibility of advanced treatment technologies combined with rigorous monitoring and public engagement [2]. Public acceptance, transparency in governance, and economic feasibility are essential for scaling wastewater reuse systems, ensuring their long-term sustainability within a circular economy framework [38].

3.17 ROLE OF WASTEWATER MANAGEMENT IN ACHIEVING SUSTAINABLE DEVELOPMENT GOALS

The achievement of Sustainable Development Goal (SDG) 6, which aims at ensuring that all people have access to water and sanitation and their sustainability, is impossible without wastewater management. It is important because it has made considerable contributions to the well-being of the population, the environment, and resource recovery, thus moving in the direction of the magnitude of the circular paradigm instead of the traditional waste disposal process. Good management addresses water shortage and water pollution, which are major issues worldwide due to population surge, urbanization, and climate change, and is the core of broader sustainable development initiatives that influence such sectors as health and economic development.

Furthermore, wastewater management is essential in meeting many other SDGs through guaranteeing environmental safety, promoting human health, and facilitating sustainable resource management; proper treatment and reuse are crucial to the maintenance of water quality, ecosystem protection, and socio-economic development. It is also applicable to sustainable cities and communities and affordable and clean energy. Without strong wastewater systems and practices in its management, any form of development toward these interdependent aims would be significantly slowed.

3.18 CONTRIBUTIONS TO WATER, SANITATION, AND HEALTH GOALS

Wastewater management has a central role in supporting the water, sanitation, and health (WASH) goals by transforming a waste product that has been traditionally seen as a waste into a resource and hence ensuring a continuous cycle of water reuse, akin to natural processes. The achievement of SDG Target 6.3, which focuses on enhancing water quality by reducing pollution, halving untreated wastewater, and increasing recycling and safe reuse, directly benefits public health [69]. Currently, around 80% of all wastewater is untreated, contributing to large-scale water contamination and health risks [47]. As an example, 2.2 billion people lack access to safely managed drinking water, and 4.2 billion lack access to safely managed sanitation services, leading to approximately 300,000 child deaths annually due to WASH-related diarrheal diseases [70].

Wastewater treatment significantly reduces the risks of waterborne diseases by preventing exposure to contaminated water sources, thereby ensuring access to safe drinking water [67]. UNICEF emphasizes the creation of environments where children can grow and learn through access to safe drinking water, sanitation, and hygiene services in homes, schools, and healthcare facilities [68]. The WHO/UNICEF Joint Monitoring Programme (JMP) plays a crucial role in tracking SDG 6 progress by monitoring safely managed drinking water, sanitation services, and access to handwashing facilities [33].

Safe sanitation encompasses not only the availability of facilities but also the appropriate treatment or disposal of excreta, either on-site or off-site, with secondary or higher treatment levels considered safely managed for SDG monitoring. The incorporation of sustainable water-management practices, including advanced wastewater treatment systems, directly benefits local communities by improving health and environmental behaviors. Applying circular-economy principles enables the recovery and reuse of water, energy, and nutrients, generating significant social and economic benefits.

3.19 CLIMATE ACTION, CLEAN ENERGY, AND ENVIRONMENTAL PROTECTION

The integration of renewable energy technologies within wastewater treatment plants (WWTPs) contributes significantly to climate action by improving energy efficiency and reducing environmental impacts. WWTPs are energy-intensive and often reliant on centralized power grids, leading to substantial greenhouse gas emissions. Renewable energy recovery technologies such as anaerobic digestion, microbial fuel cells, and sludge gasification can lower operational costs and reduce fossil fuel dependence [44, 60]. Anaerobic digestion, widely used in sewage sludge treatment, produces biogas that can meet internal energy demands and enable energy self-sufficiency [44].

Microbial fuel cells generate electricity from wastewater through bacterial electrochemical activity, offering low-energy power solutions. Sludge gasification converts sludge into syngas while reducing sludge volume, eliminating odors, and destroying pathogens, making wastewater management more sustainable [44]. The transition to renewable energy integration in WWTPs aligns with climate targets to increase renewable energy use and reduce carbon footprints [30]. These approaches enhance urban infrastructure resilience and promote sustainable water resource management.

3.20 SOCIO-ECONOMIC BENEFITS AND LIVELIHOOD GENERATION

Improved wastewater management generates substantial socio-economic benefits and livelihood opportunities. The construction and operation of modern WWTPs require skilled professionals, directly employing engineers, operators, and technical personnel. Additionally, the reuse of treated wastewater in agriculture and aquaculture stimulates local economies. Reclaimed wastewater used for irrigation provides a nutrient-rich and reliable water source, improving crop productivity and diversification [1], thereby enhancing farmer incomes and food security in water-stressed regions [70].

Decentralized wastewater treatment systems further empower local communities by fostering small-scale enterprises for system operation and maintenance, building technical capacity, and economic self-reliance. Resource recovery from wastewater, such as renewable natural gas (RNG) or biomethane production, creates new economic opportunities and employment; for example, converting human waste to RNG could generate tens of thousands of jobs while reducing fossil fuel use [1]. These projects reduce reliance on synthetic fertilizers and fossil fuels, lower operational costs, and generate additional revenue streams.

Improved environmental conditions from effective wastewater treatment reduce disease prevalence, lower healthcare costs, and increase workforce productivity [28]. Cleaner water bodies also support tourism and recreation, providing further economic benefits. Together, these outcomes highlight the broad socio-economic significance of sustainable wastewater management.

3.21 ADVANCED AND EMERGING TECHNOLOGIES FOR SUSTAINABLE WASTEWATER TREATMENT

Advanced and emerging technologies play a critical role in optimizing wastewater treatment by improving efficiency, reducing environmental impacts, and enhancing water sustainability amid global water pollution and scarcity. Innovations such as nanotechnology-based purification and smart real-time monitoring systems extend beyond conventional methods, offering safer, more efficient, and cost-effective treatment solutions. These technologies enable effective contaminant removal, ecosystem protection, and resource recovery, supporting circular economy objectives.

3.22 NANOTECHNOLOGY-BASED PURIFICATION METHODS

Nanotechnology-based purification exploits the unique properties of nanomaterials, including high surface-area-to-volume ratios and enhanced reactivity, for wastewater treatment through adsorption, filtration, and photocatalysis. Carbon nanomaterials such as graphene oxide (GO) and carbon nanotubes (CNTs) are widely used in molecular separation processes. For example, unzipped CNT–GO membranes have demonstrated high stability across pH ranges and permeability exceeding that of pure GO membranes [31].

These composite membranes achieve dye rejection rates exceeding 99% for contaminants such as methylene blue and rhodamine B [31]. CNT-modified nanofiltration membranes also exhibit improved permeability and separation efficiency [71]. Incorporation of double-walled CNT interlayers has been shown to significantly increase water flux while maintaining high sulfate rejection rates [73].

CNT-based membranes offer advantages such as high water permeability, selectivity, and antifouling properties, making them suitable for wastewater filtration [34, 65]. Incorporation of CNTs into polymeric membranes improves hydrophilicity, porosity, mechanical stability, and contaminant rejection [71]. Metal nanoparticle-based technologies further enhance adsorption, degradation, and nanoscale filtration processes, offering efficient solutions for removing diverse pollutants.

3.23 SMART SYSTEMS FOR REAL-TIME MONITORING

Smart real-time monitoring systems integrate the Internet of Things (IoT), Artificial Intelligence (AI), and Supervisory Control and Data Acquisition (SCADA) to enhance wastewater treatment efficiency and reliability. IoT sensors continuously monitor parameters such as pH, temperature, turbidity, dissolved oxygen, chemical oxygen demand, and flow rates [32, 61]. Real-time data collection enables immediate performance assessment and regulatory compliance while preventing environmental pollution [32, 49].

AI and machine-learning algorithms analyze large datasets to support predictive maintenance, fault detection, and process optimization. AI-driven optimization of aeration, chemical dosing, and sludge handling significantly reduces energy consumption, particularly in aeration-intensive operations [74]. SCADA systems provide centralized control and oversight, enabling integrated monitoring and automation [5].

These intelligent systems support proactive maintenance, minimize downtime, automate compliance reporting, and enable remote operation. Digital twin technology further enhances decision-making by simulating operational scenarios and optimizing responses. Overall, smart monitoring represents a paradigm shift toward data-driven, sustainable wastewater management.

3.24 ENERGY-EFFICIENT AND LOW-WASTE TREATMENT SOLUTIONS

Energy-efficient and low-waste treatment technologies are essential for sustainable wastewater management due to the high energy demand and sludge production of conventional systems. Anaerobic membrane bioreactors (AnMBRs) combine anaerobic treatment with membrane filtration, delivering high-quality effluent, reduced sludge production, compact footprints, and energy recovery [50]. Partial nitrification/anammox (PN/A) processes significantly reduce aeration energy requirements compared to conventional nitrogen removal.

Microalgae-based systems utilize photosynthesis to remove nutrients and organic matter while generating reusable biomass, offering eco-friendly and energy-efficient solutions. Heat recovery systems extract thermal energy from wastewater to reduce heating and cooling demands in treatment facilities and nearby buildings [42, 45]. Anaerobic digestion remains a cornerstone of energy-efficient wastewater treatment, converting organic matter into biogas and enabling energy self-sufficiency while reducing greenhouse gas emissions [11, 20].

Integrating these complementary technologies into comprehensive treatment maximizes performance, enhances resource efficiency, conserves energy, and minimizes environmental impacts. Continued research and technological advancement are required to overcome remaining technical and economic barriers and enable widespread adoption of sustainable wastewater treatment systems.

3.25 FUTURE PERSPECTIVES, POLICY IMPLICATIONS, AND CONCLUSION

Future environmental health and resource management is based on the shift to sustainable and circular sanitation systems, as the traditional methods of work are largely based on disposal and should be converted into reuse and recovery of resources. Limited financial resources, poor policy frameworks, and inappropriate governance are some of the factors that challenge this shift in that they reduce effective implementation and sustainability. In order to overcome these challenges, a strong political will, sound governance institutions, stronger stakeholder partnerships, and combined policy strategies are mandatory in ensuring sustainable and equitable sanitation services across the world.

3.26 TRANSITION TOWARDS SUSTAINABLE AND CIRCULAR SANITATION SYSTEMS

The shift to sustainable and circular sanitation systems is essential to deal with the global concern of climate change, resource depletion, and population health, and requires a thorough change in the practice of wastewater management. The change will be the shift from traditional linear disposal of waste to the integrated systems that are more concerned with resource recovery, such as nutrients and energy within the human streams of waste. These systems will need more than technological enhancements; they will need great policy change, money, and community involvement in order to break through obstacles like infrastructure expenses and social approval of new ways of sanitation.

Circular sanitation systems have significant co-benefits, and this will contribute to climate action by reducing emissions, improving food security by using recycled fertilizers, and increasing energy security by producing biogas using wastewater. Nevertheless, the unawareness of interrelationships between sanitation and climate change and the low levels of inclusion in climate policy and finance are currently hindering progress, which reinforces the need to develop strong frameworks and systematic evaluation to guide decision-making and provide increased scale to sustainable solutions worldwide.

3.27 POLICY, GOVERNANCE, AND IMPLEMENTATION CHALLENGES

It means that effective policies have to find a way across the intricate jurisdiction, incorporate a variety of interests of various stakeholders, and be able to adjust to the continuously changing scientific knowledge on the topic of wastewater contaminants and treatment efficacy. The governance frameworks that will be needed to manage the implementation of smart wastewater technologies, like the real-time monitoring systems and AI-based predictive analytics, should be strong, open, and able to promote inter-agency cooperation. Funding limitations, issues related to public acceptance, and the requirement of skilled personnel to run and maintain sophisticated systems also contribute to the implementation challenges.

To deal scientifically with these multifaceted issues, it is necessary to create adaptive policy frameworks capable of integrating the dynamic feedback mechanisms of smart technologies, establish the regulatory standards of data privacy and security in wastewater management clearly, and create incentive schemes that could incentivize the practices of sustainability and technological uptake under the different socio-economic conditions. This holistic strategy is needed in order to transform the potential of smart technologies into practical changes in sanitation and water sustainability.

3.28 KEY CONCLUSIONS AND WAY FORWARD

The successful incorporation of smart technologies in sewage and wastewater management is the key to global water sustainability and increased sanitation. A scientific conclusion that is very critical is that real-time sensor data, combined with the latest AI and machine-learning algorithms, has never offered such opportunities in the field of predictive maintenance, optimization of treatment processes, and prompt detection of contamination incidents. The energy use and chemical waste used in the treatment plants can be reduced significantly by this technological synergy, and this will have a direct impact on resource efficiency and decreased operational expenses. Moreover, the intensive quality and quantity control of effluents with the help of smart systems is also imperative, as it helps to adhere to the changing regulatory requirements and reduce the environmental effects. The way forward is to create stronger, self-learning systems that can adjust to the changing environmental actions and development trends in cities to have a resilient and sustainable wastewater system.

REFERENCES

1. ADEC ESG. *Environmental, social, and governance strategies for wastewater-based resource recovery.* Alaska Department of Environmental Conservation, 2024.
2. Aleksic, A., & Sustersic, V. Potable water reuse systems and public acceptance. *Water Policy*, 24(3), 345–360 (2022). https://doi.org/10.2166/wp.2022.053
3. Aman, H., Doost, Z. H., Hejran, A. W., Mehr, A. D., Szczepanek, R., & Gilja, G. Survey on the challenges for achieving SDG 6: Clean water and sanitation—A global insight. *Knowledge-Based Engineering and Sciences*, 5(3), 21–42 (2024).
4. Andersson, K., Dickin, S., & Rosemarin, A. Towards sustainable sanitation: Challenges and opportunities. *Sustainability*, 10(6), 1820 (2018). https://doi.org/10.3390/su10061820
5. Applied Technology & Innovation. *Smart SCADA solutions for wastewater treatment plants*, 2025.
6. Aso, S. N. Digestate: The coproduct of biofuel production in a circular economy. In *Renewable energy technologies and applications.* IntechOpen, 2020. https://doi.org/10.5772/intechopen.90546
7. AUC Group. *Advantages of decentralized water and wastewater treatment*, 2023. Available online: https://aucgroup.net/advantages-of-decentralized-water-and-wastewater-treatment/
8. Encyclopaedia Britannica. *Wastewater treatment—Primary treatment.* Available online: https://www.britannica.com/technology/wastewater-treatment/Primary-treatment
9. AUC Group. *Activated sludge process*, 2025. Available online: https://aucgroup.net/how-activated-sludge-process-works/
10. Back to Blue Initiative. *Global water pollution and wastewater solutions*, 2025.
11. Bahman, A., & Yazdian, M. Environmental sustainability of decentralized wastewater systems. *Journal of Environmental Management*, 345, 118130 (2024). https://doi.org/10.1016/j.jenvman.2023.118130
12. Batool, A., & Shahzad, K. Anaerobic digestion for renewable energy recovery from sludge. *Renewable Energy*, 168, 1105–1115 (2021). https://doi.org/10.1016/j.renene.2020.12.077
13. Berbel, J., Mesa-Jurado, M. A., & Pistón, J. M. Water reuse regulations and public health protection. *Water Research*, 231, 119135 (2023). https://doi.org/10.1016/j.watres.2023.119135
14. Bora, R. R., Wang, R., & You, S. Thermochemical conversion of sewage sludge: A review. *Renewable and Sustainable Energy Reviews*, 123, 109734 (2020). https://doi.org/10.1016/j.rser.2020.109734
15. Cambi Group. *Biogas from wastewater*, 2025. Available online: https://www.cambi.com
16. Chavhan, N., et al. Autonomous IoT monitoring of industrial wastewater. *Digital Chemical Engineering*, 14, 100217 (2025). https://doi.org/10.1016/j.dche.2025.100217
17. CleanTech Water. *Energy-efficient centralized wastewater systems*, 2025.
18. Collivignarelli, M. C., Abbà, A., & Miino, M. C. Sludge rheology and wastewater treatment optimization. *Journal of Environmental Engineering*, 148(4), 04022012 (2022). https://doi.org/10.1061/(ASCE)EE.1943-7870.0001962
19. Columbia Climate School. *Water infrastructure and climate resilience*, 2012.
20. Crone, B. C., Garland, J., & Svoronos, S. A. Methane recovery from wastewater treatment. *Environmental Science & Technology*, 50(5), 2345–2353 (2016). https://doi.org/10.1021/acs.est.5b04466
21. Dąbrowska, L., & Masłoń, A. Energy efficiency in wastewater treatment plants. *Energies*, 13(4), 875 (2020). https://doi.org/10.3390/en13040875
22. Decentralized Wastewater Systems. *Advantages and disadvantages*, 2025.
23. Dereszewska, A., & Cytawa, S. Circular bioeconomy through sewage composting. *Waste Management*, 154, 160–171 (2023). https://doi.org/10.1016/j.wasman.2022.09.028
24. Dickin, S., Bisung, E., & Schuster-Wallace, C. Water, sanitation and child health. *International Journal of Environmental Health Research*, 26(1), 1–15 (2016). https://doi.org/10.1080/09603123.2015.1020414
25. Duque, A. F., Bessa, V. S., & Reis, M. A. Valorisation of wastewater-derived resources. *Bioresource Technology*, 321, 124486 (2021). https://doi.org/10.1016/j.biortech.2020.124486
26. Environmental and Energy Study Institute. *Biogas—Converting waste to energy*, 2017. Available online: https://www.eesi.org
27. FAO. *Guide to wastewater treatment and reuse in agriculture.* Food and Agriculture Organization, 2011.
28. Fayssal, F., Mortada, A., & Ghandour, M. Wastewater Treatment: Overview, Types, Energy Consumption & Actual State in Lebanon with Proposal of Using Renewable Energy. *Acta Materialia Turcica*, 4(2), 1–9 (2020).
29. Fluence Corporation. *Biogas upgrading and biomethane technologies*, 2025.

30. Gruber, J., & Kirchmair, M. Renewable energy targets and wastewater infrastructure. *Energy Policy*, 182, 113125 (2025). https://doi.org/10.1016/j.enpol.2023.113125
31. Han, Y., Xu, Z., & Gao, C. CNT–GO membranes for water purification. *Journal of Membrane Science*, 620, 118901 (2021). https://doi.org/10.1016/j.memsci.2020.118901
32. Hasan, S. W., Liu, H., & Naddeo, V. Smart monitoring in wastewater treatment. *Journal of Environmental Management*, 260, 110005 (2020). https://doi.org/10.1016/j.jenvman.2019.110005
33. Huber, M. Monitoring SDG 6 progress through JMP. *Water International*, 46(4), 512–526 (2021). https://doi.org/10.1080/02508060.2021.1902033
34. Ihsanullah. Carbon nanotubes for water treatment. *Separation and Purification Technology*, 209, 307–337 (2019). https://doi.org/10.1016/j.seppur.2018.07.012
35. Ilyas, H., Masih, I., & Van der Hoek, W. Industrial wastewater impacts on ecosystems and public health. *Environmental Monitoring and Assessment*, 191, 536 (2019). https://doi.org/10.1007/s10661-019-7680-6
36. Jones, E. R., van Vliet, M. T. H., & McDonald, R. I. Future water quality threats and management strategies. *Nature Sustainability*, 4, 317–324 (2021). https://doi.org/10.1038/s41893-020-00628-5
37. Kathi, S., Sinha, A., & Singh, R. Nutrient recovery from wastewater. *Journal of Cleaner Production*, 385, 135604 (2023). https://doi.org/10.1016/j.jclepro.2022.135604
38. Katz, B. G., & Tennyson, J. Public perception of potable water reuse. *Water Research*, 144, 102–112 (2018). https://doi.org/10.1016/j.watres.2018.07.018
39. Kendrovics-Boda, E., et al. Agricultural reuse of treated wastewater. *Agricultural Water Management*, 206, 12–22 (2018). https://doi.org/10.1016/j.agwat.2018.04.018
40. Kerpaci, M., & Lako, A. Urban wastewater pollution mitigation strategies. *Environmental Science and Pollution Research*, 31, 4567–4580 (2024). https://doi.org/10.1007/s11356-023-29672-3
41. Kesari, K. K., et al. Health risks of untreated wastewater reuse. *Science of the Total Environment*, 755, 142417 (2021). https://doi.org/10.1016/j.scitotenv.2020.142417
42. Kowalik, P., & Bąk-Patyna, A. Heat recovery from wastewater. *Energy and Buildings*, 250, 111316 (2021). https://doi.org/10.1016/j.enbuild.2021.111316
43. Li, X., & Yang, Q. Pharmaceutical wastewater characteristics and treatment. *Journal of Hazardous Materials*, 343, 209–220 (2018). https://doi.org/10.1016/j.jhazmat.2017.09.019
44. Lima, R. M., Martins, R. C., & Quinta-Ferreira, R. M. Renewable energy integration in WWTPs. *Renewable Energy*, 224, 1–15 (2024). https://doi.org/10.1016/j.renene.2024.01.043
45. Łokietek, J., Kulczycka, J., & Pająk, M. Thermal energy recovery systems. *Sustainable Energy Technologies and Assessments*, 52, 101999 (2022). https://doi.org/10.1016/j.seta.2022.101999
46. Mannina, G., et al. Wastewater reuse in agriculture. *Water Research*, 215, 118251 (2022). https://doi.org/10.1016/j.watres.2022.118251
47. Mara, D. Sanitation and wastewater treatment challenges. *Waterlines*, 35(1), 3–20 (2016). https://doi.org/10.3362/1756-3488.2016.002
48. Mara, D., & Evans, B. Sanitation targets and financing. *Water Policy*, 20(3), 1–15 (2018). https://doi.org/10.2166/wp.2018.019
49. Matheri, A. N., et al. Digitalization in wastewater treatment. *Journal of Cleaner Production*, 345, 130945 (2022). https://doi.org/10.1016/j.jclepro.2022.130945
50. Menon, R., & Kalyanraman, V. Anaerobic membrane bioreactors. *Water Science and Technology*, 91(2), 345–356 (2025). https://doi.org/10.2166/wst.2024.513
51. Milićević, D., et al. Decentralized wastewater systems. *Sustainable Cities and Society*, 101, 104118 (2024). https://doi.org/10.1016/j.scs.2023.104118
52. National Institutes of Health. Synergies and trade-offs between sanitation and the SDGs. *Sustainability*, 13(14), 7892 (2021). https://doi.org/10.3390/su13147892
53. Obaideen, K., et al. Role of wastewater treatment in achieving SDGs. *Renewable and Sustainable Energy Reviews*, 156, 112029 (2022). https://doi.org/10.1016/j.rser.2021.112029
54. Okereke, C., et al. Industrial wastewater impacts. *Environmental Development*, 17, 1–12 (2016). https://doi.org/10.1016/j.envdev.2015.12.002
55. Our Future Water. *Wastewater treatment and resource recovery*, 2024.
56. Periyasamy, S., et al. Circular economy approaches. *Journal of Environmental Management*, 312, 114910 (2022). https://doi.org/10.1016/j.jenvman.2022.114910
57. Pfluger, A., et al. Methane mitigation in WWTPs. *Energy*, 195, 116991 (2020). https://doi.org/10.1016/j.energy.2020.116991
58. Porto, M. F., et al. Wastewater-to-energy systems. *Renewable and Sustainable Energy Reviews*, 176, 113125 (2023). https://doi.org/10.1016/j.rser.2023.113125

59. Saadatinavaz, S., et al. Urban wastewater infrastructure challenges. *Water Research*, 240, 120028 (2024). https://doi.org/10.1016/j.watres.2023.120028
60. Sharma, S., et al. Energy recovery technologies. *Renewable Energy*, 170, 130–142 (2021). https://doi.org/10.1016/j.renene.2021.01.045
61. Soetedjo, A., et al. IoT-based wastewater monitoring. *Sensors*, 24(3), 112 (2024). https://doi.org/10.3390/s24030112
62. Sonune, A. Organic matter recovery from sewage sludge. *Waste Management*, 156, 1–10 (2023). https://doi.org/10.1016/j.wasman.2022.11.017
63. Sustainable Sanitation and Water Management. *Trickling filter*, 2019. Available online: https://sswm.info/fact-sheet/trickling-filter
64. Tasker, T. L., et al. Radioactive pollutants in oil and gas wastewater. *Environmental Science & Technology*, 52(20), 11840–11849 (2018). https://doi.org/10.1021/acs.est.8b03369
65. Thamaraiselvan, C., et al. Carbon nanotube membranes. *Desalination*, 433, 129–140 (2018). https://doi.org/10.1016/j.desal.2018.01.013
66. The Nature Conservancy. *Wastewater pollution and ecosystem impacts*, 2025.
67. Tortajada, C. Water quality and public health. *International Journal of Water Resources Development*, 36(1), 1–14 (2020). https://doi.org/10.1080/07900627.2019.1708896
68. UNICEF. *Sanitation*, 2020. Available online: https://www.unicef.org/wash/sanitation
69. United Nations Environment Programme. *Indicators: Sustainable development goals*, 2023.
70. United Nations. *Goal 6: Clean water and sanitation*, 2023.
71. Vargas-Figueroa, J., et al. Nanofiltration membranes. *Journal of Membrane Science*, 675, 121132 (2023). https://doi.org/10.1016/j.memsci.2023.121132
72. Vutai, C., et al. Biogas from sewage sludge. *Renewable Energy*, 95, 18–27 (2016). https://doi.org/10.1016/j.renene.2016.03.073
73. Wang, Z., et al. CNT interlayer nanofiltration membranes. *Water Research*, 215, 118128 (2022). https://doi.org/10.1016/j.watres.2022.118128
74. Water & Wastewater. *Energy use in wastewater treatment plants*, 2025.
75. World Health Organization. *Sanitation and wastewater management*, 2023. Available online: https://www.who.int
76. Kato, S., & Kansha, Y. Comprehensive review of industrial wastewater treatment techniques. *Environmental Science and Pollution Research,* 31, 51064–51097 (2024). https://doi.org/10.1007/s11356-024-34584-0

Urban Wetlands and Green Infrastructures

4

Aravind Chauhan, Lokeshwer Sharma, Saurav Kumar Singh, Prerna Thakur, and Abhineet Anand

4.1 INTRODUCTION

Urban centers all over the world are grappling with complicated issues that include rapid urbanization, storm events getting stronger, loss of biodiversity, rising temperatures, and deteriorating water quality. It is becoming more and more apparent that the traditional "grey infrastructure" (pipes, concrete channels, storm drains) is not only very expensive but also inflexible and lacks the ability to adapt to the changing climate and the increasing population. Extreme weather events, like flash [1] floods or droughts that last for weeks, are becoming more frequent and, at the same time, they are showing how insufficient the conventional urban systems are, which were primarily designed for efficiency rather than resilience. In this scenario, nature-based solutions (NbS) have attracted worldwide attention as a viable, sustainable, and environmentally friendly alternative. To this effect, the two main categories, urban wetlands and green (and blue-green) infrastructures, are the most significant systems that mingle ecological concepts along with urban design and planning. These systems utilize the inherent potential that is already present in the vegetation, the soil, and the water bodies to reverse the hydrological cycles, purify the air and the water, regulate the temperature, and contribute to making the city more habitable overall.

Urban wetlands can be either natural or artificially created ecosystems that are vital from an ecological perspective within the urban matrix. By absorbing and slowly dispensing the water from the storm, they are the first line of defense against flooding in a given area. As biofiltration systems, they also serve by collecting the sediments and pollutants, and as carbon sinks, they aid in the fight against the cause of global warming. Besides that, they offer dwelling places for various plants and [2] animals and thus contribute to the building of urban green biodiversity networks. If one adds green infrastructure elements, e.g., parks, bioswales, vegetated roofs, and porous pavements armed with these technologies, one is assured of environmentally sound and resilient blue-green corridors that not only output the ecosystem services but also enhance human well-being in their presence. Green infrastructure is not only about planting or adding more green space. It is a well-thought-out interconnected plan for managing land and water resources that also takes into consideration the built environment. This accentuation marks a paradigm change, from treating nature as an entity separate from the cities to being aware that it is an inseparable part of the urban systems. On top of that, green infrastructure planning is consistent with

DOI: 10.1201/9781003742227-4

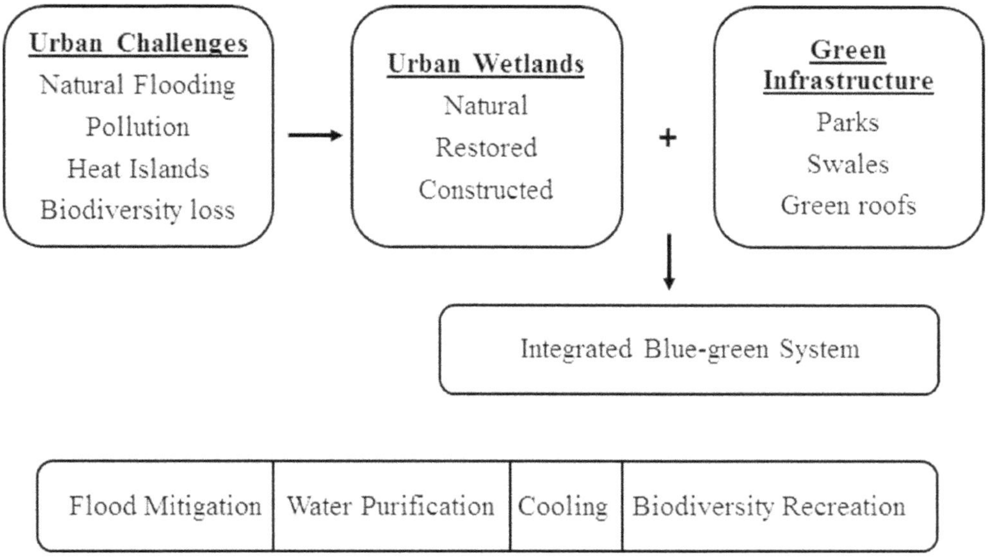

FIGURE 4.1 Conceptual framework of urban wetlands and green infrastructure.

worldwide guidelines, like the United Nations Sustainable Development [3] Goals (SDGs), particularly SDG 11 (Sustainable Cities and Communities) and SDG 13 (Climate Action). This chapter, on the one hand, presents a detailed analysis of the role those urban wetlands and green infrastructure perform and provides answers to questions such as what they are, why they matter, how they function, what benefits they deliver, and how they can be designed, implemented, and maintained in the urban context. It serves as a roadmap [4] for understanding how these technologies might be used for climate adaptation, disaster risk reduction, and the promotion of sustainable urban growth. In an attempt to address the issues, identify the best approaches, and also through illustrating a number of case studies, this chapter intends to be an all-inclusive resource for understanding the integration of nature in urban planning as a [5] means of achieving resilience, fairness, and ecological balance in future cities.

4.1.1 Green Infrastructure (GI)

Green Infrastructure (GI) is a network that is not only natural but also semi-natural and engineered green spaces. The network is strategically planned and managed in such a way as to provide the ecosystem services that are essential for urban sustainability. For instance, GI may be able to provide air purification, temperature regulation, flood control, groundwater [6] recharge, carbon sequestration, and recreational as well as aesthetic benefits. In contrast to the conventional gray infrastructure that depends on hard and engineered systems such as storm drains, pipes, and concrete channels, green infrastructure [7] still embraces natural processes to achieve environmental, social, and economic goals in tandem. Furthermore, it regards vegetation, soil, and natural hydrological systems [8] as assets rather than limitations of urban development.

Some of the typical green infrastructures are urban parks, green roofs, vegetated swales, urban forests, community gardens, rain gardens, and permeable pavements. These features are not independent; they operate as a collective and interconnected network that enhances urban resilience [9] by managing stormwater, mitigating the urban heat island effect, supporting biodiversity, and improving overall livability. On a strategic level, GI planning focuses on two aspects, namely multifunctionality and connectivity; this implies that not only should green spaces be present, but they should also be connected through ecological corridors which permit species movement and hydrological continuity. The systems-based approach understands the city as a living ecosystem made up of natural and human-made elements that need to coexist harmoniously. GI is now worldwide [10] recognized as an essential instrument for the deployment of nature-based solutions (NbS) and for making progress towards the Sustainable Development Goals (SDGs), especially SDG 11 (Sustainable Cities and Communities) and SDG 13 (Climate Action). It is the basis of the climate-resilient urban design which thus makes it [11] possible for cities to grow in a sustainable manner while at the same time increasing the well-being of their inhabitants.

4.1.2 Blue-Green Infrastructure (BGI)

Blue-Green Infrastructure (BGI) is an extension of the green infrastructure idea that deliberately incorporates the water systems (blue elements), such as rivers, streams, ponds, lakes, and wetlands, with the vegetation and soil systems (green elements) in the urban landscape. The integration recognizes that water and vegetation cannot be separated in the functioning of ecosystems and that managing them together gives urban sustainability yields of synergistic benefits. They also integrate a range of purely natural, nature-based, and hybrid origin infrastructure solutions as part of the BGI system, including wetlands, bioswales, retention and detention ponds, constructed wetlands, green roofs, rain gardens, permeable pavements, and restored waterways. Collectively, these features manage stormwater at its source, enhance water quality through natural filtration, and increase infiltration, thus lessening the load on drainage networks and mitigating urban flooding. The "blue-green" approach is aimed at imitating the natural water cycle whereby rainwater is allowed to infiltrate, evaporate, and be reused within the urban system instead of being rapidly channeled away. Besides, it supports urban cooling, improves microclimatic conditions, and creates aesthetic and recreational opportunities for city residents.

Additionally, BGI entails less carbon emissions, higher biodiversity, increased property values, and the mental and physical health of people through more access to natural spaces besides the hydrological functions it has. In contemporary urban planning, BGI is getting more and more acknowledged as the foundation of climate adaptation and water-sensitive urban design (WSUD) systems, thus helping cities to make a shift from reactive drainage management to resilience building that is proactive.

4.1.3 Urban Wetlands

An urban wetland is a natural or restored or even a constructed wetland ecosystem that is located either within or adjacent to the urban or peri-urban areas, and thus it carries out essential ecological and hydrological functions even though it is surrounded by a heavily modified environment. Wetlands are those areas that are defined by the presence of water, either permanently or seasonally, and by distinctive vegetation adapted to saturated soil conditions. Urban wetlands might be any of the following: marshes, swamps, mangroves, riparian buffers, ponds, and constructed wetlands. By their origin and management, they might be differentiated as follows:

- Natural wetlands are ecosystems that existed before and have become part of the urban landscape as cities expanded.
- Restored wetlands are those wetlands that were degraded or drained and have now been rehabilitated to regain ecological and hydrological functions.
- Constructed wetlands are engineered systems that are specifically designed to treat wastewater, manage stormwater, or provide ecological and recreational services.

Within the urban environment, these wetlands represent the vital elements of blue-green infrastructure. Among other things, they are natural water retention and purification facilities that clean filtered sediments, nutrients, and pollutants from the runoff, which then goes to the rivers or aquifers. In addition to these, they serve as flood buffers, which on the one hand do the temporary storage of excess stormwater during heavy rains and on the other hand release it slowly, thus reducing the downstream flood risks.

Besides their essential hydrological role, urban wetlands can be considered biodiversity hotspots because they can offer safe habitats for birds, fish, amphibians, and insects, which live within the built environment. Also, they help in carbon sequestration, microclimate regulation, and promoting human well-being through the provision of educational, cultural, and recreational opportunities. Urban wetlands, influenced by the multiple advantages, are at the moment acknowledged as multifunctional urban assets instead of being treated as marginal or leftover spaces. Besides, looking after them and incorporating them into the city plan is not only a matter of protecting nature but also a [12] condition for obtaining sustainable, resilient, and livable urban futures. By integrating the wetland hydrological functions with green infrastructure principles, cities have the opportunity, through nature-based solutions, not only to build resilience but also to promote well-being and conserve biodiversity.

4.2 FUNCTION AND VALUE OF URBAN WETLANDS AND GREEN INFRASTRUCTURE

4.2.1 Hydrological and Water Management Functions

Urban wetlands and the related green infrastructures are not only the lungs of the cities but also the main actors; they play a role in the drama of stormwater and urban runoff management:

- By their interception, storage, and slowing down of surface runoff, they thus lower peak flows, mitigate flood risk to an extent, and lessen the load on drainage systems.
- Vegetation in the wetland and soils are, in fact, biofilters, enabling water purification through the elimination of nutrients, sediment, and pollutants. At the same time, they maintain groundwater recharge and hydrological connectivity where their presence is eco-friendly and sustainable.
- Through their evaporation, shading, and transpiration, wetlands also help regulate the local microclimate, thus contributing to the reduction of the urban heat island effect.

4.2.2 Ecological and Biodiversity Benefits

Urban wetlands allow many types of plants, insects, birds, amphibians, and small mammals to live there; they become, in many cases, the main biodiversity that survives in the urban matrix. Moreover, they can become a necessity for the continuity of natural habitats, thus connecting the green spaces or water

TABLE 4.1 Comparison between Grey and Green Infrastructure

PARAMETER	GREY INFRASTRUCTURE	GREEN INFRASTRUCTURE
Design approach	Engineered, rigid, single-function (drainage, flood control)	Nature-based, flexible, multi-functional
Examples	Drains, culverts, concrete channels	Parks, wetlands, bioswales, green roofs
Environmental impact	High carbon footprint, disrupts natural systems	Enhances biodiversity, improves air and water quality
Cost and maintenance	High operational cost, aging infrastructure	Lower long-term cost, natural regeneration
Climate resilience	Limited adaptability	High adaptability to climate extremes
Social benefits	Minimal recreational value	Recreation, aesthetics, wellbeing, education

bodies, which are fragmented in the urban landscape. Additionally, they are involved in absorbing carbon from the atmosphere (although their contribution is usually much less than that of large forests) and in regulating natural processes of ecosystems.

4.2.3 Social, Aesthetic, and Recreational Value

- The natural areas such as wetlands along with green infrastructure are able to deliver spaces of amenity and recreation: walking on boardwalks of wetland, parks, nature [13] trails improves the physical and mental health of people, creates new recreational opportunities, and promotes social interaction.
- They also serve as sources of education and inspiration (e.g., learning in nature, involvement of the local community in restoration activities).
- Their existence enhances the look of the cities, the feeling of belonging, and frequently leads to the rise of property values (however, also consider the possibility of green gentrification).

4.2.4 Climate Adaptation and Mitigation Roles

As climate change continues to increase the severity of rainfall, raise flood risk, and cause heatwaves and droughts, wetlands and green infrastructure still offer the adaptive capacity needed to cope with such events by their very nature. They store water, cool urban areas, and act as a buffer against extreme events. Being a component of a nature-based solution strategy, they are consistent with the principles of sustainable urban development and the circular economy (e.g., water reuse, nutrient cycles, ecosystem service delivery).

4.3 TYPES AND DESIGN CONSIDERATIONS

Urban wetlands and green infrastructures are natural, restored, or constructed systems with the aim of water management, habitat creation, and recreation. An [14] efficient design should take into account hydrology, water quality, integration of the area, and native vegetation in order to assure ecological performance. The choice of a site should encourage connectivity within the urban green-blue network while at the same time allowing human access and addressing the needs of biodiversity. Upkeep, surveillance, and governance are, however, as important as design features to ensure that the system is sustainable and able to adapt to the changing urban and climatic conditions in the long run.

4.3.1 Types of Urban Wetlands/Green Infrastructure

Urban wetlands exist in various forms—natural, restored, and constructed—each having different origins, structures, and functions of the ecosystem. These classes are illustrated in Figure 4.2, which shows the typology of urban wetlands and the respective subtypes, thus indicating the system diversity that is the basis for urban resilience and ecological balance.

- **Constructed Wetlands**: Man-made environmentally friendly systems used for water purification, stormwater management, or leisure purposes.
- **Restored/Naturalized Wetlands**: Once degraded or drained wetlands that have been given a new lease on life and are now part of the urban fabric.

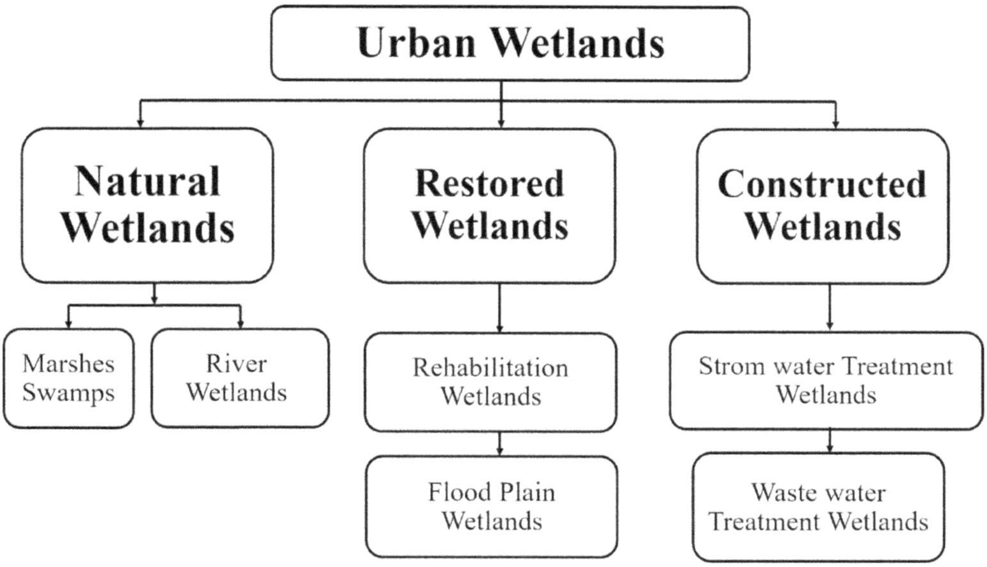

FIGURE 4.2 Typology of urban wetlands.

- **Bioswales, Rain Gardens, Permeable Pavements**: These GI elements, although not wetlands, usually provide the source water to or support the wetland systems in the urban environment.
- **Wetland Parks/Ecological Corridors**: Urban parks with the elements of open water, riparian zones, ponds, and wetlands for recreation and habitat.

4.3.2 Key Design and Implementation Considerations

The effective design of urban wetlands as well as the green infrastructure needs to consider the following points:

- **Catchment and Hydrology**: Grasping inflows (stormwater, runoff, wastewater), storage demands, retention/detention volumes, overflow paths, groundwater levels.
- **Water Quality**: Determine pollutant loads, nutrient inputs, sediment loads, and accordingly design vegetation/soils for filtration, settling and uptake.
- **Spatial Integration**: integration into urban structure, link to [15] other green/blue elements, safe access for people, synergies with urban design.
- **Vegetation and Habitat Structure**: Dependence on native species, zonation (emergent vegetation, open water, riparian buffer), habitat enhancement.
- **Maintenance and Management**: Sediment accumulation, invasive species, water levels, vegetation health, monitoring; maintenance, good cooperation.
- **Socio, Ecological Considerations**: Access, safety, programming of space, avoiding exclusion, balancing ecological function and human use.
- **Cost and Benefits**: By the way, even though green infrastructure may have a lower operational cost than gray infrastructure in most cases, one should not disregard the upfront cost, land availability, and long, term maintenance.
- **Policy, Governance, and Stakeholder Engagement**: Urban wetlands/green infrastructure typically necessitate coordination across departments (water, planning, parks), policy support, community involvement.

Basically, these design and implementation considerations for an urban blue-green ecosystem, when combined, constitute the core principles of a holistic system.

4.4 URBAN WETLANDS AND GI IN PRACTICE: BENEFITS AND CHALLENGES

4.4.1 Benefits in Practice

According to a single study, urban wetlands may be called "the core unit of urban ecosystems" because they deliver a wide range of positive effects like water purification, lessening of urban runoff, climate stabilization as well as giving psychological and social benefits.

- Another review states that installation of wetlands in cities is a green infrastructure solution that has been confirmed to be effective in water management, habitat creation, recreation, and landscaping.
- The presence of wetlands in the urban environment enhances human health, motivates people to be physically active in nature, and furthers social life by increasing the number of contacts and the availability of green/blue space.

The flowchart conveys the chain of events leading to the positive changes in the environment and society of urban wetlands. It illustrates the role of wetlands as Earth's water regulators, pollutant filters, and habitat providers for biodiversity. These features then lead to larger impacts such as reduction of floods, improvement of water quality, and carbon storage. Moreover, wetlands beautify cities, provide leisure activities, and strengthen the general climate resilience. All these mutually linked benefits, which are at the core of the existence of nature, highlight the indispensable function of wetlands in the formation of sustainable and livable urban ecosystems.

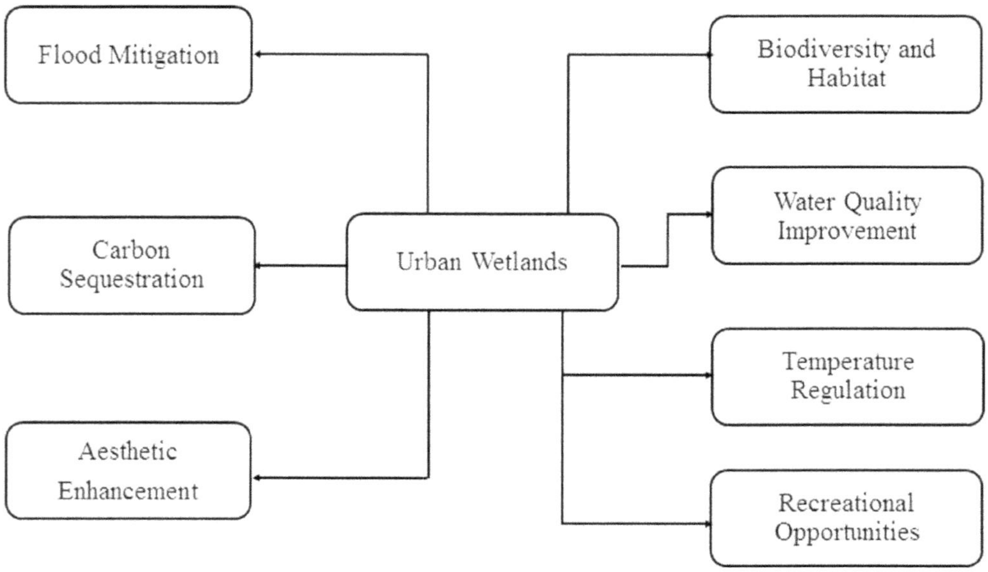

FIGURE 4.3 Benefits flowchart of urban wetlands.

4.4.2 Challenges and Pitfalls

Land Availability and Competing Uses: Urban land tends to be [16] pricey; therefore, there might be competition for the space that is to be set apart for wetland and green infrastructure (GI).
Maintenance Burden: Pollutant load, invasive species, water level management, safety issues (mosquitoes) can require a certain amount of money and attention in the long run.
Design Mismatch: Inefficiently designed systems may deliver less performance than expected (e.g., too short retention time, too heavy pollutant loads).
Perception and Acceptance: Some stakeholder groups might consider wetlands as "leftover" spaces, waterlogged, dangerous, or less utilized unless thoroughly integrated.
Equity Issues: The green infrastructure can cause green gentrification, i.e., where the enhancements lead to an increase in the value of local properties, thus displacing lower-income residents.
Policy and Governance Fragmentation: Infrastructure may cross several jurisdictions/departments; the absence of clear governance can be a barrier to the realization.
Climate Change Uncertainties: Design features (rainfall intensity, storm frequency) that have been relied on may not be valid anymore; the ability to adapt has to be incorporated.

4.5 PLANNING AND IMPLEMENTATION FRAMEWORK

4.5.1 Step-by-Step Numerical Framework for Cities

Assessment and Baseline

Identify existing resources and constraints:

$$A_w = \sum A_{wetlands}, Q_r = C \cdot I \cdot AS_e = f\left(B_d, L_u, I_c\right)$$

where

A_w= wetland area (ha),
Q_r= runoff (m³/s),
C= runoff coefficient,
I= rainfall intensity (mm/hr),
A= catchment area (ha),
S_e= socio-economic index,
B_d= biodiversity,
L_u= land-use conflict index,
I_c= income/community dependence.

Vision and Objectives

Define target performance values:

$$Q_{f,red} \geq 30\%, WQ_{imp} \geq 40\%, B_{gain} \geq 20\%$$

where

$Q_{f,red}$= flood peak reduction,
WQ_{imp}= water quality improvement,
B_{gain}= biodiversity gain.

Design Selection

Select wetland/GI type based on suitability index:

$$SI = w_1 H + w_2 E + w_3 U$$

where

SI = suitability index,
H = hydrologic compatibility,
E = ecological potential,
U = urban integration score,
$w_1 + w_2 + w_3 = 1$.

4.5.2 Policy and Institutional Levers

- Embed wetlands and green infrastructure in the urban master plans, stormwater management plans, land use zoning, and development regulations.
- Offer incentives (for example, green infrastructure credits, stormwater fee reductions) to developers to encourage the integration of wetlands/GI.
- Create maintenance funds or dedicated governance units to ensure the continued operation and avoid the degradation of the facilities.
- Encourage community stewardship and education so that wetland/GI locations become the residents' common property and they take care of them.
- Connect climate resilience targets (flood risk, heat mitigation) with GI/wetland planning.
- Implement performance metrics, such as retention volume, pollutant removal, biodiversity indicators, and user access, to assess the efficiency of projects.

4.6 CASE STUDIES AND LESSONS FOR EMERGING CITIES

4.6.1 Example: Wetlands as Part of Urban Blue-green Infrastructure

Urban wetlands comprise three major categories—natural, restored, and constructed—each of which performs distinct ecological and hydrological functions in urban areas. Natural wetlands are those that have existed before urban development and thus have become a part of the expanding cityscape [17]. They usually have a high level of biodiversity and can serve as ecological refuges in densely built-up areas. Restored wetlands are those areas of wetlands that were once degraded or drained and have now been rehabilitated to regain their natural functions [18]. By managing water flow, vegetation, and soil [19] conditions carefully, these ecosystems become capable of filtering pollutants, storing stormwater, and providing habitats for wildlife again. On the other hand, constructed wetlands are artificially created systems that closely resemble natural wetland processes to achieve specific objectives such as stormwater treatment, flood control, and recreational use. They are a viable [20] solution to the problem of how to integrate ecology with infrastructure by using the principles of both engineering and natural design. These three types of wetlands, in combination, constitute a network that helps urban areas become resilient by ensuring the availability and proper management of water resources, flood protection, purification of water, and increasing biodiversity. The incorporation of these wetlands in urban planning is an example of the shift from traditional, engineered drainage systems to nature-based solutions that contribute to sustainability and human well-being, as well as strengthening ecological integrity in the urban fabric.

Example: Constructed wetlands for multiple purposes

Constructed wetlands technology is a well-known environmentally friendly multi-purpose solution for water management and wastewater treatment that brings in a lot of environmental and economic advantages.

4.6.2 Particular Relevance for Cities in India and Similar Contexts

Indian cities (as well as other emerging cities) facing rapid urbanization, increased flood risk, water pollution, and heat stress can immensely benefit from the integration of urban wetlands/green infrastructure:

- Creation of wetlands through the use of degraded urban land or stormwater detention basins.
- Designing for multi-use amenities: flood control + recreation + biodiversity + education.
- Local ecology and climate (monsoon intensity, high rainfall variability) consideration.
- Compliance with national/regional policy mechanisms (urban water body protection laws, GI standards). For instance: in the UK, wetlands have become a part of the [21] national green infrastructure standards.

4.6.3 Key Lessons

- Involving them early in the integration planning helps to avoid gray conversions (e.g., wetlands that are drained and built over).
- Hybrid (mixing gray and green) usually gives the most resilient result.
- Upkeep is not optional; the long-term success will depend on money and institutional clarity.
- Connecting ecological and social value can facilitate community support (parks, trails, wetlands).
- Think about fairness: make sure that any urban resident has access to wetland/GI facilities and that no one is displaced due to value uplift.
- Keep track of and adjust to: climate change is such that even the assumptions have to be revisited regularly.

4.7 EMERGING TRENDS AND RESEARCH FRONTIERS

- There is a growing global effort in measuring and quantifying ecosystem services from green urban infrastructure (including wetlands); however, heterogeneity remains a significant challenge.
- Urban Nature-based Solutions (NbS) conceptual models are increasingly being used, which focus on multifunctionality, resilience, and ecosystem services.
- Innovative sensor/IoT devices enable continuous recording of the efficiency of green infrastructure (water levels, infiltration, and runoff reduction), thus providing flexible management.
- The emphasis on social justice in the provision of green-blue infrastructure has increased considerably: The research shows that the distribution of nature in the cities is highly uneven and thus may worsen the inequalities if taken for granted.

4.8 CONCLUSIONS

Urban wetlands and green infrastructures are a complete turnaround from the conventional methods by which cities react to water, biodiversity, climate, and social wellbeing problems. As cities shift from using only grey, system approaches to mixing nature with urban design [22] and infrastructure, they can make the landscapes multifunctional, which are resilient, livable, and sustainable. Yet, the promise will only be fulfilled if implementation is supported by good design, strong governance, community involvement, continuous maintenance, and monitoring, and if issues of equity are dealt with proactively. The embedding of urban wetlands and green infrastructures early on is, therefore, a huge opportunity for emerging cities. However, it is equally a big chance for the incorrect execution of such a strategy for these cities where land pressures and climate stresses are severe.

REFERENCES

1. P. Indrayani et al., "Geographic information system-based assessment of environmental challenges and green infrastructure planning in the tallo river watershed, makassar city," *Indonesian Journal of Urban and Environmental Technology*, vol. 8, no. 2, pp. 613–631, 2025, doi: 10.25105/urbanenvirotech.v8i2.23083.
2. G. Hu et al., "Exploring the decentralized blue-green infrastructure network layouts for urban flood risk management," *Landscape and Ecological Engineering*, vol. 21, no. 4, pp. 757–771, 2025, doi: 10.1007/s11355-025-00670-9.
3. C. Wang et al., "Urban soil health improves with green infrastructure and reduced pathogen risks," *Journal of Environmental Management*, vol. 393, 2025, doi: 10.1016/j.jenvman.2025.127136.
4. Y. Yan et al., "Multifunctional mof-based superhydrophobic quartz sand: interfacial interactions for durable and efficient microplastic removal in stormwater," *Chemical Engineering Journal*, vol. 522, 2025, doi: 10.1016/j.cej.2025.167692.
5. A. Motlagh, "Urban stormwater and groundwater quality: pathways, risks, and green infrastructure solutions," *Environments*, vol. 12, no. 11, 2025, doi: 10.3390/environments12110446.
6. V. A. K. Dewi et al., "Integrated flood control strategy in new capital city of nusantara (ikn) indonesia," E3S Web of Conferences, 2025, doi: 10.1051/e3sconf/202565211005.
7. H. Chen et al., "Balancing urbanisation and habitat conservation on lingkun island, china," *Marine and Freshwater Research*, vol. 76, no. 15, 2025, doi: 10.1071/mf25042.
8. H. Harish and V. Jegatheesan, "A review of sources, worldwide legislative measures and the factors influencing the treatment technologies for contaminants of emerging concern (cecs)," *Current Pollution Reports*, vol. 11, no. 1, 2025, doi: 10.1007/s40726-025-00357-7.
9. N. K. Berger, l. Porter, l. H. Teixeira, and J. Kollmann, "Testing native plants for multifunctional infiltration swales: effects of substrate amendments and flooding," *Urban Forestry & Urban Greening*, vol. 114, 2025, doi: 10.1016/j.ufug.2025.129177.
10. I. Vera-puerto, D. Istenič, P. Molle, C. Hernández-crespo, and C. A. Arias, "Constructed wetlands for combined sewer overflow treatment: a 30-year review, current trends and future projections," *Current Opinion in Environmental Science & Health*, vol. 48, 2025, doi: 10.1016/j.coesh.2025.100681.
11. K. Zakharova, R. Rodela, and K. Lehtilä, "Indicators of visual and ecological complexity of urban wetlands and lakes in stockholm," *Ecological Indicators*, vol. 181, 2025, doi: 10.1016/j.ecolind.2025.114362.
12. M. H. Morgado, D. V Vastardi, F.-E. Baudot Almeida, and H. Dahy, "'Spongetown' christiania as an urban living lab: nature-based solutions for resilient, circular, symbiotic, and regenerative transitions in urban waters," *Nature-based Solutions*, vol. 8, 2025, doi: 10.1016/j.nbsj.2025.100237.
13. J. A. Arteaga-zambrano, O. S. Areu-rangel, J. E. Rosero-moran, and l. Lizárraga-mendiola, "Multicriteria evaluation of green infrastructure for public spaces in urban areas: hydrological functionality and ecosystem services," *Environmental Management*, vol. 75, no. 12, pp. 3635–3657, 2025, doi: 10.1007/s00267-025-02287-y.
14. I. Oluk, S. D. Rey, M. E. Foltz, J. C. Schussler, and D. M. Byrne, "Characterizing sustainability of green stormwater infrastructure technologies using life cycle assessment and life cycle costing: a systematic review," *Water Research*, vol. 288, 2026, doi: 10.1016/j.watres.2025.124586.
15. A. M. James and P. K. Amritha, "Exploring the significance of social-ecological systems approach in urban green infrastructure planning for sustainable climate resilient cities," in *Lecture Notes in Civil Engineering*, 2026, pp. 101–113, doi: 10.1007/978-981-95-0241-7_9.
16. A. Fatima and A. Khare, "Urban wetlands as green infrastructure for sustainable cities," in *Sustainable Development Goals Series*, vol. Part f1171, 2026, pp. 103–123, doi: 10.1007/978-3-032-06881-1_6.

17. S. Dutta and D. Roy, "An integrated approach to the planning and management of a peri-urban wetland: a case study of chatra beel, english bazar, malda district, west bengal," in *Sustainable Development Goals Series*, vol. Part f1171, 2026, pp. 357–373, doi: 10.1007/978-3-032-06881-1_19.
18. R. Baa, "Urban wetland unveiling the socioeconomic benefit and sociocultural significance: reference to five different case studies," in *Sustainable Development Goals Series*, vol. Part f1171, 2026, pp. 235–253, doi: 10.1007/978-3-032-06881-1_13.
19. F. Zhao, Y. Wang, D. Zhang, Y. Yang, X. Li, and Y. Liang, "Patterns and drivers of beta diversity in bird communities associated with blue-green infrastructure in rapidly urbanizing landscapes," *Avian Research*, vol. 17, no. 1, 2026, doi: 10.1016/j.avrs.2025.100316.
20. J. Sánchez, K. Corada, J. Furlong, C. Nash, S. Connop, and E. S. J. Carreras, "Blue-green infrastructure and socio-spatial changes: a study of urban wetlands restoration, housing development and gentrification in london using census data," *Cities*, vol. 169, 2026, doi: 10.1016/j.cities.2025.106500.
21. S. Bhattacharya, B. K. Sachdev, and A. Seth, "Policy regulation and innovative strategies for overcoming contemporary challenges to combat climate crisis by sustainable approach toward urban wetland ecosystem," in *Sustainable Development Goals Series*, vol. Part f1171, pp. 145–161, 2026, doi: 10.1007/978-3-032-06881-1_8.
22. T. Xu, L. Huang, A. M. Aini, and N. A. Nordin, "Characterizing the critical role of older people's overall satisfaction with green spaces for their well-being using machine learning methods: feature extraction and predictive modeling," *Acta Psychologica (Amsterdam)*, vol. 262, p. 106001, 2026, doi: 10.1016/j.actpsy.2025.106001.

Smart Water Metering and Leak Detection

5

Shuchi Tiwari, Bhausaheb Ashok Botre, B. Arundhathi, S. Sridhar, and Caroliny M. da Silva Santos

5.1 INTRODUCTION

Smart water meters and leak detection technologies have significantly advanced water management by combining precise measurement with automated monitoring systems to identify leaks promptly and enhance conservation efforts. Smart water meters leveraging Automated Meter Reading (AMR) technologies provide critical consumption data, enabling utilities to monitor water use quantitatively and detect leaks rapidly [1].

These devices facilitate improved feedback to consumers for better control over water consumption and household expenditure, establishing a foundational step toward efficient water use and conservation [2]. The deployment of smart water meters is influenced not only by engineering and cost considerations but also by regulatory, social, and privacy concerns, which can influence user acceptance and the pace of adoption in various regions [2]. As these technologies mature and costs decrease, broader adoption is anticipated, supported by thoughtful communication strategies that address consumer concerns and regulatory compliance.

Water leak detection methods integrated into smart meters and networks are diverse, employing both software and hardware-based solutions. These include acoustic sensing technologies such as hydrophone sensors, vibration sensors, and pressure monitoring systems for real-time leak identification [3, 4]. Acoustic methods using hydrophones stand out for their capability in detecting leakage through sound waves propagating in the pipeline, although challenges such as false alarms, sensitivity to ambient noise, and accurate leak localization remain [5]. Advanced signal processing and data analytics algorithms apply to hydrophone and vibration sensor data to pinpoint leak locations accurately and reduce false-positive rates [3].

Complementing acoustic sensors, wireless communication technologies play a pivotal role in the timely transmission of leak-related data and enable near real-time monitoring. Low Power Wide Area Network (LPWAN) technologies, including Narrowband IoT (NB-IoT), LoRaWAN, and SigFox, have been evaluated for their effectiveness in smart water grid applications. These technologies offer long communication range with low energy consumption, supporting extended device operational lifetimes (exceeding five years) without external power sources, crucial for meters installed in remote areas [6, 7]. Comparative analyses show NB-IoT has superior scalability and lower packet error rates, making it highly suitable for large-scale smart water metering networks [7].

DOI: 10.1201/9781003742227-5

From a system perspective, smart water metering systems integrate multiple sensing modalities and communication technologies to optimize leak detection sensitivity, robustness, and operational cost-effectiveness. The use of piezoelectric sensors that harvest energy from pipe vibrations exemplifies innovative self-powered solutions enabling easier deployment and continuous monitoring without external power requirements [4]. Moreover, combining sensor data with intelligent models within smart buildings or water distribution networks facilitates real-time leak detection and automatic alerts with minimal human intervention, an emerging research and development area crucial for future water management [8].

Comprehensive reviews of water leak detection methods highlight that while traditional approaches such as manual sounding and periodic inspections remain in use, modern methodologies based on sensor networks and smart analytics bring greater accuracy and faster response times [9]. Nonetheless, challenges such as leak sensitivity thresholds, cost of long-term monitoring, and false alarm management require ongoing research. Future directions emphasize the development of integrated smart water technologies that combine leak detection, evaluation, and control into cohesive systems for enhancing water distribution safety and sustainability [10].

In summary, smart water meters equipped with advanced leak detection technologies including acoustic sensors, vibration-based self-powered devices, and supported by LPWAN communication are revolutionizing water network management. These technologies enable prompt leak detection, precise localization, improved resource conservation, and cost savings. Adoption is shaped by technical capabilities as well as regulatory and social factors, with increasing emphasis on scalable, real-time monitoring that supports sustainable water infrastructure for the future [2, 6, 9, 11].

Water leakage remains a major contributor to non-revenue water (NRW), which can reach 30–50% in some regions [9, 10]. Traditional mechanical meters and manual inspection techniques are inadequate for modern water networks. The transition toward smart water metering enables automated monitoring, rapid leak detection, and improved operational efficiency [1, 2].

Recent advances in IoT, machine learning, and low-power communication technologies have transformed conventional water distribution systems into intelligent monitoring networks [12, 13]. Smart water meters now function as distributed sensing nodes capable of continuous data acquisition and anomaly detection.

5.2 EVOLUTION OF SMART WATER METERING

The evolution of smart water metering reflects the broader transformation of water distribution systems from manually monitored infrastructure to intelligent, data-driven networks. Over the past decade and a half, advancements in sensing technologies, communication protocols, and artificial intelligence have progressively enhanced monitoring accuracy, operational efficiency, and leak-detection capability. This evolution can be categorized chronologically into four major phases. Table 5.1 summarizes the technological progression of smart water metering from basic automated meter reading (AMR) systems to AI-enabled edge intelligence (TinyML). It highlights the evolution in communication infrastructure, intelligence level, and system capabilities over the past decade.

5.2.1 Early Automation and Automated Meter Reading (2010–2014)

The initial stage of smart water metering focused primarily on automation of meter reading and reduction of manual intervention. During this period, Automated Meter Reading (AMR) systems were introduced to replace conventional mechanical meters that required on-site data collection. AMR technologies enabled

TABLE 5.1 Evolution of Smart Water Metering Technologies

PHASE (YEAR RANGE)	TECHNOLOGY FOCUS	KEY COMPONENTS	COMMUNICATION TYPE	INTELLIGENCE LEVEL
2010–2013	Automation (AMR)	Digital meters, basic data loggers	GSM/Wired	Low
2013–2016	Analytics Integration	Centralized databases, statistical tools	Cellular/GPRS	Moderate
2016–2019	IoT-Cloud Integration	Smart sensors, LPWAN, cloud storage	LoRaWAN/NB-IoT	High
2019–2022	Machine Learning Era	Predictive models, anomaly detection	LPWAN + Cloud APIs	Very High
2022–2025	AI-Edge (TinyML)	Smart meters with embedded ML	Edge + Hybrid Cloud	Intelligent and Adaptive

one-way communication, where consumption data was periodically transmitted to utilities through radio frequency (RF) modules or wired networks. This transition significantly improved billing accuracy, minimized human error, and reduced operational costs [11].

Wireless Sensor Networks (WSNs) also began to emerge during this phase. These networks comprised distributed sensing nodes capable of collecting flow and pressure data across localized sections of water distribution systems. Although early WSN deployments were limited by battery life and communication range, they marked a critical shift toward remote monitoring and early-stage anomaly detection. The primary objective during this phase was operational efficiency rather than predictive analytics. Leak detection remained largely reactive, based on threshold-based alarms and abnormal consumption patterns. Despite these limitations, the automation phase laid the foundation for digital transformation in water utilities by demonstrating the feasibility of remote monitoring and networked infrastructure.

5.2.2 Integration of Data Analytics (2015–2017)

The second phase witnessed the incorporation of statistical analysis and early machine learning (ML) techniques into smart water systems. With increasing data availability from AMR systems, utilities began applying analytical tools to extract insights from consumption patterns. Statistical process control, time-series analysis, and regression-based anomaly detection became common approaches for identifying irregular flow behavior.

During this period, research expanded toward predictive leak detection rather than simple threshold-based alerts. Early ML models such as k-Nearest Neighbors (kNN), Support Vector Machines (SVM), and decision trees were applied to classify abnormal water usage events. These approaches improved detection sensitivity and reduced false alarms compared to conventional rule-based systems.

Another key advancement was the use of pressure transient analysis and pattern recognition techniques to detect small leaks before catastrophic failure. Although computational resources were primarily centralized, this stage marked the beginning of data-driven water management.

However, challenges persisted, including limited scalability, insufficient labeled datasets, and computational inefficiencies. Most systems still relied on periodic data transmission rather than real-time streaming, constraining rapid response capabilities.

5.2.3 IoT and Cloud Platforms (2018–2020)

The widespread adoption of Internet of Things (IoT) technologies between 2018 and 2020 significantly accelerated the evolution of smart water metering. Smart meters transitioned from standalone data collection devices to fully networked IoT nodes capable of real-time communication.

Low-Power Wide-Area Network (LPWAN) technologies such as LoRaWAN and Narrowband IoT (NB-IoT) enabled long-range, energy-efficient communication suitable for large-scale deployments [7, 14]. These protocols reduced energy consumption while ensuring reliable data transmission across urban and rural areas. Consequently, utilities could monitor distributed assets continuously with minimal maintenance overhead.

Cloud computing platforms further transformed water management by enabling centralized data storage, scalable processing, and advanced analytics. Cloud-based dashboards provided utilities with real-time visualization tools, automated alerts, and consumer usage insights [15]. Consumers also benefited from mobile applications that promoted water conservation through usage transparency and feedback mechanisms.

During this phase, system architectures evolved toward multi-layered designs integrating sensing, communication, cloud storage, and analytics layers. Scalability improved substantially, and smart water metering became integral to broader smart city initiatives. Nevertheless, cybersecurity concerns and interoperability challenges emerged due to heterogeneous IoT ecosystems and proprietary communication standards.

5.2.4 AI and Edge Computing (2021–2025)

The most recent phase represents the integration of advanced artificial intelligence (AI) and edge computing into smart water infrastructure. Unlike earlier centralized models, modern systems incorporate on-device intelligence capable of performing real-time analytics at the network edge.

Deep learning models, including Convolutional Neural Networks (CNNs) and Long Short-Term Memory (LSTM) networks, have demonstrated high accuracy in detecting leaks from flow and pressure time-series data [16]. These models capture spatial-temporal dependencies and complex nonlinear patterns that traditional algorithms fail to detect.

Graph Neural Networks (GNNs) further enhance leak localization by incorporating network topology information into predictive models [17]. By modeling water distribution systems as graphs, GNNs can identify leak locations more precisely, even in large-scale and complex pipeline networks.

Additionally, Tiny Machine Learning (TinyML) techniques enable lightweight models to operate on microcontrollers embedded within smart meters [18]. This reduces latency, lowers bandwidth requirements, and enhances system resilience by minimizing dependence on cloud connectivity.

Edge-cloud hybrid architectures now dominate system design. Edge devices perform preliminary anomaly detection, while cloud platforms handle long-term analytics, model training, and visualization. This approach balances computational efficiency with scalability.

Despite these advancements, challenges remain in terms of model generalization across diverse hydraulic conditions, data privacy protection, and integration with legacy infrastructure. However, the convergence of AI, IoT, and edge computing signals a paradigm shift from reactive maintenance toward predictive and prescriptive water management. The evolution of smart water metering reflects a progression from simple automated billing systems to intelligent, self-learning infrastructures capable of real-time leak detection and predictive maintenance. Each phase built upon previous technological advancements.

The automation phase significantly enhanced operational efficiency, while the analytics phase introduced predictive capabilities that improved system planning and performance. Subsequently, the IoT–cloud phase enabled scalable and real-time monitoring, and the AI edge phase advanced intelligent, decentralized decision-making for optimized operations. This chronological transformation demonstrates

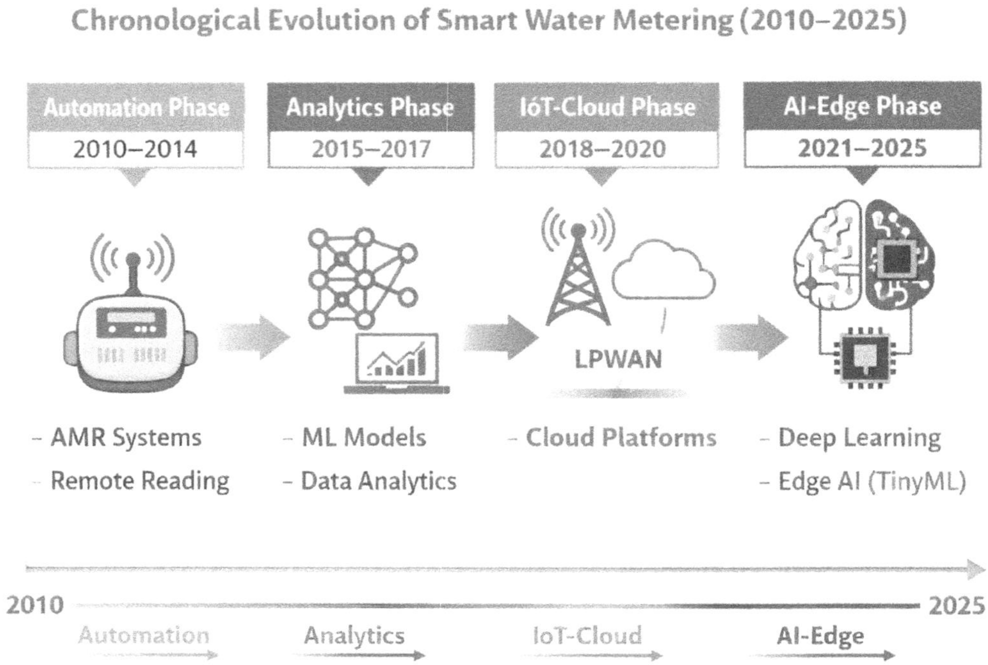

FIGURE 5.1 Evolution of smart water metering.

how interdisciplinary integration combining sensing technologies, communication networks, cloud computing, and artificial intelligence has reshaped water distribution management into a data-driven, sustainable infrastructure system. Figure 5.1 illustrates the progressive transformation of smart water metering from basic automation (AMR) to AI-enabled edge intelligence systems. It highlights the shift from centralized data logging to adaptive, real-time decision-making architectures.

5.3 SYSTEM ARCHITECTURE OF SMART WATER NETWORKS

5.3.1 Layered Architecture of Modern Smart Water Monitoring Systems

Modern smart water monitoring systems are designed using a multi-layered architecture that enables seamless data acquisition, transmission, processing, and decision-making. This structured framework enhances system reliability, scalability, and real-time responsiveness. By integrating sensing technologies, embedded computing, wireless communication, cloud analytics, and user-facing applications, smart water infrastructure has evolved into an intelligent cyber-physical system. IoT-enabled frameworks support continuous monitoring and automated alerts, significantly improving operational efficiency and leak management [19, 20].

The layered architecture typically consists of five interconnected components: the sensing layer, edge layer, communication layer, cloud layer, and application layer.

5.3.1.1 Sensing Layer

The sensing layer forms the physical foundation of the smart water monitoring system. It comprises distributed sensors installed across water distribution networks to collect real-time hydraulic and environmental parameters. Commonly deployed sensors include:

- **Flow sensors** to measure volumetric water consumption and detect abnormal flow variations.
- **Pressure sensors** to monitor network stability and identify pressure drops associated with leaks.
- **Acoustic sensors** and hydrophones to detect leak-induced sound waves.
- **Vibration sensors** to capture structural disturbances in pipelines.
- **Ultrasonic sensors** for non-invasive flow measurement and leak localization.

These sensors convert physical parameters into digital signals, enabling continuous data acquisition. High-resolution sensing enhances detection sensitivity but increases data volume and energy consumption. Therefore, sensor placement optimization and calibration are critical to ensure accuracy and cost-effectiveness.

The sensing layer is often battery-powered, necessitating low-power operation and energy-efficient sampling strategies. In large-scale networks, sensors are geographically distributed, forming Wireless Sensor Networks (WSNs) capable of transmitting data to centralized or edge processing units.

5.3.1.2 Edge Layer

The edge layer consists of embedded microcontrollers or single-board computing devices installed near or within smart meters. This layer performs local data preprocessing, filtering, and preliminary analytics before transmitting information to higher layers.

Edge computing reduces system latency by enabling on-site anomaly detection. Instead of continuously transmitting raw data, edge devices can execute lightweight machine learning algorithms to identify suspicious patterns, such as sudden pressure drops or abnormal consumption spikes. When anomalies are detected, only relevant data or alerts are forwarded to the cloud, reducing bandwidth usage and conserving energy.

Hybrid edge-cloud architectures enhance system efficiency by distributing computational tasks appropriately [21]. Real-time decision-making at the edge improves responsiveness in critical situations, such as burst pipe events. Furthermore, localized processing enhances resilience in areas with unstable internet connectivity.

Recent advancements in microcontroller performance and TinyML frameworks allow the implementation of compact neural networks directly on embedded hardware. This enables intelligent leak detection without relying entirely on cloud-based infrastructure.

5.3.1.3 Communication Layer

The communication layer connects sensing and edge components to centralized platforms. Reliable and energy-efficient data transmission is crucial for real-time monitoring.

Several wireless technologies are widely used:

- **LoRaWAN (Long Range Wide Area Network)**: Provides long-range communication with low power consumption, making it suitable for rural and large-area deployments.
- **NB-IoT (Narrowband Internet of Things)**: Offers improved scalability and secure cellular connectivity for dense urban networks.
- **GSM and Cellular Networks**: Provide widespread coverage but may consume more power compared to LPWAN technologies.

The choice of communication protocol depends on network size, geographic coverage, power availability, and data transmission frequency. LPWAN technologies are particularly advantageous for battery-operated smart meters due to extended operational lifetimes.

However, communication reliability remains a challenge in underground or densely constructed environments. Signal attenuation, packet loss, and latency can impact real-time performance. Therefore, adaptive transmission strategies and redundancy mechanisms are often integrated to maintain system robustness.

5.3.1.4 Cloud Layer

The cloud layer functions as the centralized intelligence hub of the smart water monitoring architecture. It is responsible for large-scale data storage, advanced analytics, model training, and system-wide coordination.

Cloud platforms aggregate data from distributed sensors and edge devices, enabling utilities to analyze long-term consumption trends and network performance. Machine learning models, including anomaly detection algorithms and predictive maintenance frameworks, are typically trained in the cloud due to higher computational capacity.

Visualization tools and dashboards hosted on cloud platforms provide interactive interfaces for utility operators. These dashboards display real-time flow maps, pressure distribution, historical trends, and leak alerts. Advanced analytics can also support demand forecasting and infrastructure planning.

IoT-enabled cloud systems facilitate automated alert mechanisms that notify maintenance teams immediately when anomalies occur [19, 20]. Cloud scalability ensures that the system can accommodate expanding sensor networks without significant infrastructure upgrades.

Nevertheless, reliance on centralized cloud systems introduces concerns related to data privacy, cybersecurity, and latency in mission-critical scenarios.

5.3.1.5 Application Layer

The application layer represents the user interface and decision-support component of the system. It includes utility control dashboards, web-based monitoring portals, mobile applications for consumers and maintenance teams, and automated notification systems (SMS/email alerts). This layer transforms complex data into actionable insights. Utility managers can prioritize maintenance activities, optimize pressure regulation, and reduce non-revenue water losses. Consumers benefit from usage transparency and leak alerts, promoting water conservation. Integration with Geographic Information Systems (GIS) further enhances spatial visualization of network conditions. Application-level analytics support performance benchmarking and regulatory reporting. Figure 5.2 presents the multi-layer system architecture comprising sensing, edge, communication, cloud, and application layers. It demonstrates how data flows from field sensors to intelligent analytics and user dashboards.

5.3.2 Hybrid Architecture and Interoperability Challenges

Modern smart water monitoring systems increasingly adopt hybrid edge-cloud architectures to balance scalability and responsiveness. Edge devices handle immediate anomaly detection, while cloud systems manage long-term analytics and strategic planning. This distributed intelligence approach reduces latency and bandwidth consumption while maintaining centralized oversight [21]. Table 5.2 presents the multi-layer architecture of the smart water monitoring framework, describing the components and core functions at each layer. It clarifies how sensing, edge processing, communication, cloud analytics, and application interfaces interact within the integrated system.

However, interoperability challenges remain significant. Smart water networks often integrate heterogeneous devices from multiple vendors using proprietary communication standards. Lack of standardized

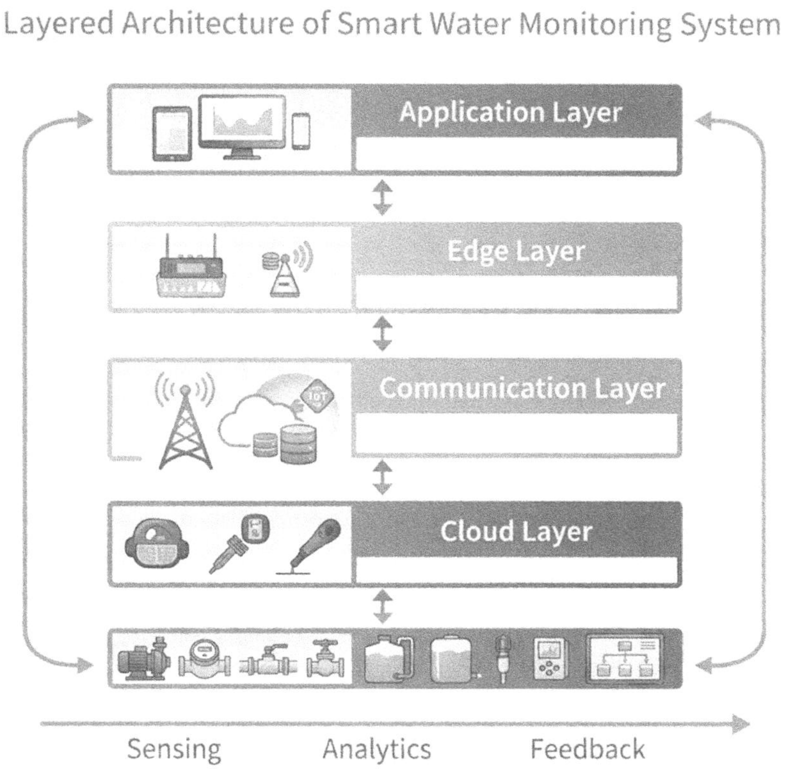

FIGURE 5.2 Layered architecture of smart water monitoring system.

TABLE 5.2 Layer-wise Functional Description of Smart Water Monitoring System

LAYER	MAIN COMPONENTS	KEY FUNCTIONS	TECHNOLOGIES USED
Sensing Layer	Flow, Pressure, Acoustic Sensors	Data acquisition	MEMS, Ultrasonic
Edge Layer	Microcontrollers, TinyML	Local anomaly detection	Edge AI, TinyML
Communication Layer	Gateways, Routers	Data transmission	LoRaWAN, Wi-Fi, NB-IoT
Cloud Layer	Servers, Databases	Data storage and model training	AWS, Azure, GCP
Application Layer	Dashboard, Alerts	Visualization and decision support	Web Apps, Mobile Apps

protocols complicates data integration and system expansion [13]. Ensuring compatibility among sensing devices, communication modules, and cloud platforms is critical for sustainable deployment.

Standardization efforts, open communication frameworks, and modular system design are increasingly recognized as essential to overcome these barriers.

The layered architecture of modern smart water monitoring systems reflects a sophisticated integration of sensing technologies, embedded intelligence, wireless communication, cloud analytics, and user-centric applications. Each layer performs a specialized function while contributing to a cohesive, intelligent water management ecosystem.

IoT-enabled frameworks enable real-time monitoring, automated alerts, and predictive maintenance, significantly enhancing operational efficiency [19, 20]. Hybrid edge-cloud models further optimize system performance by distributing computational tasks effectively.

Despite technological progress, addressing interoperability, cybersecurity, and scalability challenges remains crucial for achieving fully integrated, resilient, and sustainable smart water infrastructure systems.

5.4 LEAK DETECTION TECHNOLOGIES

Leak detection approaches are broadly classified into acoustic, hydraulic, data-driven, and hybrid methods.

5.4.1 Acoustic-Based Detection

Acoustic sensors, including hydrophones and MEMS devices, detect leak-induced vibrations and sound waves [5]. These methods are effective for underground pipelines but are sensitive to environmental noise. Machine learning models such as Random Forest and CNN improve acoustic signal classification and reduce false alarms [18, 22].

5.4.2 Flow and Pressure-Based Detection

Pressure and flow deviations are commonly analyzed for anomaly detection [23, 24]. These approaches are cost-effective and compatible with existing smart meters. However, demand variability and hydraulic uncertainties affect reliability [25].

5.4.3 Ultrasonic and Vibration Sensors

Ultrasonic sensors provide non-destructive leak detection [26]. Piezoelectric vibration sensors enable self-powered monitoring systems [4]. These technologies improve sensitivity but increase deployment costs and maintenance complexity [27].

5.5 WIRELESS COMMUNICATION TECHNOLOGIES

Reliable data transmission is essential for enabling real-time leak detection and alert generation in smart water distribution systems. Among available communication technologies, *LoRaWAN* offers a low-power, long-range solution that is particularly suitable for rural and geographically dispersed deployments, where infrastructure density is limited [28]. In contrast, *NB-IoT* provides higher network scalability, improved coverage in dense urban environments, and lower packet error rates, making it well-suited for city-scale smart water networks [7]. Conventional communication options such as *GSM* and *Wi-Fi* remain common in industrial and legacy monitoring contexts; however, these technologies are comparatively more energy-intensive and may not be optimal for battery-powered smart meters [29]. Therefore, the adoption of energy-efficient transmission strategies such as adaptive data rates, duty cycling, edge-based preprocessing, and event-driven communication is critical to extending device lifespan and ensuring sustainable long-term operation in battery-powered deployments [30, 31].

5.6 MACHINE LEARNING FOR LEAK DETECTION

Machine learning (ML) has become a transformative tool in smart water monitoring systems, significantly enhancing the accuracy, speed, and reliability of leak detection. Unlike traditional rule-based or threshold-driven approaches, ML models can identify subtle and complex patterns in flow, pressure, and acoustic data that indicate the presence of leaks. By learning from historical datasets, these algorithms improve detection performance over time and reduce false alarms, which are common in conventional monitoring systems.

5.6.1 Supervised Machine Learning Models

Supervised learning techniques are widely used in leak detection applications where labeled datasets indicating normal and abnormal conditions are available. Common algorithms include Random Forest (RF), Support Vector Machines (SVM), and k-Nearest Neighbors (kNN). These models classify system states based on extracted features such as pressure gradients, flow deviations, and consumption anomalies.

Random Forest models are particularly effective due to their ensemble structure, which combines multiple decision trees to improve robustness and reduce overfitting. Support Vector Machines provide strong performance in high-dimensional feature spaces and are well-suited for distinguishing subtle leak signatures. Similarly, kNN offers a simple yet effective method for anomaly classification based on similarity measures.

Studies report detection accuracies exceeding 90% under controlled experimental conditions [32, 33]. However, supervised models rely heavily on high-quality labeled datasets, which can be difficult and costly to obtain in real-world water distribution networks.

5.7 DEEP LEARNING APPROACHES

Deep learning methods have further advanced leak detection by capturing nonlinear and spatial-temporal relationships in complex datasets. Convolutional Neural Networks (CNNs) are effective in processing structured data such as pressure maps and acoustic spectrograms. They automatically extract hierarchical features, reducing the need for manual feature engineering.

Long Short-Term Memory (LSTM) networks, a type of recurrent neural network, are particularly valuable for analyzing time-series data. LSTMs can model temporal dependencies in flow and pressure signals, enabling early detection of small leaks that gradually develop over time [16, 24].

Graph Neural Networks (GNNs) represent a more recent innovation in leak detection. By modeling water distribution systems as interconnected graphs, GNNs incorporate network topology into predictive analysis. This approach significantly enhances leak localization accuracy, especially in large-scale pipeline systems [17, 34]. Deep learning models generally outperform traditional ML techniques in complex scenarios but require substantial computational resources and large training datasets.

5.8 UNSUPERVISED AND SEMI-SUPERVISED METHODS

In many practical scenarios, labeled leak data is scarce. To address this limitation, unsupervised learning techniques such as autoencoders and clustering algorithms are increasingly applied. Autoencoders learn normal operating patterns and identify anomalies based on reconstruction error. When incoming data deviates significantly from learned patterns, a potential leak is flagged [35]. Clustering methods group similar consumption or pressure patterns to detect outliers without predefined labels [36]. These approaches are particularly useful for early-stage deployments where historical failure data is limited.

5.9 MULTI-SENSOR DATA FUSION

Multi-sensor data fusion has emerged as a powerful strategy to enhance the reliability and precision of leak detection in smart water distribution systems. Instead of relying on a single parameter, modern monitoring frameworks integrate multiple sensing modalities such as pressure, flow rate, acoustic signals, and vibration data to generate a comprehensive representation of pipeline behavior. This multi-parameter approach significantly improves leak localization accuracy and reduces false positives compared to single-sensor systems [37, 38].

Each sensing modality contributes unique information. Pressure sensors detect transient drops or fluctuations that may indicate leakage. Flow meters identify discrepancies between expected and actual consumption patterns. Acoustic sensors capture sound waves generated by escaping water, while vibration sensors detect structural disturbances along pipeline walls. Individually, these signals may be influenced by environmental noise, demand variability, or hydraulic transients. However, when combined through data fusion techniques, complementary information strengthens anomaly detection and improves diagnostic confidence.

Data fusion can be implemented at different levels:

- **Data-level fusion**, where raw sensor signals are integrated before processing.
- **Feature-level fusion**, where extracted features from different sensors are combined into a unified model.
- **Decision-level fusion**, where independent models generate predictions that are subsequently aggregated. Figure 5.3 depicts the integration of pressure, flow, acoustic, and vibration sensors for enhanced leak detection. The framework shows how multi-sensor data fusion improves reliability and reduces false positives.

Feature-level fusion is particularly common in leak detection applications, as it balances computational efficiency with improved classification performance. By integrating diverse features such as pressure gradients, acoustic frequency components, and flow variance, machine learning models achieve higher sensitivity and robustness.

Graph-based and spatial-temporal modeling techniques further enhance multi-sensor frameworks. Water distribution networks can be represented as interconnected graphs where nodes correspond to junctions and edges represent pipes. Graph Neural Networks (GNNs) utilize this topology to propagate information across the network, enabling more precise leak localization [17, 39]. Spatial-temporal models also account for time-dependent changes in demand and hydraulic conditions, reducing misclassification due to routine consumption fluctuations.

Despite these advancements, optimal sensor placement remains a significant research challenge. Installing sensors at every possible location is economically infeasible. Therefore, determining the

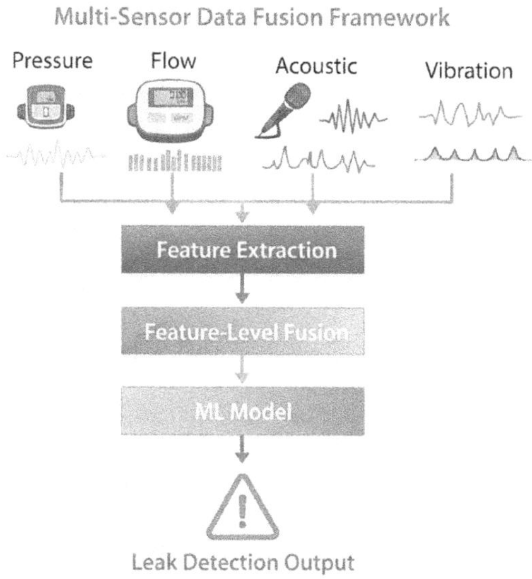

FIGURE 5.3 Multi-sensor data fusion framework.

TABLE 5.3 Sensor Characteristics for Multi-Sensor Data Fusion

SENSOR TYPE	MEASURED PARAMETER	LEAK SENSITIVITY	DATA TYPE	ADVANTAGES
Pressure sensor	Pressure drop	Medium	Continuous	Early pressure anomaly detection
Flow sensor	Flow rate variation	High	Time-series	Detects abnormal consumption
Acoustic sensor	Leak sound	Very high	Frequency spectrum	Detects underground leaks
Vibration sensor	Pipe vibration	High	Signal waveform	Structural leak indication

minimum number and optimal positions of sensors required to maximize detection accuracy is critical [25]. Poor placement can result in blind spots, delayed detection, or ambiguous localization results. Table 5.3 compares different sensor types used for leak detection and data fusion, outlining their measured parameters and sensitivity levels. The table emphasizes the complementary advantages of pressure, flow, acoustic, and vibration sensors in improving detection accuracy.

5.10 HYBRID DATA-DRIVEN AND HYDRAULIC

Hybrid modeling approaches have gained increasing attention in smart water monitoring because neither purely data-driven nor purely hydraulic models is sufficient to address the complexities of real-world water distribution systems. Data-driven models, particularly machine learning algorithms, excel at identifying patterns in historical flow and pressure data. However, they often struggle when network conditions change due to seasonal demand variation, pipe aging, valve operations, or infrastructure modifications. Without retraining or recalibration, their predictive accuracy may degrade significantly [40].

On the other hand, hydraulic models are based on physical principles such as mass conservation, energy balance, and pipe friction laws. Tools such as EPANET-based simulations allow utilities to model

TABLE 5.4 Comparison of Hydraulic and Machine Learning Models

CRITERIA	HYDRAULIC MODEL (EPANET)	MACHINE LEARNING MODEL	HYBRID MODEL
Approach	Physics-based	Data-driven	Physics + data
Data requirement	Network parameters	Historical data	Both
Accuracy	Moderate	High (data dependent)	Very high
Real-time capability	Limited	Good	Excellent
Adaptability	Low	High	Very high

pressure distribution, flow dynamics, and system responses under different operating scenarios. These physics-based models provide interpretability and theoretical consistency. However, they require precise calibration using accurate pipe parameters, roughness coefficients, boundary conditions, and demand patterns. In practice, obtaining such detailed and continuously updated parameters are challenging. Small discrepancies in calibration can result in significant modeling errors.

Hybrid approaches aim to integrate the strengths of both paradigms. In these systems, hydraulic simulations provide physically meaningful baseline predictions, while machine learning algorithms capture nonlinear residual patterns or anomalies that the hydraulic model cannot explain. For example, ML models may analyze deviations between simulated and measured pressure values to detect potential leaks. By combining physical knowledge with data-driven adaptability, hybrid models improve robustness and reduce false alarms [23, 41].

One common strategy involves using hydraulic models to generate synthetic training datasets under various leak scenarios, which are then used to train supervised ML algorithms. Another approach integrates ML within the calibration process itself, enabling continuous parameter adjustment based on real-time sensor data. This dynamic calibration enhances model reliability under varying demand conditions.

Despite their advantages, hybrid systems face computational and implementation challenges. Running hydraulic simulations alongside real-time ML inference requires substantial computational resources, particularly in large urban networks. Integration between simulation platforms and IoT-based data streams can also be technically complex. Additionally, ensuring synchronization between real-time sensor data and model predictions is essential for accurate leak localization.

Scalability remains a critical concern. Deploying hybrid frameworks across extensive distribution systems demands efficient model reduction techniques and optimized algorithms. Future research is focused on lightweight hybrid architectures, cloud-edge coordination strategies, and adaptive calibration mechanisms to enable cost-effective large-scale deployment.

Overall, hybrid data-driven and hydraulic models represent a promising direction for achieving accurate, interpretable, and resilient leak detection in modern smart water infrastructure. Table 5.4 provides a comparative evaluation of hydraulic modeling (e.g., EPANET), machine learning approaches, and hybrid frameworks.

It demonstrates how integrating physics-based and data-driven models enhances accuracy, adaptability, and real-time performance. Figure 5.4 illustrates the hybrid modeling approach combining hydraulic simulations (e.g., EPANET) with machine learning algorithms. The integration enables improved prediction accuracy, adaptability, and real-time anomaly detection.

5.11 EDGE AND CLOUD COMPUTING

Edge–cloud computing architectures are fundamental to intelligent and scalable smart water monitoring systems, as they enable efficient processing of large volumes of real-time sensor data while minimizing latency and bandwidth consumption. Edge computing processes data close to the source within smart

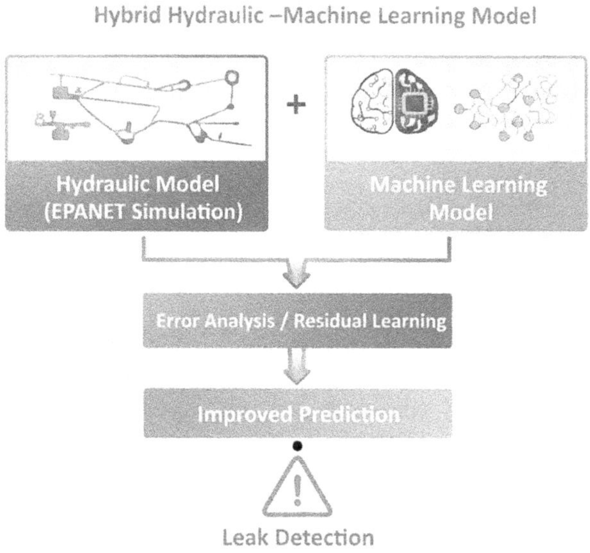

FIGURE 5.4 Hybrid hydraulic–machine learning model.

meters or gateways allowing local preprocessing and anomaly detection. This reduces communication overhead and enables immediate response to events such as pressure drops or abnormal flow variations without relying solely on centralized servers [21]. The emergence of TinyML further strengthens edge intelligence by enabling lightweight machine learning models to run on low-power microcontrollers, supporting real-time inference with minimal energy consumption particularly valuable for battery-operated and remote deployments [18]. Figure 5.5 demonstrates the edge–cloud collaborative framework where anomaly detection is performed at the smart meter level and advanced model training occurs in the cloud. The architecture ensures low-latency response while maintaining scalable analytics capabilities.

However, cloud computing remains essential for large-scale analytics, long-term storage, and model training. Cloud platforms aggregate distributed sensor data to support predictive maintenance, demand forecasting, and system-wide performance monitoring through interactive dashboards [15].

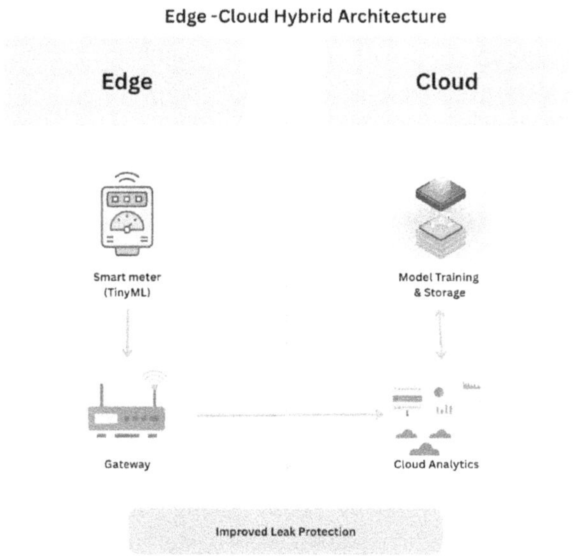

FIGURE 5.5 Edge–cloud hybrid architecture.

TABLE 5.5 Edge vs Cloud Processing Capabilities

PARAMETER	EDGE PROCESSING	CLOUD PROCESSING
Location	On-device (smart meter)	Remote servers
Latency	Very low	Moderate
Computation power	Limited	High
Use case	Real-time anomaly detection	Model training and storage
Energy consumption	Low	High
Internet dependency	Not mandatory	Mandatory

Hybrid architectures combine the strengths of both paradigms: edge devices perform real-time anomaly detection, while the cloud manages advanced analytics and periodic model updates, ensuring scalability, adaptability, and optimized bandwidth usage [20]. Despite challenges related to synchronization, interoperability, and cybersecurity, edge–cloud integration represents a critical advancement toward resilient and energy-efficient smart water infrastructures. Table 5.5 contrasts edge and cloud computing paradigms in smart water systems based on latency, computational power, and operational dependency. The comparison highlights the advantages of a hybrid edge–cloud architecture for real-time anomaly detection and scalable model training.

5.12 ENERGY EFFICIENCY AND SUSTAINABILITY

Energy constraints are a major limitation in large-scale smart water metering deployments, as most devices are battery-powered and installed in locations where maintenance is costly or impractical. Long operational lifetimes (5–15 years) require ultra-low power hardware and intelligent power management. Modern microcontrollers support multiple sleep modes and optimized memory usage, enabling continuous sensing with minimal energy consumption, while efficient firmware design reduces unnecessary processing and sampling [42].

Energy harvesting further enhances sustainability by capturing power from water flow, pressure variations, vibrations, or small solar sources, though harvested energy is often intermittent and requires hybrid storage systems. Since communication modules consume the most power, adaptive transmission strategies such as event-driven reporting, data compression, and batch transmission significantly reduce energy use [19]. However, a trade-off exists between sampling frequency and battery life, as higher data resolution improves leak detection but increases power consumption [31]. Therefore, integrating ultra-low power electronics, adaptive communication, and renewable energy harvesting into energy-aware architectures is essential for reliable and scalable smart water monitoring systems.

5.13 PRACTICAL IMPLEMENTATION AND SCALABILITY

Pilot deployments of smart water metering systems confirm their technical feasibility and operational benefits in urban and residential settings. IoT-enabled meters with real-time analytics reduce non-revenue water, improve billing accuracy, and enable early leak detection through continuous monitoring and automated alerts [19, 43]. Although initial installation costs are high, lifecycle analyses show long-term savings through reduced water losses, predictive maintenance, and lower manual labor, improving overall return on investment [28].

However, large-scale adoption faces challenges including integration with legacy infrastructure, aging pipeline networks, and interoperability with existing SCADA and billing systems. Data privacy and cybersecurity concerns are critical, as granular consumption data require secure transmission and regulatory compliance [13]. Communication reliability in underground or dense urban environments and the lack of standardized IoT governance frameworks further hinder implementation. Addressing these issues through standardized protocols, interoperable platforms, and modular system design is essential for scalable and sustainable smart water monitoring.

5.14 CONCLUSION

Smart water metering and leak detection systems have evolved into highly sophisticated technological frameworks that integrate Internet of Things (IoT) architectures, multi-sensor networks, advanced machine learning algorithms, and energy-efficient communication technologies. Together, these innovations are transforming traditional water distribution systems into intelligent, data-driven infrastructures capable of real-time monitoring and predictive maintenance. By continuously capturing flow, pressure, acoustic, and vibration signals, smart systems provide early leak detection, reduce non-revenue water losses, and enhance operational transparency. The integration of analytics and automated alert mechanisms enables utilities to respond proactively rather than reactively, significantly improving network reliability and service quality.

The convergence of edge computing and cloud analytics has further strengthened system responsiveness and scalability. Local anomaly detection reduces latency, while centralized platforms enable large-scale data processing and strategic planning. Multi-sensor data fusion and graph-based modeling enhance leak localization accuracy, and hybrid AI-hydraulic models improve robustness under varying demand conditions. Additionally, energy-efficient hardware and adaptive communication protocols extend device lifetime, supporting sustainable long-term deployment.

Despite these advancements, several technical and practical challenges persist. Scalability remains a major concern, particularly in large and aging urban networks where integration with legacy infrastructure can be complex. Energy optimization continues to be critical for battery-powered smart meters deployed in remote or underground environments. Interoperability issues caused by heterogeneous devices and proprietary communication standards limit seamless integration across vendors. Moreover, many proposed models require further real-world validation under diverse hydraulic and environmental conditions to ensure generalizability and reliability.

Ultimately, the successful deployment of intelligent smart water systems will depend on a balanced integration of technological innovation, economic feasibility, regulatory alignment, and sustainable design principles. With continued interdisciplinary research and coordinated implementation efforts, smart water metering and leak detection technologies have the potential to play a pivotal role in ensuring efficient, resilient, and sustainable water resource management worldwide.

5.15 FUTURE RESEARCH DIRECTIONS

The rapid evolution of smart water meters highlights the need for continued research and innovation in this field. Future research should prioritize the development of hybrid AI-hydraulic frameworks that combine physical modeling with adaptive machine learning to enhance accuracy and interpretability. Lightweight and adaptive learning models capable of operating on edge devices will be essential for reducing computational overhead and improving resilience. Standardized, interoperable IoT architectures must also be established to facilitate scalable implementation and cross-platform compatibility.

Leak detection systems have created strong opportunities to improve reliability, scalability, and sustainability in water distribution networks. Future research must focus on overcoming current limitations by developing advanced multi-leak detection algorithms capable of handling complex hydraulic interactions, optimizing sensor placement for cost-effective monitoring, and designing lightweight edge AI models (e.g., TinyML) that balance accuracy with low power and computational constraints. In addition, hybrid hydraulic–machine learning calibration frameworks can enhance model interpretability and adaptability under dynamic operating conditions.

Equally important are system-level advancements, including IoT standardization and interoperability to enable seamless large-scale deployment, along with privacy-preserving analytics and secure communication mechanisms. Emerging technologies such as digital twins and graph neural networks are expected to further strengthen predictive maintenance and leak localization accuracy. Collectively, these innovations will accelerate the transition toward intelligent, resilient, and self-adaptive water infrastructure systems.

REFERENCES

1. Farah, E.; Shahrour, I. Leakage Detection Using Smart Water System: Combination of Water Balance and Automated Minimum Night Flow. *Water Resour. Manag.*, **2017**, *31* (15), 4821–4833. https://doi.org/10.1007/s11269-017-1780-9.
2. Koech, R.; Cardell-Oliver, R.; Syme, G. Smart Water Metering: Adoption, Regulatory and Social Considerations. *Aust. J. Water Resour.*, **2021**, *25* (2), 173–182. https://doi.org/10.1080/13241583.2021.1983968.
3. Phua, W. K.; Rabeek, S. M.; Han, B.; Njihof, E.; Huang, T. T.; Chai, K. T. C.; Yeo, J. H. H.; Lim, S. T. AIN-Based MEMS (Micro-Electro-Mechanical System) Hydrophone Sensors for IoT Water Leakage Detection System. *Water*, **2020**, *12* (11), 2966. https://doi.org/10.3390/w12112966.
4. Okosun, F.; Cahill, P.; Hazra, B.; Pakrashi, V. Vibration-Based Leak Detection and Monitoring of Water Pipes Using Output-Only Piezoelectric Sensors. *Eur. Phys. J. Spec. Top.*, **2019**, *228* (7), 1659–1675. https://doi.org/10.1140/epjst/e2019-800150-6.
5. Bakhtawar, B.; Zayed, T. Review of Water Leak Detection and Localization Methods through Hydrophone Technology. *J. Pipeline Syst. Eng. Pract.*, **2021**, *12* (4), 03121002. https://doi.org/10.1061/(asce)ps.1949-1204.0000574.
6. Pointl, M.; Fuchs-Hanusch, D. Assessing the Potential of LPWAN Communication Technologies for Near Real-Time Leak Detection in Water Distribution Systems. *Sensors*, **2021**, *21* (1), 293. https://doi.org/10.3390/s21010293.
7. Lalle, Y.; Fourati, L. C.; Fourati, M.; Barraca, J. P. A Comparative Study of LoRaWAN, SigFox, and NB-IoT for Smart Water Grid. *2019 Glob. Inf. Infrastruct. Netw. Symp. GIIS*, **2019**. https://doi.org/10.1109/GIIS48668.2019.9044961.
8. Yussof, N. A. M.; Ho, H. W. Review of Water Leak Detection Methods in Smart Building Applications. *Build.*, **2022**, *12* (10), 1535. https://doi.org/10.3390/buildings12101535.
9. Farah, E.; Shahrour, I. Water Leak Detection: A Comprehensive Review of Methods, Challenges, and Future Directions. *Water*, **2024**, *16* (20), 2975. https://doi.org/10.3390/w16202975.
10. Joseph, K.; Sharma, A. K.; van Staden, R.; Wasantha, P. L. P.; Cotton, J.; Small, S. Application of Software and Hardware-Based Technologies in Leaks and Burst Detection in Water Pipe Networks: A Literature Review. *Water*, **2023**, *15* (11), 2046. https://doi.org/10.3390/w15112046.
11. Farah, E.; Shahrour, I. Smart Water for Leakage Detection: Feedback about the Use of Automated Meter Reading Technology. *2017 Sensors Networks Smart Emerg. Technol. SENSET*, **2017**, January, 1–4. https://doi.org/10.1109/SENSET.2017.8125061.
12. Guezouli, L.; Guezouli, L.; Eddine Djeghaba, M. B.; Bentahrour, A. IoT and AI for Real-Time Water Monitoring and Leak Detection. *J. Renew. Energies*, **2024**, *27* (2), 243–281. https://doi.org/10.54966/jreen.v27i2.1210.

13. Alvisi, S.; Tudon-Martinez, L. C. ; Lozoya-Santos, J. C. ; On, J. D. J.; Water, S.; Quintana, D.; Felix-Herran, L. C.; Tudon-Martinez, J. C.; De, J.; Lozoya-Santos, J. On Smart Water System Developments: A Systematic Review. *Water*, **2025**, *17* (17), 2571. https://doi.org/10.3390/w17172571.
14. Mohammed, E.; Jamal, E. M.; Abdelilah, J. Advanced IoT Techniques for Detecting Water Leaks in Supply Networks with LoRaWAN. *Int. J. Adv. Comput. Sci. Appl.*, **2024**, *15* (6), 107–115. https://doi.org/10.14569/IJACSA.2024.0150613.
15. Ray, A.; Goswami, S. IoT and Cloud Computing Based Smart Water Metering System. 2020 Proceedings of the 2020 International Conference on Power Electronics, IoT Applications, Renewable Energy and Its Control (PARC), **2020**, 308–313. https://doi.org/10.1109/PARC49193.2020.236616.
16. Syed, T. A.; Muhammad, M. A.; AlShahrani, A. A.; Hammad, M.; Naqash, M. T. Smart Water Management with Digital Twins and Multimodal Transformers: A Predictive Approach to Usage and Leakage Detection. *Water*, **2024**, *16* (23), 3410. https://doi.org/10.3390/w16233410.
17. Zhang, Z.; Fink, O. Algorithm-Informed Graph Neural Networks for Leakage Detection and Localization in Water Distribution Networks. *Reliab. Eng. Syst. Saf.*, **2026**, *265*. https://doi.org/10.1016/j.ress.2025.111494.
18. Atanane, O.; Mourhir, A.; Benamar, N.; Zennaro, M. Smart Buildings: Water Leakage Detection Using TinyML. *Sensors*, **2023**, *23* (22), 9210. https://doi.org/10.3390/s23229210.
19. Oberascher, M.; Maussner, C.; Hinteregger, P.; Knapp, J.; Halm, A.; Kaiser, M.; Gruber, W.; Truppe, D.; Eggeling, E.; Sitzenfrei, R. Experiences from a Large-Scale Implementation of Digital Water Meters Used for Improved Leakage Management. *EGU24*, **2024**. https://doi.org/10.5194/egusphere-egu24-8150.
20. Bharudin, M. S. A.; Zainuddin, A. A.; Roslee, A. M.; Razak, N. A. A.; Abd Halim, M. H. F.; Saifudin, N. N. M. Comprehensive IoT-Based Water Management System: Cloud Integration for Residential Conservation. Proceedings of the IEEE Region 10 Humanitarian Technology Conference (R10-HTC), **2024**. https://doi.org/10.1109/R10-HTC59322.2024.10778751.
21. Garlisi, D.; Restuccia, G.; Tinnirello, I.; Cuomo, F.; Chatzigiannakis, I. Leakage Detection via Edge Processing in LoRaWAN-Based Smart Water Distribution Networks. Proceedings of the 2022 18th International Conference on Mobility, Sensing and Networking (MSN), **2022**, 223–230. https://doi.org/10.1109/MSN57253.2022.00047.
22. Bakhtawar, B.; Fares, A.; Zayed, T. AIoT-Driven Leak Detection in Real Water Networks Using Hydrophones. *Water Resour. Manag.*, **2024**, *39* (6), 2551–2566. https://doi.org/10.1007/s11269-024-04077-3.
23. Tyagi, V.; Pandey, P.; Jain, S.; Ramachandran, P. A Two-Stage Model for Data-Driven Leakage Detection and Localization in Water Distribution Networks. *Water*, **2023**, *15* (15), 2710. https://doi.org/10.3390/w15152710.
24. Dong, B.; Shu, S.; Li, D. A Unified Spatial-Pressure Sensitivity Partitioning and Leakage Detection Method within a Deep Learning Framework. *Water (Switzerland)*, **2024**, *16* (4), 542. https://doi.org/10.3390/w16040542.
25. Steffelbauer, D. B.; Fuchs-Hanusch, D.; Steffelbauer, D. B.; Fuchs-Hanusch, D. Efficient Sensor Placement for Leak Localization Considering Uncertainties. *Water Resour. Manag.*, **2016**, *30* (14), 5517–5533. https://doi.org/10.1007/s11269-016-1504-6.
26. Mohammed, E. G.; Zeleke, B.; Abebe, S. L.; Girma, E.; Surafel, M.; Abebe, L.; Zeleke, E. B. Water Leakage Detection and Localization Using Hydraulic Modeling and Classification. *J. Hydroinformatics*, **2021**, *23* (4), 782–794. https://doi.org/10.2166/hydro.2021.164.
27. Bakhtawar, B.; Zayed, T. State-of-the-Art Review of Leak Diagnostic Experiments: Toward a Smart Water Network. *Wiley Interdiscip. Rev. Water*, **2023**, *10* (5), e1667. https://doi.org/10.1002/wat2.1667.
28. Murdan, A. P.; Hardin, A. Exploring the Economic Viability of LoRaWAN Based Smart Water Meters for Sustainable Water Management. *Proc. 9th IRC Conf. Sci. Eng. Technol.*, **2023**, 216–225. https://doi.org/10.1007/978-981-99-8369-8_21.
29. Dickson, T. K.; Mduma, N.; K.tarus, J. Industrial-Based GSM Water Leakage Detection, Monitoring and Controlling System: A Case of North Rift Valley Water Agency in Kenya. *Transdiscipl. J. Eng. Sci.*, **2024**, *15*, 39–67. https://doi.org/10.22545/2024/00256.
30. Athinarayanan, H. P.; Selvam, M. Multi Energy-Harvesting Smart Water Meter Design for Underground Water Pipeline Leakage Detection. *Lect. Notes Networks Syst.*, **2023**, *672 LNNS*, 95–111. https://doi.org/10.1007/978-981-99-1624-5_7.
31. Gorawski, M.; Marjasz, R.; Grochla, K.; Frankiewicz, A. Comparative Analysis of Energy Consumption in Simulated LoRa Water Meter Reconfiguration vs. Real-World Readings. Proceedings of the 2024 IFIP Networking Conference (IFIP Networking 2024), **2024**, 696–701. https://doi.org/10.23919/IFIPNetworking62109.2024.10619861.
32. Kanyama, M. N.; Bhunu Shava, F.; Gamundani, A. M.; Hartmann, A. Machine Learning Applications for Anomaly Detection in Smart Water Metering Networks: A Systematic Review. *Phys. Chem. Earth, Parts A/B/C*, **2024**, *134*, 103558. https://doi.org/10.1016/j.pce.2024.103558.

33. Karimanzira, D. Simultaneous Pipe Leak Detection and Localization Using Attention-Based Deep Learning Autoencoder. *Electron.*, **2023**, *12* (22), 4665. https://doi.org/10.3390/electronics12224665.
34. Li, X.; Wu, Y. A Convolutional Graph Neural Network Model for Water Distribution Network Leakage Detection Based on Segment Feature Fusion Strategy. *Water*, **2024**, *16* (24), 3555. https://doi.org/10.3390/w16243555.
35. Zhao, F.; Xia, W.; Zhang, Z.; Zhang, S.; Tao, L.; Wang, J.; Li, N. Unsupervised Water Pipe Network Leakage Detection Method Based on Discrete-Time Dynamic Graph Network. Proceedings of the 2025 IEEE 2nd International Conference on Energy and Electrical Engineering (EEE 2025), **2025**. https://doi.org/10.1109/EEE64897.2025.11162765.
36. Oluyomi, A.; Abedzadeh, S.; Bhattacharjee, S.; Das, S. K. Unsafe Events Detection in Smart Water Meter Infrastructure via Noise-Resilient Learning. Proceedings of the 15th ACM/IEEE International Conference on Cyber-Physical Systems (ICCPS 2024), **2024**, 259–270. https://doi.org/10.1109/ICCPS61052.2024.00030.
37. Khalifeh, A.; Aweisi, N. An Overview on Advancements in Water Leakage Detection Integrating Cutting-Edge Solutions. Proceedings of the 22nd International Conference on Research and Education in Mechatronics (REM 2024), **2024**, 364–368. https://doi.org/10.1109/REM63063.2024.10735620.
38. Martinez-Rios, E. A.; Barrientos, D.; Bustamante, R. Water Leakage Classification With Acceleration, Pressure, and Acoustic Data: Leveraging the Wavelet Scattering Transform, Unimodal Classifiers, and Late Fusion. *IEEE Access*, **2024**, *12*, 84923–84951. https://doi.org/10.1109/ACCESS.2024.3416056.
39. Örn Garðarsson, G.; Boem, F.; Toni, L. Graph-Based Learning for Leak Detection and Localisation in Water Distribution Networks*. *IFAC-PapersOnLine*, **2022**, *55* (6), 661–666. https://doi.org/10.1016/j.ifacol.2022.07.203.
40. Mazaev, G.; Weyns, M.; Vancoillie, F.; Vaes, G.; Ongenae, F.; Hoecke, S. Van. Data for Replication of the Publication: Probabilistic Leak Localization in Water Distribution Networks Using a Hybrid Data-Driven and Model-Based Approach. https://doi.org/10.5281/ZENODO.7255403.
41. Momeni, A.; Piratla, K. R. Model-Based Leakage Detection for Large-Scale Water Pipeline Networks. Pipelines 2023: Condition Assessment, Utility Engineering, and Surveying—Multidisciplinary Sessions, Proceedings of the Pipelines 2023 Conference, **2023**, 120–127. https://doi.org/10.1061/9780784485033.014.
42. Chakir, Y.; Aaroud, A. Innovative Solution for Enhancing Mechanical Water Meters: A Smart Water Meter Add-On. Proceedings of the 2024 International Conference on Circuits, Systems and Communications (ICCSC 2024), **2024**. https://doi.org/10.1109/ICCSC62074.2024.10616942.
43. Akimu, O.; Rashayi, E.; Gumindoga, W.; Gotora, G. M.; Zulu, G. B. Optimising Harare's Water Dissemination Network: Development of an IoT-Driven Leak Monitoring System for Sustainable Urban Water Management. *East African J. Eng.*, **2025**, *8* (1), 219–230. https://doi.org/10.37284/eaje.8.1.3344.

6 Privacy-Preserving IoT for Smart Water Systems Using Homomorphic Encryption

Anupam Tiwari, Neha Sharma, and Shilpi Harnal

6.1 INTRODUCTION

Water is an essential resource, and sustainable water management [1] is increasingly important amid population growth and global warming. The concept of the Internet of Things (IoT) has opened up the possibility of monitoring the usage of water in real time [2]. This can be utilized to ensure efficient management of water and prevent wastage. However, the collection of high-resolution data in the context of the usage of water has opened up a major privacy issue, as the data collected can be utilized to derive sensitive information.

The chapter aims to present a solution to ensure the sustainability of water management in a manner that ensures the privacy of the data. This is achieved through homomorphic encryption [HE][3]. It is a technique for performing computation in a direct manner. The proposed solution can be utilized to ensure data confidentiality. This is essential for the sustainable management of water. The chapter presents a small-scale implementation using Python. This can be utilized to understand the efficiency of the proposed solution.

The subsequent subsections present the information in the context of IoT, the significance of sustainable management of water, and the necessity of ensuring computation in a manner that provides privacy.

DOI: 10.1201/9781003742227-6

6.1.1 IoT in Modern Water Systems

The use of IoT technology in water management has revolutionized water monitoring. Sensors are being used for monitoring water flow, pressure, and quality. This data can be transmitted for analysis in the cloud or edge computing systems.

Components of the IoT technology used in water management systems:

- Sensors: These are used for monitoring water flow, pressure, and quality.
- Gateways: These are used to integrate data from various sensors and to securely send data for analysis in the cloud or edge computing systems.
- Cloud/Edge Analytics: This is used for calculations and predictions for decision-making.

The use of IoT technology in water management [4]:

The use of IoT technology has helped in the efficient use of water, has reduced water wastage, and has been useful for decision-making. However, a new challenge has arisen with regard to the transmission and storage of water consumption data, and the associated issues of privacy, especially if a third-party cloud service is used.

6.1.2 Water Sustainability Challenges

Sustainable water management is being challenged by various factors, including an increasing population, climate change, aging water infrastructure, and inefficient water distribution systems. Inefficient water management systems, such as traditional water meter readings and periodic water inspections, cannot sufficiently identify water leaks, overuse, and contamination.

Water management systems that utilize IoT technology can provide precise data to ensure optimal water distribution and reduce water wastage. For example, predictive analytics can be used to forecast peak water demand, identify leaks in the water distribution system, and optimize water reservoirs. However, precise water usage data collection may result in privacy issues, as it may provide insights into the activities of individuals or organizations.

To overcome these problems, a framework must be developed to ensure water sustainability without compromising data privacy [5]. HE can provide an efficient solution to ensure that water sustainability is achieved without compromising the privacy of individuals or organizations.

6.1.3 Privacy Concerns in Water IoT Systems

The data gathered through IoT technology in the management of water systems has the potential to reveal sensitive information. For instance, continuous flow monitoring can reveal information about household occupancy and even industrial operation schedules. This information, if not properly managed, can compromise security and privacy.

Traditional approaches to data protection, like anonymization and aggregation [6], cannot be relied upon. Aggregated data can be de-anonymized using auxiliary information. In addition, other forms of encryption require data to be decrypted before analysis, making it prone to data breaches. The other challenge to cloud analytics in water management is trust. Third-party cloud providers will have access to sensitive information regarding water consumption.

To address the limitations of traditional data protection approaches, privacy-preserving computation approaches are significant. HE [7] has the capacity to carry out arithmetic and statistical analysis on encrypted data. This gives the potential for water authorities and providers to employ IoT technology in their sustainable management.

6.1.4 Homomorphic Encryption: Enabling Privacy-Preserving Analytics

Homomorphic encryption is a cryptographic technique that allows operations to be performed on the encrypted data without the need for decrypting the data. This is the biggest advantage of HE, which can be used to maintain data confidentiality, and this is an important requirement for IoT systems. The types of HE schemes are as follows:

- Partially Homomorphic Encryption (PHE): The operations of either addition or multiplication can be performed on the ciphertext. However, it is not possible to perform both operations. Examples for this kind of HE are the RSA and Paillier encryption schemes [8].
- Somewhat Homomorphic Encryption (SHE): A few operations of addition and multiplication can be performed on the ciphertext, but only up to a certain level of noise [9, 10].
- Fully Homomorphic Encryption (FHE): Any operation can be performed on the ciphertext. However, this is a computationally expensive operation. The newly developed Cheon–Kim–Kim–Song (CKKS) and BFV schemes are examples of this kind of HE [11, 12].

A list of commonly used HE schemes is presented in Table 6.1. Additive HE schemes, such as Paillier encryption [13], are appropriate for simple aggregation operations, whereas fully HE schemes, such as BFV and BGV [14, 15], are appropriate for more complex computations. The CKKS scheme is especially appropriate for IoT-based water analytics, as it supports approximate arithmetic for real numbers, which is appropriate for aggregation and statistical computation. Emerging bitwise-based schemes such as TFHE and FHEw are appropriate for more complex computations but are computationally expensive and therefore not appropriate for use in typical IoT-based systems.

In an IoT-based system for water analysis, HE allows for aggregation, averaging, and statistical analysis of sensor readings without exposing the actual sensor readings. For example, the flow rate from multiple households can be encrypted and sent to a central server for aggregation and analysis.

However, the key challenge with the use of HE in IoT-based systems is the computational overhead, which is inherent in most IoT-based systems. Edge-based encryption [16] and lightweight HE schemes [17] help alleviate some of these challenges.

TABLE 6.1 Overview of Homomorphic Encryption (HE) Schemes

SCHEME	*COMPUTATION TYPE*	*KEY FEATURES*	*ADVANTAGES*	*LIMITATIONS*
Paillier	Additive	Supports addition of encrypted values	Simple, efficient for sums	Cannot perform multiplication directly
BFV	Fully Homomorphic (Integer)	Supports addition and multiplication on integers	Exact arithmetic for integers	Larger ciphertexts, slower than additive HE
CKKS	Approximate (Real/Float)	Supports addition & multiplication on real numbers	Efficient for statistics, ML, IoT data	Approximation error due to floating-point arithmetic
BGV	Fully Homomorphic (Integer)	Supports batching (SIMD operations)	Parallel operations for vectors	Complex parameter selection, slower performance
TFHE / FHEW	Bitwise operations	Fast bootstrapping, supports Boolean circuits	Fine-grained operations	High computational cost, mainly for experimental use

6.1.5 Motivation and Scope of This Chapter

The rise in IoT-based water monitoring systems reveals the trade-off between the utility of the data collected and the need for preserving the privacy of the collected data. Although water management is made efficient with the help of accurate sensors, leaks are detected, and predictions are made, there is a possibility of compromising the privacy of household or industrial water use details. The traditional methods of preserving the privacy of collected data, such as anonymization or aggregation, are no longer effective against inference or cloud-based attacks.

HE is a strong solution for maintaining the privacy of collected water use data without sacrificing its effectiveness. Although HE is a strong solution, its use in IoT-based water monitoring systems is limited due to the computational overhead involved in HE, its difficulty in integration with IoT systems, and the absence of real-world usage of HE in IoT-based water monitoring systems.

This chapter fills this gap by:

- Proposing a privacy-preserving IoT framework for smart water systems using HE;
- Developing a proof-of-concept using a small-scale Python implementation;
- Investigating the trade-offs in terms of computational costs, accuracy, and scalability;
- Exploring the potential benefits for sustainable water management and its implications for smart water systems, emphasizing the feasibility of privacy-preserving IoT systems.

The scope of this research covers aggregation and analytics for smart water systems. Predictive modeling and/or large-scale smart cities remain future research directions, while this research emphasizes the feasibility of using HE in resource-constrained IoT systems.

6.2 LITERATURE REVIEW

6.2.1 IoT in Water Systems

In the past ten years, there has been tremendous growth in the adoption of IoT technologies for water management. For example, smart meters, flow sensors, pressure transducers, and water quality sensors can be used to monitor water consumption in real time. Several studies have shown that the adoption of IoT for water management can result in a reduction in water loss. For instance, water leaks can be detected in the water pipelines using anomaly detection algorithms. In a similar manner, water consumption patterns can be used to better manage water reservoirs to avert water shortages.

Edge computing [18] has been proposed as a solution to solve some of the problems associated with the use of IoT for water management. However, most studies on IoT for water management seem to focus more on functionality than on privacy and security issues, which are paramount when water consumption patterns for each household or industry are being monitored.

6.2.2 Privacy Concerns in IoT Data

IoT systems used in water systems produce high-resolution time series data, which may reveal detailed patterns of individual or organizational behavior. It has been found that detailed consumption data may be used to infer information regarding occupancy, appliance usage, or industrial operation schedules. If this data falls into the wrong hands, there may be serious privacy risks, including profiling, attacks, and unauthorized surveillance.

Traditional privacy preservation techniques, including anonymization, aggregation, and access control, may be insufficient. It has been found that aggregated data can be de-anonymized using auxiliary information, and traditional encryption techniques require decryption before computation, which may lead to exposure. Additionally, cloud computing may introduce new trust issues because third-party providers may have access to sensitive data.

Recent research has highlighted the need to use privacy preservation computation techniques, including differential privacy, secure multiparty computation, and HE, to support data-driven insights for sustainable water management without compromising sensitive data.

6.2.3 HE in IoT Applications

HE has been recognized as one of the major techniques for secure computation on encrypted data in IoT systems [19–21]. Several research studies have been carried out on its feasibility for resource-constrained devices such as smart meters, industrial sensors, and healthcare-based IoT systems. HE supports statistical analysis and certain machine learning operations without leaking any information.

For example, additive HE schemes such as Paillier are used for secure data aggregation in smart metering systems, which allows utility companies to calculate aggregate or average consumption without access to individual customer information. More advanced schemes, such as variants of fully HE like CKKS and BFV [22, 23], have also shown promise for the ability to perform approximate arithmetic and more advanced analytics while maintaining confidentiality. Table 6.2 provides an overview of some of the key IoT domains where HE has been applied. HE is an important tool for analytics on sensitive information from sensors while maintaining confidentiality and is particularly important for healthcare, smart water, and smart grid applications. However, there are some challenges with the use of HE in these systems:

- Computational Overhead: Performing HE is much more computationally expensive than performing computations in the clear.
- Communication Latency: The sizes of the encrypted data are much larger than the original sensor values.
- Integration Complexity: Integrating HE into existing IoT systems is complex.

TABLE 6.2 HE Applications in IoT

IOT DOMAIN	*HE USE CASE*	*KEY BENEFIT*	*KEY CHALLENGE*
Smart Water Systems	Aggregated water consumption, leak detection	Preserves individual sensor privacy; enables analytics	Approximate computation (CKKS) or overhead (BFV)
Smart Grids / Energy IoT	Encrypted energy usage aggregation	Confidential monitoring of household consumption	Real-time computation constraints
Healthcare IoT	Encrypted vital sign analysis	Protects patient privacy during remote monitoring	Resource-limited devices, latency
Industrial IoT (IIoT)	Predictive maintenance on encrypted sensor data	Prevents leakage of sensitive operational data	Computational overhead; integration complexity
Smart Cities / Environmental Sensors	Pollution or traffic data aggregation	Allows policy-level analytics without exposing locations	Large-scale deployment and bandwidth constraints

But recent research has shown the promise of using HE in some small-scale proofs-of-concept for demonstrating the viability of HE for use in some of these smart systems, which is then extended for smart water systems where sensitive information must be protected without sacrificing the ability to perform analytics.

6.2.4 Research Gaps

From the available literature on IoT-based water systems and privacy-preserving computation techniques, several research gaps have been identified that motivated the present research effort as follows:

- Limited Adoption of HE in Water IoT: Although HE has been successfully implemented in smart grid and healthcare IoT applications, there is a significant lack of research on smart water system implementations.
- Lack of Proof-of-Concept Implementations: There are a limited number of papers available that present a proof of concept of HE-based IoT systems with actual IoT devices and/or simulation environments.
- Computational and Resource Constraints: IoT devices are often associated with strict computational and resource constraints. Although HE-based IoT applications have been implemented in several cases, these implementations have mostly ignored the computational and resource constraints associated with IoT devices.
- Policy and Sustainability Integration: The available literature on HE-based IoT implementations has mostly addressed the technical viability of HE-based IoT systems without considering the policy and sustainability implications of such implementations.
- Trade-off Analysis: The trade-off between privacy and computational efficiency and analytical accuracy has not been addressed in the available literature on smart water systems.

In order to overcome the identified research gaps and to make a significant contribution to the field of smart water systems and privacy-preserving computation techniques, a HE-based IoT system framework for smart water systems is presented in the chapter along with a Python-based proof of concept and trade-off analysis.

6.3 METHODOLOGY

6.3.1 System Architecture

The proposed privacy-preserving IoT architecture for smart water management aims to facilitate the collection, transmission, and analysis of water consumption data in a secure manner. The architecture of the proposed IoT architecture comprises a sensing layer, aggregation layer, and analytics layer, as depicted in Figure 6.1.

- Sensing Layer: In the sensing layer, various IoT devices are used, and these devices are installed in various parts of the water supply, for instance, through water pipelines and reservoirs, and at the consumer end, i.e., households and industries. In this layer, the data on water consumption is collected by IoT devices at periodic intervals. In this architecture, data is encrypted using HE, and no data leakage occurs during the transmission of the data.
- Aggregation Layer (Gateways / Edge Devices): In the aggregation layer, various gateways and edge devices are used, and these devices collect data from various IoT devices. In the

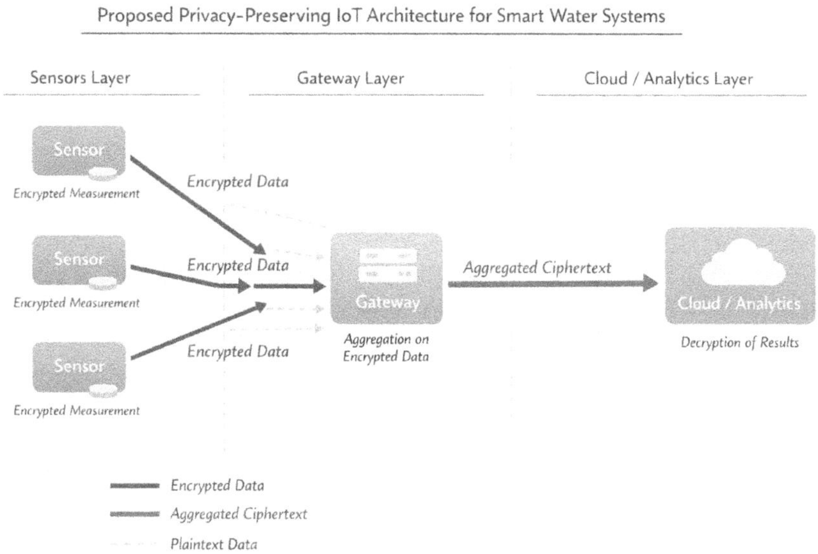

FIGURE 6.1 Proposed privacy-preserving IoT architecture for smart water systems.

aggregation layer, various aggregation operations, for instance, summation and averaging, are performed on the data using HE. After performing the aggregation operations, the data is transmitted to the cloud or the central server for processing, and data privacy is ensured.

- Analytics Layer (Cloud / Central Server): In the analytics layer, a central server is used, and the aggregated data from the gateways and edge devices is transmitted to the central server for processing. In the analytics layer, computations are performed on aggregated data, and data is decrypted only at the end of the computations, ensuring data privacy.

6.3.1.1 *Key Features of the Architecture*

- End-to-End Privacy: The sensor-level encryption of data ensures that the data remains private.
- Scalability: The aggregation of data at the edge of the network helps to reduce the overall bandwidth and computational cost.
- Flexibility: The various operations of HE ensure that analytics can be performed while still maintaining end-to-end privacy.

Figure 6.1 shows the architecture of the end-to-end privacy-preserving smart water IoT analytics system. In this architecture, the sensors will first locally encrypt the water consumption data and then send it to the gateway. The gateway will then aggregate the data and send it to the cloud for decryption and analysis. The dashed gray arrow shows the hypothetical data that could have been sent to the cloud had it not been private, indicating that the data was indeed exposed.

6.3.2 Data Flow and Privacy Design

This system will ensure end-to-end privacy and will also allow for the analysis of the data coming from the IoT water sensors. The IoT water sensors will encrypt the data locally before sending it using HE. The gateways will then aggregate the data, which could be a sum or an average, directly from the encrypted data without exposing any individual data. This aggregated encrypted data will then be sent to the central server, and only the aggregated data will be decrypted for analysis while maintaining end-to-end privacy for individual data.

The framework proposed here guarantees end-to-end privacy, as the individual data from the sensors remains encrypted while being sent and aggregated. It also maintains secure key management, ensuring that only authorized entities have access to decrypt the data, thereby ensuring that unauthorized entities do not have access to the sensitive information. Furthermore, the aggregation of data at the edge also reduces the amount of data being sent to the central server.

6.3.3 HE Implementation

The framework uses a CKKS-type [24] HE Scheme, which offers the ability to perform approximate arithmetic on the ciphertext of floating-point numbers and is best for IoT devices. The devices send the ciphertext of the aggregated values to the gateway. Only the aggregated values are decrypted on the server.

A proof-of-concept implementation using Python for a small-scale system was done using the Pyfhel library and shows the following steps:

- Key generation and distribution.
- Sensor-level encryption of flow values.
- Aggregation of the ciphertext values.
- Decryption of the aggregated values.

6.3.4 Small-Scale Python Proof-of-Concept

A small-scale proof of concept, implemented using the Python programming language, was used to verify the applicability of the proposed privacy-preserving IoT architecture. The Pyfhel library was used to simulate the process of sensor-level encryption, edge-level aggregation, and decryption at the central server. The water flow data, simulated from various sensors, were encrypted at the sensor level, aggregated at the gateway, and then decrypted after aggregation at the central server.

Initialize HE context and generate keys

Encrypt the sensor readings
Send the readings to the gateway

At the gateway:

Sum (or average) the readings directly
Send the aggregated ciphertext to the central server

At the central server:

Decrypt the aggregated data
Perform the final analytics (e.g., average consumption)

The prototype validates the possibility of privacy-preserving aggregation on small-scale IoT data with minimal computational overhead, thus proving the applicability of HE in smart water systems.

TABLE 6.3 Plaintext vs Encrypted Aggregation Result

METRIC	*PLAINTEXT VALUE*	*ENCRYPTED RESULT*
Sum	42.5	42.50000
Average	14.1666667	14.1666682
Absolute Error	–	~1.5 × 10^{-6}

6.4 RESULTS

This section will validate the proposed privacy-preserving IoT framework's applicability through a small-scale IoT proof of concept using the CKKS HE scheme in Python. The results of this experiment are provided in this section. It will particularly focus on the accuracy of the proposed framework because it is a crucial consideration in smart water systems.

6.4.1 Correctness of Encrypted Aggregation

To validate the accuracy of the results, this experiment utilized three different readings from water flow sensors in the simulation, which are 12.5, 15.3, and 14.7 units. The results of the encrypted computation are:

Average consumption: 14.166668193793678

Table 6.3 shows the aggregation of water consumption values calculated from plaintext data and homomorphically encrypted data. The results show that the sum and average values calculated from the encrypted data are almost similar to the plaintext data with an absolute error of 10^{-6}, which is due to the approximate arithmetic in CKKS HE.

6.4.2 Computational Overhead

The computational cost of the HE approach was also analyzed based on the execution time for the key operations, which include encryption, aggregation, and decryption. The experiments were carried out in a typical desktop environment, which simulates a low-resource scenario for the IoT application. Table 6.4 shows the average execution time for the key operations involved in the HE approach. The computational cost of the encryption operation is the highest, taking about 0.012 s for a single vector, while aggregation over the ciphertext is efficient, taking about 0.002 s. Decryption also has a moderate cost, taking about 0.006 s, and the plaintext computation has a negligible cost, taking about <0.001 s.

There are certain computational overheads involved in the process of encryption and decryption, though aggregation over encrypted data is efficient, which shows that homomorphic computations are feasible for low-frequency IoT analytics, for example, periodic reporting of water usage.

TABLE 6.4 Computation Time for HE Operations

OPERATION	*AVERAGE TIME (SECONDS)*
Encryption (vector)	~0.012
Aggregation (ciphertext)	~0.002
Decryption	~0.006
Plaintext average	<0.001

TABLE 6.5 Approximation Error Metrics for Encrypted Aggregation

METRIC	*VALUE*
Max Error	~1.5×10^{-6}
Relative Error	< 0.00001%
Impact on Policy Decisions	Negligible

6.4.3 Accuracy and Approximation Error

CKKS HE uses approximate arithmetic for computation. This introduces some numerical error in the encryption and decryption process.

In the experiment carried out:

- The error in approximation was found to be below 10^{-5}.
- There was no significant loss of utility.
- The error is negligible for water sustainability because the decision is based on trends and not precise per-unit values.

Table 6.5 shows a summary of the degree of errors introduced by CKKS HE in the aggregation of water consumption. We see that the maximum absolute error is 1.5×10^-6, while the relative error is less than 0.00001%. Such minor differences do not affect policy-making or sustainability analysis, thus validating that privacy-preserving aggregation is sufficiently accurate. These results validate that CKKS-based aggregation is sufficiently accurate for smart water monitoring.

Figure 6.2 demonstrates a comparative analysis of the average water consumption calculated using the plaintext data and the results of the homomorphic encryption. The results of the encrypted data are similar to the plaintext results, and the difference is due to the approximation of the CKKS scheme. This demonstrates the use of homomorphic encryption for the purpose of smart water monitoring.

The execution time for the major operations in the HE process is depicted in Figure 6.3. The encryption process has the maximum execution time, followed by the aggregation process. The aggregation process is performed efficiently on the encrypted data. Although the execution time for the homomorphic process is high compared to the plaintext process, the time is acceptable for low-frequency IoT data analytics in the smart water system.

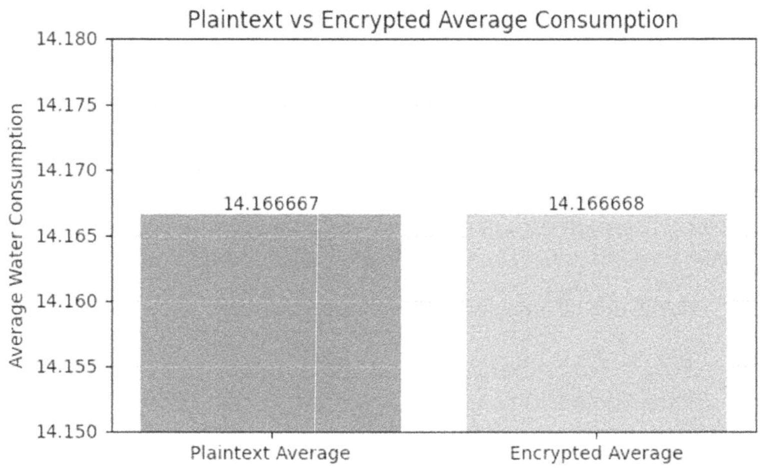

FIGURE 6.2 Plaintext vs encrypted average consumption.

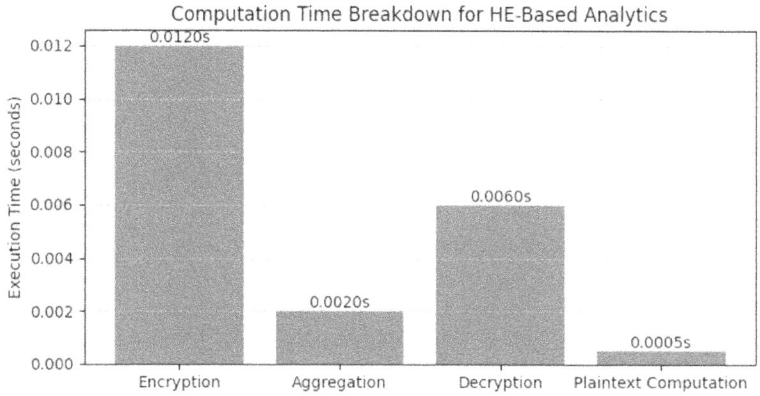

FIGURE 6.3 Computational time breakdown for HE-based analytics.

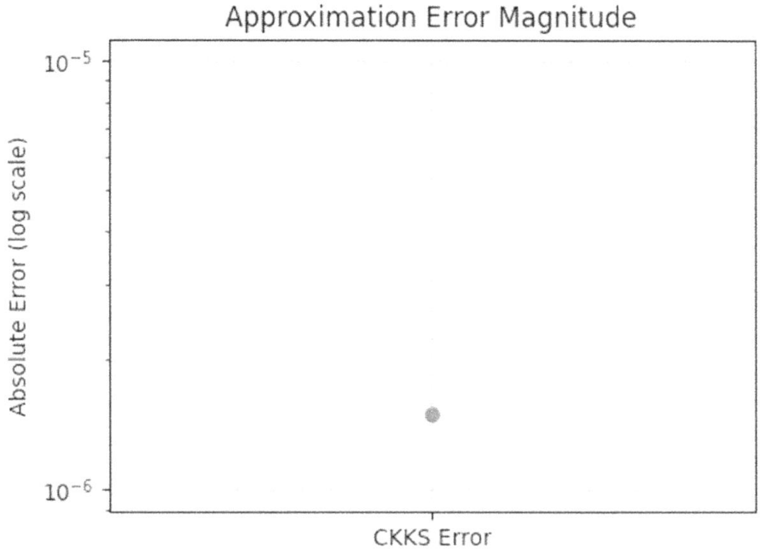

FIGURE 6.4 Approximation error magnitude.

As depicted in Figure 6.4, the absolute approximation error due to the HE using the CKKS algorithm is plotted on a logarithmic scale. As can be observed, the absolute approximation error is still on the order of 10^-6, validating the fact that the numerical precision is compromised to a very small extent. Thus, the approximation error is very small and can be ignored for smart water analytics, as decisions are based on aggregated trends rather than exact sensor readings.

6.5 DISCUSSION

6.5.1 Privacy-Utility Trade-Off

It is evident from the results that the proposed approach based on HE supports analytics over IoT-based water consumption data while maintaining confidentiality. Although CKKS proposed in this chapter only supports approximate arithmetic, the numerical error is zero in the aggregate data of water consumption, like total and average. This implies that the proposed approach does not compromise the usefulness of data, which is required to be adopted in a sustainable way.

6.5.2 Feasibility for Smart Water IoT Systems

The computational results of the proposed approach for the smart water IoT system indicate that there is an overhead in the encryption/decryption process compared to the plaintext approach, with the encryption process being the main contributor to the overhead. The aggregation process is efficient in the proposed approach. This indicates that the proposed approach is viable for the smart water IoT system, which is used for low- or moderate-frequency analytics.

6.5.3 Implications for Water Sustainability and Policy

The importance of privacy-preserving analytics in facilitating data-driven water sustainability without compromising the privacy of individuals or organizations can be highlighted. By ensuring that the raw data of water consumption is not revealed, the presented approach can facilitate the concept of privacy by design, which can further improve the trust of the public in water infrastructure built on the concept of smart technologies.

6.5.4 Limitations

In spite of the benefits of the presented approach, there are limitations to the approach. For example, the presented approach for the proof-of-concept does not consider the requirements of the city-wide water infrastructure in terms of real-time requirements. In addition, the HE approach is computationally expensive for the sensor devices of the IoT-based water infrastructure.

From the discussion, it is clear that the HE approach can provide a balance between the requirements of ensuring the privacy of individuals and the requirements of analytics in the IoT-based water infrastructure.

6.6 CONCLUSION

The contribution of this study is a proposed privacy-preserving solution for smart water systems based on IoT data and HE. The study has demonstrated that the proposed CKKS-based encrypted aggregation solution is capable of efficiently computing the most critical metrics, i.e., total and average water consumption, without compromising the privacy of the data. The study used a small-scale Python-based POC to demonstrate the efficiency of the proposed solution in terms of numerical approximation errors, which were found to be minimal, while the computational cost was acceptable for IoT data analytics.

The proposed solution is considered appropriate for sustainable water management because HE has the potential to provide a foundation for developing privacy-preserving IoT architectures [25, 26] that can support data-driven policy-making for water management, which can foster trust in smart water systems [30].

6.7 FUTURE WORK

Although the present structure proves the applicability and feasibility of HE for privacy preservation in IoT-based water analysis, there are a number of ways to increase scalability, security, and functionality.

6.7.1 Integration with Differential Privacy

In order to further increase security and privacy in IoT-based water analysis, HE can be integrated with differential privacy (DP) [27]. By integrating a privacy budget [28] mechanism, policymakers can easily track and limit privacy leakage for effective compliance with regulatory requirements while still allowing for meaningful analysis.

6.7.2 Post-Quantum Cryptography (PQC) Readiness

As IoT networks continue to expand and become ubiquitous, it is expected that future IoT networks will require quantum-resistant cryptographic techniques. By incorporating PQC [29] into the framework, future IoT networks will be secured against potential quantum attacks, even if and when such technology becomes available.

6.7.3 Edge-Side Optimization and Scalability

In order to further increase scalability and applicability in IoT networks, future work should be directed toward edge-side aggregation and HE.

6.7.4 Hybrid Encrypted Analytics and AI/ML Integration

The next step is to integrate HE with AI/ML techniques for predictive water demand, leak detection, or anomaly detection while maintaining privacy. HE may be used along with differential privacy [31] or secure multiparty computation for privacy-preserving machine learning on IoT data.

6.7.5 Exploration of Emerging Technologies

The framework may also leverage the exploration of emerging technology niches for the security, efficiency, and utility of smart water systems. Universal privacy-preserving computation frameworks may also be used for unified support of HE, secure multiparty computation, and differential privacy. This would offer versatility and strong privacy guarantees. Light-weight HE may also be used to reduce the computation costs of encrypted analytics for IoT sensors. This would offer a solution for encrypted analytics for smart water systems. In addition, secure sensor-to-cloud communication may also be used for autonomous and secure communication for the transmission of data, thus providing end-to-end data confidentiality while maintaining system scalability and reliability.

REFERENCES

1. S. Narendran, P. Pradeep and M. V. Ramesh, "An Internet of Things (IoT) Based Sustainable Water Management," 2017 IEEE Global Humanitarian Technology Conference (GHTC), San Jose, CA, 2017, pp. 1–6, doi: 10.1109/GHTC.2017.8239320.
2. V. B. Niranjane, A. Bhagat, K. M. Punwatkar, K. Chaurey and P. Niranjane, "Sustainable Water Management using Internet of Things," 2022 4th International Conference on Smart Systems and Inventive Technology (ICSSIT), Tirunelveli, India, 2022, pp. 72–75, doi: 10.1109/ICSSIT53264.2022.9716267.
3. N. N. Kucherov, M. A. Deryabin and M. G. Babenko, "Homomorphic Encryption Methods Review," 2020 IEEE Conference of Russian Young Researchers in Electrical and Electronic Engineering (EIConRus), St. Petersburg and Moscow, Russia, 2020, pp. 370–373, doi: 10.1109/EIConRus49466.2020.9039110.
4. S. Wadekar, V. Vakare, R. Prajapati, S. Yadav and V. Yadav, "Smart Water Management Using IOT," 2016 5th International Conference on Wireless Networks and Embedded Systems (WECON), Rajpura, India, 2016, pp. 1–4, doi: 10.1109/WECON.2016.7993425.
5. L. Nandakumar, A. S. Joseph, A. Shaji, A. S. Antony, A. J. A. Surendran and C. Jose, "AQUA - Water Management System at Home for a Sustainable World Using IoT," 2024 IEEE Recent Advances in Intelligent Computational Systems (RAICS), Kothamangalam, Kerala, India, 2024, pp. 1–6, doi: 10.1109/RAICS61201.2024.10689916.
6. H. Li, Q. Chen, H. Zhu, D. Ma, H. Wen and X. S. Shen, "Privacy Leakage via De-Anonymization and Aggregation in Heterogeneous Social Networks," *IEEE Transactions on Dependable and Secure Computing*, vol. 17, no. 2, pp. 350–362, 1 March–April 2020, doi: 10.1109/TDSC.2017.2754249.
7. S. Aditya G, K. B, S. M. Rajagopal and C. R. Kavitha, "Homomorphic Encryption for Secure Data Analysis: A Hybrid Approach using PKCS1_OAEP Padding," 2024 2nd International Conference on Intelligent Data Communication Technologies and Internet of Things (IDCIoT), Bengaluru, India, 2024, pp. 481–485, doi: 10.1109/IDCIoT59759.2024.10468008.
8. J. Ryu, K. Kim and D. Won, "A Study on Partially Homomorphic Encryption," 2023 17th International Conference on Ubiquitous Information Management and Communication (IMCOM), Seoul, Korea, Republic of, 2023, pp. 1–4, doi: 10.1109/IMCOM56909.2023.10035630.
9. R. R. Sahani, S. Krishnaswamy and G. Trivedi, "Implementation of Somewhat Homomorphic Encryption (SHE) using RLWE based Post Quantum Cryptography (PQC)," 2025 IEEE Space, Aerospace and Defence Conference (SPACE), Bangalore, India, 2025, pp. 1–6, doi: 10.1109/SPACE65882.2025.11170983.
10. P. S. Pisa, M. Abdalla and O. C. M. B. Duarte, "Somewhat Homomorphic Encryption Scheme for Arithmetic Operations on Large Integers," 2012 Global Information Infrastructure and Networking Symposium (GIIS), Choroni, Venezuela, 2012, pp. 1–8, doi: 10.1109/GIIS.2012.6466769.
11. A. Viand, P. Jattke and A. Hithnawi, "SoK: Fully Homomorphic Encryption Compilers," 2021 IEEE Symposium on Security and Privacy (SP), San Francisco, CA, 2021, pp. 1092–1108, doi: 10.1109/SP40001.2021.00068.
12. K. G. Kogos, K. S. Filippova and A. V. Epishkina, "Fully Homomorphic Encryption Schemes: The State of the Art," 2017 IEEE Conference of Russian Young Researchers in Electrical and Electronic Engineering (EIConRus), St. Petersburg and Moscow, Russia, 2017, pp. 463–466, doi: 10.1109/EIConRus.2017.7910591.
13. M. Nassar, A. Erradi and Q. M. Malluhi, "Paillier's encryption: Implementation and Cloud Applications," 2015 International Conference on Applied Research in Computer Science and Engineering (ICAR), Beiriut, Lebanon, 2015, pp. 1–5, doi: 10.1109/ARCSE.2015.7338149.
14. A. Satriawan, R. Mareta and H. Lee, "Exploring Possibilities of BFV-based Homomorphic Encryption for Privacy-Preserving Image Processing," 2025 IEEE International Symposium on Circuits and Systems (ISCAS), London, 2025, pp. 1–5, doi: 10.1109/ISCAS56072.2025.11044185.
15. A. Almaini, J. Folz, D. Woelfl, A. Al-Dubai, M. Schramm and M. Heigl, "A New Scalable Distributed Homomorphic Encryption Scheme for High Computational Complexity Models," 2023 International Wireless Communications and Mobile Computing (IWCMC), Marrakesh, Morocco, 2023, pp. 890–897, doi: 10.1109/IWCMC58020.2023.10183131.
16. V. Vanitha, G. Saritha, K. Hemalatha, P. Sathish and P. G. Devi, "Securing Edge Computing Environments with a Novel Chaos Theory-Based Lightweight Encryption Algorithm for Real-Time Data Protection," 2024 International Conference on Integrated Intelligence and Communication Systems (ICIICS), Kalaburagi, India, 2024, pp. 1–8, doi: 10.1109/ICIICS63763.2024.10860006.

17. M. R. Baharon, Q. Shi and D. Llewellyn-Jones, "A New Lightweight Homomorphic Encryption Scheme for Mobile Cloud Computing," 2015 IEEE International Conference on Computer and Information Technology; Ubiquitous Computing and Communications; Dependable, Autonomic and Secure Computing; Pervasive Intelligence and Computing, Liverpool, 2015, pp. 618–625, doi: 10.1109/CIT/IUCC/DASC/PICOM.2015.88.
18. L. Fiondella, L. Hogewood, A. Ligo and I. Linkov, "Edge Computing as an Enabler of Energy and Water System Resilience," *IEEE Engineering Management Review*, vol. 52, no. 1, pp. 28–42, February 2024, doi: 10.1109/EMR.2023.3320876.
19. S. Ullah et al., "Homomorphic Encryption Applications for IoT and Light-Weighted Environments: A Review," *IEEE Internet of Things Journal*, vol. 12, no. 2, pp. 1222–1246, January 15, 2025, doi: 10.1109/JIOT.2024.3472029.
20. M. Matsumoto and M. Oguchi, "IoT Device Friendly Leveled Homomorphic Encryption Protocols," 2022 IEEE International Conferences on Internet of Things (iThings) and IEEE Green Computing & Communications (GreenCom) and IEEE Cyber, Physical & Social Computing (CPSCom) and IEEE Smart Data (SmartData) and IEEE Congress on Cybermatics (Cybermatics), Espoo, Finland, 2022, pp. 525–532, doi: 10.1109/iThings-GreenCom-CPSCom-SmartData-Cybermatics55523.2022.00104.
21. H. M. Reddy, S. P C and S. Sankaran, "On the Feasibility of Homomorphic Encryption for Internet of Things," 2022 IEEE 8th World Forum on Internet of Things (WF-IoT), Yokohama, Japan, 2022, pp. 1–6, doi: 10.1109/WF-IoT54382.2022.10152214.
22. J. Barbosa and P. Zhang, "Encrypted Control Systems with Fully Homomorphic Encryption," 2025 European Control Conference (ECC), Thessaloniki, Greece, 2025, pp. 161–166, doi: 10.23919/ECC65951.2025.11186994.
23. H. Zhu, T. Suzuki and H. Yamana, "Private Function Evaluation using CKKS-based Homomorphic Encrypted LookUp Tables," 2025 22nd Annual International Conference on Privacy, Security, and Trust (PST), Fredericton, NB, Canada, 2025, pp. 1–10, doi: 10.1109/PST65910.2025.11268867.
24. B. S. Dakhare and L. L. Ragha, "CKKS Homomorphic Encryption Scheme for Financial Dataset," 2025 IEEE 14th International Conference on Communication Systems and Network Technologies (CSNT), Bhopal, India, 2025, pp. 353–357, doi: 10.1109/CSNT64827.2025.10968654.
25. D. Torre, A. Chennamaneni and A. Rodriguez, "Privacy-Preservation Techniques for IoT Devices: A Systematic Mapping Study," *IEEE Access*, vol. 11, pp. 16323–16345, 2023, doi: 10.1109/ACCESS.2023.3245524.
26. A. Almohaimeed, S. Gampa and G. Singh, "Privacy-Preserving IoT Devices," 2019 IEEE Long Island Systems, Applications and Technology Conference (LISAT), Farmingdale, NY, 2019, pp. 1–5, doi: 10.1109/LISAT.2019.8817349.
27. A. G. Sébert, M. Checri, O. Stan, R. Sirdey and C. Gouy-Pailler, "Combining Homomorphic Encryption and Differential Privacy in Federated Learning," 2023 20th Annual International Conference on Privacy, Security and Trust (PST), Copenhagen, Denmark, 2023, pp. 1–7, doi: 10.1109/PST58708.2023.10320195.
28. Z. Chen, G. Liao, Q. Ma and X. Chen, "Adaptive Privacy Budget Allocation in Federated Learning: A Multi-Agent Reinforcement Learning Approach," ICC 2024 - IEEE International Conference on Communications, Denver, CO, 2024, pp. 5166–5171, doi: 10.1109/ICC51166.2024.10622685.
29. A. Sharma and S. Rani, "Post-Quantum Cryptography (PQC) for IoT-Consumer Electronics Devices Integrated With Deep Learning," *IEEE Transactions on Consumer Electronics*, vol. 71, no. 2, pp. 4925–4933, May 2025, doi: 10.1109/TCE.2025.3569904.
30. B. S. Rao, S. Chattopadhyay, P. Singh, B. Hazela, G. Sabarinathan and K. Yamini, "Privacy-Aware Artificial Intelligence with Homomorphic Encryption using Machine Learning," 2023 International Conference on Sustainable Computing and Smart Systems (ICSCSS), Coimbatore, India, 2023, pp. 259–265, doi: 10.1109/ICSCSS57650.2023.10169776.
31. R. Aziz, S. Banerjee, S. Bouzefrane and T. Le Vinh, "Exploring Homomorphic Encryption and Differential Privacy Techniques towards Secure Federated Learning Paradigm," *Future Internet*, vol. 15, no. 9, p. 310, 2023. https://doi.org/10.3390/fi15090310

IoT-Enabled Smart Water Management

7

Advanced Sensor Technologies and Data-Driven Optimisation

R. Saroja Devi, J. Sudarvel, and Ravi Thirumalaisamy

7.1 INTRODUCTION

Water is an indispensable resource and the basis of human life, economic activities, industrialisation, and well-being of natural ecosystems. The increasing pressures of population increase, the rapid urbanisation process, and climatic changes have placed more pressure on the world's water requirements, with trends indicating that by 2030, the water demands could exceed supply levels by up to 40%. This looming water scarcity highlights the importance of new strategies in managing the water resources so as to enable them to be exploited fairly, efficiently, and in a sustainable manner [1].

The water management methods which have been largely based on periodic manual checks, fixed flow metres, and corrective action are no longer adequate in dispensing the water issues of the modern world. Looking at these traditional ways, it creates inefficiency like water leaks, excessive use, and the probability of contamination, and takes a long time to find out that something is not normal. Such imperfections underscore the importance of the immediate need to have technologies that can provide real-time, continuous, and accurate information about the water systems [2].

The emergence of the Internet of Things (IoT) has changed the water management strategy [3]. The IoT water sensors offer opportunities to control vital parameters such as the level of water, the rate of flow, the pressure, and the water quality. Such sensors can be utilised to acquire real-time data and transmit it to the cloud-based systems where sophisticated analytics and machine learning algorithms can potentially process the raw data into actionable information [4–6]. Through remote monitoring and automated control systems, the occurrence of leaks can be detected on time, the irrigation programmes can be streamlined, the usage of water by the industries can be well controlled, and the quality of water in the environment can be ensured.

The IoT usage in water management does not merely enhance efficiency in the working process but also leads towards sustainable working practices by reducing wastages, lowering costs, and conserving resources. Besides, IoT-based systems facilitate the application of data in policy making, which enables

DOI: 10.1201/9781003742227-7

municipalities, industries, and farming stakeholders to make informed decisions that will contribute to the goal of environmental sustainability and resource optimisation.

This chapter entails the discussion of the IoT-based water sensor systems, their design, architecture, and functioning. It describes the latest sensor technologies, real-life applications, approaches to its implementation, and challenges associated with its implementation. Other than that, future research directions that the chapter reveals are the integration of artificial intelligence and scalable solutions for smart cities, along with innovative methods for the sustainable management of water resources. The IoT water management system is a groundbreaking approach to managing the world water crisis, since it can achieve a balance between technology and sustainability.

7.2 IOT WATER MANAGEMENT CONCEPTUAL FRAMEWORK

The IoT-controlled water is a revolution in the old, reactionary water management methods against the new, proactive, information-driven water resources optimisation. It is based on the integration of cutting-edge sensor networks, edge computing, cloud analytics, and automated control systems that enable a full-time surveillance system, timely intervention, and sustainable utilisation of water. The IoT-based water management conceptual framework is organised in several interdependent layers that contain respective roles and that, when combined with each other, lead to effective and resilient operation.

Data Acquisition Layer: This layer is essential and has a number of elaborate sensors that are positioned in strategic places within the water systems. These sensors monitor the most significant parameters, such as the level of water, the rate of water flow, the pressure, the temperature, and the chemical quality measurements, which could be pH, turbidity, and dissolved oxygen measurements. The contemporary gathering of information (real-time) permits predicting the water consumption patterns and promptly detecting water leaks or contamination and making decisions. The sensors are significant to downstream analytics and control processes, as the accuracy and reliability of their functioning is key to their functioning [7].

Edge Computing Layer: Edge computing devices (microcontrollers or embedded processors) are deployed at the location of the data source to offer processing to a large volume of data generated by sensor networks. These machines pre-process and filter sensor raw data and lastly do local calculations and only transmit what is required to minimise network congestion and latency. This layer will make the system more responsive, offer a real-time decision-making process, and reduce the requirement of constant access to clouds which makes the system more sustainable in remote or resource-constrained settings [8].

Communication Layer: The security and reliable data transfer are a crucial aspect of water management through IoT. The communication layer facilitates the free flow of information between the sensors, edge devices, and cloud servers using any one of the numerous wireless protocols, including Wi-Fi, LoRaWAN, GSM, NB-IoT, and any other low-power wide-area network (LPWANs). The technology of communication that will be adopted will depend on the area to be covered, the volume of the data to be transferred, the energy efficiency, and the environment in which it will be used. At this level, security and encryption are applied so that the water management information is not accessed or corrupted by a person who is not supposed to have access to such information [9].

Cloud Analytics Layer: Data stored here will be processed and analysed to store data on cloud environments and utilise the power of big data analytics, machine learning algorithms, and predictive modelling to process data obtained. Cloud analytics can be used to identify anomalies and trends, predict water demand, as well as performance optimisation. The information generated in this layer is utilised in making strategic plans, resource assignment, and preventive maintenance process which ultimately lead to increased reliability and efficiency of the system [10].

Decision and Control Layer: It will be the final layer representing analytical knowledge in the action. The automated control systems are dynamically controlled valves, pumps, irrigation systems, and treatment units. This makes it possible to irrigate the agricultural system specifically, to swiftly control the leaks in the urban water system, and to reduce the use of water in industries, thereby curbing wastage and wasting resources.

The layers also allow IoT-based water management systems to be scalable, resilient, and adaptable to a broad spectrum of applications. This conceptual framework is the foundation for intelligent, sustainable, and efficient management of water resources in the digital transformation era in the municipal water distribution networks, agricultural irrigation, and industrial water monitoring [11].

7.3 WATER MANAGEMENT SENSOR TECHNOLOGIES

The accuracy, dependability, and flexibility of the sensors placed in the water systems are a great determinant of the success of the IoT-based water management. With the help of electronic sensor systems, the real-time information about such crucial parameters as the water level, flow, and quality is provided, and the appropriate management and efficient utilisation of the resources are possible. The sensors form the basis of smart water management systems, which are utilised in the urban as well as the rural water management, industry, and the protection of the environment [12].

Water Level Sensors: Water level sensors will be required to monitor the quantity of water used in tanks, reservoirs, and rivers, among other storage or distribution facilities. The most widespread of them are ultrasonic sensors since they work on the principle of non-contact, thereby reducing wear and tear as well as maintenance of the sensors. Ultrasonic sensors are sources of sound waves, and they capture the time the sound would take to bounce back, providing an extremely accurate reading of the surfaces of the water. They can be used in the large-scale water bodies, flood supervision, and computerised tank care [13].

Larger applications or applications in which the cost is a consideration can be substituted with float sensors and capacitive sensors. The float sensors operate under the principle of floating body that follows the level of water and sends switches or transmits signals to the monitoring systems. The variation in the capacitance when there is a change in the water level is however measured using capacitive sensors. In spite of the fact that they are less expensive and simple to install, such kind of sensors require maintenance at least periodically to ensure accuracy and prevent dirtiness and mechanical failures. Generally, the water level sensors play a crucial role in enabling predictive water management, overflow management, and adequate water availability to the domestic, agricultural, and industrial operations.

Flow Sensors: Flow sensors measure the volumetric or mass rate of water flow through pipes and channels, which is highly significant information, used to track consumption behaviour in real time and detect leakage and efficiency of the system. The effective flow measurement is essential to municipal and water supply services, industrial water usage, and precision agriculture, in which streamlined irrigation is needed to conserve water and give optimal crop yield [14].

Flow sensing has a number of technologies, which include electromagnetic, ultrasonic, turbine, and differential pressure sensors. Electromagnetic flow sensors are appropriate when dealing with conductive liquids, and they are highly precise and free of moving components. The ultrasonic flow metres utilise the Doppler or transit-time concept to measure non-invasive flow and give minimal disturbance and maintenance to the system. The turbine flow sensors also possess rotating blades, which are economical at small-scale use, but they require calibration on a periodic basis. By connecting flow sensors to IoT networks, water managers will be in a position to determine abnormal patterns of usage, leak fast, and take automated control measures, thereby saving a lot of water and operational costs [15].

Water Quality Sensors: In order to ensure clean and safe water, a number of quality parameters of water must continuously be checked. PH, turbidity, conductivity, sensors of dissolved oxygen, and chemical contaminants will provide comprehensive data on the welfare of water to the human body as well as to the protection of the environment. As an example, the high turbidity conditions can indicate the presence of sediment or contamination, but not an abnormal pH or conductivity values can be indicative of chemical pollution.

The further evolution of IoT water systems is likely to develop one device or a net of devices, and one can measure several parameters concurrently. This type of multi-sensor approach enhances the accuracy and reliability of the water quality cheque and allows effective corrective actions to be taken in time such as altering the treatment programme, contamination warnings, or even a change of source. In addition to this, the continued accumulation of the water quality information aids in checking in cheque regulatory compliance, environmental conservation modules, and the prudent decision-making process in the urban, industrial, and agricultural water systems [16].

With these advanced sensor technologies using the IoT, water management systems offer a suitable and data-focused foreground to simplify water use, maintain the population healthy, and support the water resources management practices conservation.

7.4 IOT SMART WATER MANAGEMENT ARCHITECTURE

It is believed that IoT water management systems should be data-driven, scalable, and modular to be able to constantly monitor water resources, predictive information, and control them automatically. The architecture also uses different layers, which possess specialised functionalities in order to promote the efficient collection, processing, transmission, and decision-making of data. This hierarchical design helps water administrators to optimise the operations, reduce the wastage, detect an anomaly in real time, and take proactive measures in the municipal, agricultural, and industrial water systems.

Sensor Layer: The IoT water management system is built on the sensor layer and is used to measure physical, chemical, and environmental properties of water. The sensors are utilised to measure the water level, flow rate, pressure, temperature, pH, turbidity, dissolved oxygen, and other relevant quality indicators. The sensors present all the details regarding the health status and performances of the water system. These two attributes are the critical ones, as they define the quality of the predictive analytics and AI-based decision-making, namely the accuracy and reliability of the sensors.

Edge Layer: The edge layer consists of microcontrollers and embedded processors (ESP32 board or Arduino or Raspberry Pi board). These gadgets process the crude sensor information at the local location, and they initially compute, filter, and smooth the data and forward the information to the cloud. Local process decreases network load, minimises latency, and serves to respond instantly to such emergent states as an unexpected leakage, a drop in pressure, or the development of contamination. The edge computing enhances system resilience, particularly in adjoined zones or in resource-scarce environments where there might not be a consistent cloud connexion.

Network Layer: Network layer enables effective and secure transfer of the edge device data to the cloud platforms. There are a number of various communication protocols which are applied depending on the coverage, power requirements, and the quantity of data: Wi-Fi, LoRaWAN, GSM, NB-IoT, and other low-power wide-area networks (LPWANs). The encryption mechanism and security available have guaranteed that the sensitive water management data cannot be intercepted by any hacking or cyber threat, and its integrity, confidentiality, and reliability will be maintained across the entire system.

Cloud Layer: Cloud layer will provide centralised storage, visualisation, and more advanced analytics of large-scale data obtained using distributed sensors. Big data platforms, real-time data, and historical data can be used to identify anomalies and predict demand due to machine learning algorithms, predictive demand, and resource optimisation. Cloud analytics enables the managers of the water to foresee system failures, develop maintenance, and make sound decisions that would enhance efficiency and sustainability. Other strategies that this layer supports are long-term trend analysis, regulatory compliance, and urban water management.

Application Layer: The analytics would be converted into actionable insights by the application layer that are accessible to the end-users through dashboards, mobile applications, and web interfaces. The users will be able to observe the real-time performance of a water system, receive notifications regarding leakage or quality issues, and operate automated devices, such as valves, pumps, and irrigation systems. Its friendly interface will be an assurance that the field technicians, decision-makers, and stakeholders can respond promptly and effectively to the information in order to bridge the gap between the information and the practise.

Integrated Benefits: Once such layers are incorporated into one system, predictive maintenance and anomaly detection, along with the ability to make dynamic decisions, can be also applied to the IoT-enabled water management systems. They make the most out of the water, reduce the amount of money used to operate the activities, eradicate wastage, and offer sustainable operation of the water resources. It is also scalable with a flexible design and thus can be connected to some of the emerging technologies, such as artificial intelligence, machine learning, and smart city platforms, to meet the future demands of water management.

7.5 APPLICATIONS IN THE REAL WORLD AND CASE STUDIES

The current water management systems with the use of the IoT are transforming the existing trends of monitoring, controlling, and preserving water resources. They have real-life applications in the farming field, urban water distribution, flood control, and water quality. By combining sensor networks, edge computing, and cloud analytics, these systems enable making decisions based on data, predicting on the basis of the intervention, and using water in a sustainable way. The most important uses and findings of field studies are illustrated in the subsequent subsections.

Smart Irrigation Systems: One of the biggest threats to crop productivity and sustainability in the agricultural industry is water shortages and poor irrigation systems. The smart irrigation systems proposed by the IoT provide solutions to these issues by measuring the soil moisture, tank capacity, and weather conditions constantly. The automated irrigation controllers make use of sensors to feed real-time data to the controller and apply the data to optimise the water application based on crop requirements. It has been shown that smart irrigation can cut water consumption by 25–30%, and the crop yield is not affected by it, and predictive analytics can even better optimise the irrigation schedule and prevent over-utilisation and waste of resources. It is also these systems that enable the monitoring and control of remotely, allowing a farmer to respond fast to the fluctuating field environment and climatic changes.

Urban Water Distribution: The municipalities usually have a high water wastage due to leaks in the water supply networks, bursting of the pipes, and inefficient channels of distribution. The IoT and smart metres placed on pipelines determine the flow rate, pressure, and consumption habits in real time. This will enable the detection of leaks and unnatural utilisation at an earlier stage to enable the administrators to reduce on the loss of water by 15-25%. Moreover, evidence-based information is helpful in the optimisation of resource allocation, infrastructure maintenance, and policy planning. The IoT and the introduction of technologies into the city water systems enhance the efficiency of its work, reduce the maintenance cost, and promote sustainable management of the city water.

Flood Management and Flood Monitoring: The importance of flooding as a risk to the city infrastructure, agriculture, as well as the safety of the people. IoT-enabled water level sensors in rivers, canals, and drainage channels give the opportunity to monitor the volume of water and channel flow all the time. Real-time data supports early flood warning systems, thereby increasing the responsiveness of such systems, and the level of destruction of infrastructure and property can be reduced through proactive action taken. There is also a greater rate of accuracy in using predictive analytics in flood monitoring systems since authorities would be able to give timely warnings and successfully plan how to deal with the emergency.

Water Quality Monitoring: The drinking water safety and quality are quite a significant component in water management. The parameters identify the anomalies in the parameters, such as pH, turbidity, conductivity, dissolved oxygen, and chemical contaminants in the water early; therefore, the parameters make sure that the anomalies, which reveal the water quality, are identified early. The IoT sensors provide real-time alerts to the operators so as to allow preventive action before contamination. Such systems are invaluable in municipal water supply as well as in industrial effluents, environmental monitoring, which aids in ensuring that the health of the population is not endangered, and that regulations are adhered to.

Experimental Insights: The pilot projects and field tests prove the usefulness of the systems of water management in the context of IoT. The farmers who utilised the IoT-based irrigation reported that the saved water was approximately 28%, but production was the same, indicative of the efficacy and precision of smart water management. Introduction of sensor-based monitoring in water networks in cities resulted in the reduction of cost in running the water networks, reliability and fast response to leakages and anomalies. These experimental results prove the applicability of the IoT technologies in terms of water efficiency, sustainable management, and actionable information to the decision-makers.

The solution to all these applications, which is based on the IoT, is not only improving the efficiency of operations but also assisting in the management of sustainable water resources, climate resiliency, and environmental preservation. IoT systems give a feasible route in the world water issue with actual time checks, predictive analytics, and automatic control.

7.6 CHALLENGES AND LIMITATIONS

Despite the high volume of benefits that are linked with the implementation and application of the IoT-based water management systems, there are several challenges and limitations that come along with it. These are the issues that should be learned to create reliable, scalable, and cost-effective systems.

High Initial Investment: One of the key reasons why the large-scale use of the IoT water management system is not possible is its high cost of installation. The use of advanced sensors, edge gadgets, communication systems, and cloud computing can be expensive, particularly to localities, small-scale farmers, or developing regions. Even though these expenses can be covered by long-term savings from operation and optimisation of the resources, initially the initial cost is a significant consideration in the planning and implementation.

Network Reliability: The monitors and the system of remote controls over the water systems also require effective network connexion. The internet and cellular connexions may be ineffective, spotty, or fail at remote, rural, or unmapped areas. Issues on the reliability of network may contribute to the sluggish flow of information, data sets, and inefficient automated control systems. The solutions that can reduce the challenges to a certain extent and require careful planning and investment are low-power wide-area networks (LPWANs) or hybrid communication systems.

Sensor Maintenance: The data collection is founded on relevant and precise data collection upon well-calibrated and well-kept sensors. Sensors may be contaminated over time with sediment, biofilm, or chemistry and the measurement error. To ensure that the system is by the book, sensors should be cleaned, calibrated, and replaced on a regular basis. The maintenance problems can increase the price of operation and require staff training, particularly on large systems of water networks or distributed water networks.

Cybersecurity Risks: The water management systems which are IoT-enabled are closely connected with the digital networks, and this makes them vulnerable to cyber-attacks such as hacking, loss of data, as well as unauthorised control systems. The impact of weakened security can be drastic because there is a possibility of water pollution, outage of services, or destruction of infrastructure. This is to be achieved through the implementation of robust cybersecurity controls like encryption, authentication, intrusion detection, and regular auditing of the system to make sure that data integrity and system continuity is not affected.

Integration Complexity: The IoT water management is a set of different components that include types of sensors, microcontrollers, communication units, cloud services, and application interfaces. Such components require a superior system design, interoperability guidelines, and effective data administration plans to synchronise them to be functioning in unison. The intricacy of integration might bring about technical problems, extended implementation periods, and greater maintenance needs, especially when carrying out the activity of modifying the current infrastructure or connecting the IoT networks to the current water management systems.

Additional Considerations: Others like energy consumption of sensor nodes that are not in-gird, data traffic and data processing of large volumes, and regulatory/policy obstacles in the implementation of new technologies, among others. In order to defeat these limitations, there is a necessity to choose the comprehensive approach that involves technological innovation, planning of the strategy, and involvement of the stakeholders and their building of the capacity. Nonetheless, this does not reduce the significance that the IoT-based water management is a game-changer in water management in the world since the long-term benefits include optimised water utilisation, improved service delivery, improved sustainability, and resistance to climatic changes.

7.7 PERSPECTIVES AND TRENDS OF FUTURE OF THE IOT WATER MANAGEMENT

The future evolution of the IoT-based water management is anticipated to evolve at a rapid pace because of the enhancement of sensor technologies, artificial intelligence, edge analytics, and connexion to smart cities. Due to continuously rising global water problems due to population, urbanisation, and climate change, the future is on the aspect of emerging innovations that will enable more intelligent, adaptive, and sustainable systems. Its future requirements encompass greater automation, interoperability, and intelligence of the data so that it can support long-term environmental and economic sustainability.

Artificial Intelligence and Machine Learning Assessment: One of the radically new directions of the evolution of the IoT-based water systems development is artificial intelligence (AI) and machine learning (ML). With large volumes of data collected by sensors, it is possible to use the AI algorithms to identify latent patterns, predict system anomalies, as well as optimise the allocation of resources. The machine learning frameworks can forecast water demands, detect leaks prior to outburst, and propose alterations in response to control to utilise real-time and historical data. Proactive and not reactive management will therefore be implemented with the help of predictive analytics that will promote efficiency and reduce the costs of operations. In the long term, with automation, there will be more precise decisions made due to AI, which will lead to a more resilient and self-sustaining water system.

The System Will Be Combined with the Smart City Infrastructure: The water management systems will be integrated into the overall system of smart city and will be integrated with the water distribution, energy grids, waste management, and transportation systems. The IoT will also enable smart water systems to be linked to the municipal control stations, and this will enable them to possess a single dashboard where they can access the consumption of the resources and the environmental conditions. This type of integration helps in proper urban planning, equal sharing of water, and sustainable development. Data sharing across sectors in decision-making processes between related systems will continue to encourage the achievement of the Sustainable Development Goals (SDGs) of clean water, sustainable cities, and climate action.

Emerging Sensor and Communication Technology: New developments in sensors and protocols will result in a significant improvement of system effectiveness, precision, and availability. Nano-sensors and bio-sensors will enable the detection of chemical and biological pollutants at the micro-levels in real time that will enhance the quality of water and the safety of people. The 5G, LoRaWAN, and NB-IoT technologies will assist in transmitting data faster, more reliably, and efficiently, even in the most problematic and remote areas on the communication front. Besides this, the energy problem will be resolved by designing solar or kinetic-powered low-power and self-sustaining sensor nodes, and the IoT will be disseminated to rural and developing locations.

Real-time Decisions to Be Made Using Edge and Fog Computing: As the IoT networks are spread, the volume of data generated by the networks increases exponentially. This information will require edge and fog computing to compute efficiently and bring real-time analytics at the data source. The computing paradigms reduce the latency, enhance the responsiveness of the system, and reduce the dependence on cloud connectivity. The future water systems will incorporate edge-enabled gadgets to execute the localised analytical depiction of the water applications such as leakage identification, pumping, and irrigation of the water systems which will be more dependable and faster.

Blockchain of Data Security and Transparency: Another significant problem of the IoT-based water systems is the integrity and transparency of data. Blockchain technology offers a promising solution by providing a decentralized and tamper-resistant framework for data management, thereby enhancing data integrity, transparency, and security. Blockchain is capable of ensuring that the data is genuine through its impartial and non-malicious records, that it can be billed easily, and that it cannot be altered by unwarranted parties. This will come in handy, particularly in the water utilities in the general populace and management of the water resources across boundaries, where responsibility and trust are critical.

Policy, Governance, and Capacity Building: Besides technological innovation, good policy frameworks, institution governance, and capacity building programmes will also be necessary in the future. Governments and regulatory bodies ought to develop efficient standards in the internet of things interoperability, internet security, and data sharing. This will be made possible by the investment in training programmes and community awareness so that the operators, engineers, and policymakers at the time can harness the IoT technologies effectively. The links with the academic, industry, and government can also be developed to accelerate the innovation process and offer equal opportunities to smart water solutions.

To Sustainable and Adaptive Water Systems: The last vision is to come up with self-adaptive, sustainable, and resilient water management systems that are capable of dynamically responding to changes in the environment and society. Such systems will make use of the AI, IoT, and renewable energy integration to make sure that its performance will be optimal and to conserve resources. The sustainable Internet of Things water systems will also be directly aimed at achieving the world goals, such as SDG 6 (Clean Water and Sanitation), SDG 11 (Sustainable Cities and Communities), and SDG 13 (Climate Action), which will also be useful to create and establish the ecological balance and human welfare.

7.8 CONCLUSION

The use of the Internet of Things (IoT) in water management is a game-changer in the process of ensuring sustainability, efficiency, as well as resilience in the water resources systems. The introduction of interconnected sensors, communication network, and intelligent data analytics makes it possible to monitor in real-time, predictive maintenance, and make informed decisions. The IoT systems operated on the principle of sensing and feedback to maximise the use of water, reduce the wastes due to leakages, and give equitable distribution of water among the domestic, agricultural, and industrial sectors.

As established in the chapter, the IoT architecture (comprising sensor, edge, network, cloud, and application layers) forms the basis of smart water systems since smart water systems can easily accomplish operational smartness through the efficient flow of data, uninterrupted. However, problems such as steep start-up costs, sensor calibration, connexion issues, and cybersecurity risks should be addressed to ensure that implementation and scaling become stable. The solution to counter these barriers is the presence of effective governance structures, interoperability standards, and capacity-building programmes.

The possibilities of the IoT systems are shifting due to the new technologies, among which are artificial intelligence (AI), machine learning (ML), blockchain, and edge computing. Besides the accuracy of data and speed of data processing, these innovations have been making the system more autonomous and transparent. Also, the addition of IoT-based water systems to an overall smart city system leads to the overall sustainability of the city and the correspondence to the global agenda to develop the United Nations Sustainable Development Goals (SDGs), namely to SDG 6 (Clean Water and Sanitation) and SDG 13 (Climate Action).

In conclusion, the use of IoT-based water management as a technology is a very important aspect of the convergence of sustainability and governance. Its future lies in the chances of leveraging smart automation, strong infrastructures, and the possibility of forming a policy formulation that is non-discriminative to offer water security to all. The IoT technologies will continue to play a role in making the world sustainable, data-driven, and water-efficient by being innovative, collaborative, and adaptive to manage.

REFERENCES

1. Gubbi, J., Buyya, R., Marusic, S., & Palaniswami, M. (2013). Internet of Things (IoT): A vision, architectural elements, and future directions. *Future Generation Computer Systems,* 29(7), 1645–1660. https://doi.org/10.1016/j.future.2013.01.010
2. Sharma, N., Chakraborty, C., & Kumar, R. (2023). Optimized multimedia data through computationally intelligent algorithms. *Multimedia Systems*, 29(5), 2961–2977.
3. Srivastava, D., Sharma, N., Sinwar, D., Yousif, J. H., & Gupta, H. P. (Eds.). (2023). *Intelligent Internet of Things for Smart Healthcare Systems.* CRC Press.
4. Sharma, N., & Batra, U. (2021). An enhanced Huffman-PSO based image optimization algorithm for image steganography. *Genetic Programming and Evolvable Machines*, 22(2), 189–205.
5. Jain, A., & Rani, A. (2020). IoT-based smart water management system: An overview. *International Journal of Advanced Science and Technology,* 29(6), 4550–4563.
6. Sharma, N., Soni, M., Kumar, S., Kumar, R., Deb, N., & Shrivastava, A. (2023). Supervised machine learning method for ontology-based financial decisions in the stock market. *ACM Transactions on Asian and Low-Resource Language Information Processing*, 22(5), 1–24.
7. Sharma, N., & Batra, U. (2017, October). A review on spatial domain technique based on image steganography. In *2017 International Conference on Computing and Communication Technologies for Smart Nation (IC3TSN)* (pp. 24–27). IEEE.
8. Sharma, N., & Batra, U. (2017, October). A study on integrating crypto-stego techniques to minimize the distortion. In *International Conference on Recent Developments in Science, Engineering and Technology* (pp. 608–615). Springer Singapore.
9. Kumar, A., & Singh, P. (2021). IoT-based water quality monitoring system. *Journal of Environmental Management*, 293, 112–120. https://doi.org/10.1016/j.jenvman.2021.112120
10. Rani, R., & Sharma, S. (2022). Smart irrigation using IoT: A review. *Agricultural Water Management*, 261, 107–115. https://doi.org/10.1016/j.agwat.2022.107115
11. Gill, S., Sharma, N., Gupta, C., & Samanta, A. (2021, November). Attendance management system using facial recognition and image augmentation technique. In *2021 International Conference on Intelligent Technology, System and Service for Internet of Everything (ITSS-IoE)* (pp. 1–6). IEEE.
12. Beri, M., Kumar, B., Tiwari, S., Sharma, N., Vashishtya, H., & Chaudhary, P. (2022, April). IoT based health monitoring system built on ESP32. In *2022 2nd International Conference on Advance Computing and Innovative Technologies in Engineering (ICACITE)* (pp. 454–458). IEEE.
13. Sharma, N., Srivastava, D., & Sinwar, D. (Eds.). (2024). *Artificial Intelligence Technology in Healthcare: Security and Privacy Issues.* CRC Press.
14. United Nations. (2023). *Water Scarcity: Facts and Trends.* UN Water Reports. https://www.unwater.org/publications

15. Zanella, A., Bui, N., Castellani, A., Vangelista, L., & Zorzi, M. (2014). Internet of Things for smart cities. *IEEE Internet of Things Journal, 1*(1), 22–32. https://doi.org/10.1109/JIOT.2014.2306328
16. Kumari, M., Sharma, N., Chauhan, R., Joshi, K., Madaan, V., & Sunil, G. (2024, February). Investigation of OWC system incorporating different filters. In *2024 4th International Conference on Innovative Practices in Technology and Management (ICIPTM)* (pp. 1–4). IEEE.

Smart Water Governance Using Blockchain

8

Deepak Joshi

ABBREVIATIONS

IoT: Internet of Things
PBFT: Practical Byzantine Fault Tolerance
NGO: Non-Governmental Organization
WASH: Water, Sanitation, and Hygiene
DLT: Distributed Ledger Technology
ECDSA: Elliptic Curve Digital Signature Algorithm
BWSSB: Bangalore Water Supply and Sewerage Board
WARMS: Water Resources Management System
ZK-SNARKs: Zero-Knowledge Succinct Non-Interactive Argument of Knowledge
AI: Artificial Intelligence
GIS: Geographic Information System
ICS: Industrial Control Systems
OPC UA: Open Platform Communications Unified Architecture
ISO: International Organization for Standardization
MQTT: Message Queuing Telemetry Transport
CoAP: Constrained Application Protocol
RESTful APIs: Representational State Transfer Application Programming Interfaces
GDPR: General Data Protection Regulation
LGDSAX: LGDSAX
ZKPs: Zero-Knowledge Proofs
RSA: Rivest–Shamir–Adleman
ECDSA: Elliptic Curve Digital Signature Algorithm
PQC: Post-Quantum Cryptography
NIST: National Institute of Standards and Technology
FIPS: Federal Information Processing Standards
ML-KEM: Module-Lattice-Based Key Encapsulation Mechanism

DOI: 10.1201/9781003742227-8

ML-DSA: Module-Lattice-Based Digital Signature Algorithm
SLH-DSA: Stateless Hash-Based Digital Signature Algorithm
SPHINCS+: Stateless Practical Hash-Based Incredibly Nice Collision-Resistant Signatures
DPI: Digital Public Infrastructure
NID: National Identity
PPP: Public–Private Partnership
UN SDG 6: United Nations Sustainable Development Goal 6.

8.1 INTRODUCTION

The twenty-first century represents a critical turning point for water governance as mounting pressures from rapid population growth, accelerated urbanization, climate change, and competing sectoral demands increasingly strain limited water resources [9]. Traditional water management frameworks are struggling to cope with these stresses due to fragmented institutional arrangements, siloed data systems, limited transparency, and declining trust among stakeholders [19]. These challenges are particularly acute in regions experiencing chronic water scarcity, where inefficiencies in allocation, monitoring, and enforcement exacerbate social and economic inequities. As water scarcity intensifies, the need for robust, technology-driven governance [17] mechanisms becomes both urgent and unavoidable.

Conventional water governance systems are largely centralized and institution-driven, relying on manual reporting, opaque decision-making processes, and retrospective audits [20]. Such systems often fail to ensure accountability, timely access to reliable data, and equitable participation of stakeholders. Partitioned data repositories prevent holistic visibility across the water value chain, while weak verification mechanisms undermine confidence in reported usage, quality metrics, and entitlement records [3]. This erosion of trust limits cooperation among governments, utilities, industries, farmers, and citizens, thereby reducing the effectiveness of policy interventions and conflict resolution mechanisms in water-stressed environments [11].

Blockchain technology [9] offers a paradigm shift by addressing fundamental governance gaps that traditional systems are unable to resolve. As a form of distributed Ledger [1] technology, blockchain introduces immutability, decentralization, and verifiable [25] transparency into record-keeping and transactional processes. In the context of water governance [17], blockchain enables the creation of tamper-resistant records of water extraction, distribution [20], quality measurements, and financial transactions without reliance on centralized intermediaries. By embedding accountability directly into the system architecture, blockchain shifts trust away from institutions toward cryptographic and consensus-based mechanisms.

When integrated with complementary technologies such as the Internet of Things (IoT) and Smart Contracts [14], blockchain-enabled water governance systems can support real-time monitoring, automated compliance enforcement, and transparent water disbursement. IoT sensors can continuously capture data on flow rates, quality parameters, and infrastructure performance, while Smart Contracts [14] can automatically execute regulatory rules, tariffs, subsidies, or trading arrangements based on predefined conditions. This convergence enables responsive, data-driven governance that enhances operational efficiency, reduces administrative overhead, and minimizes opportunities [7] for manipulation or fraud.

This chapter examines the role of Blockchain technology [9] in transforming water governance through the lens of data-driven decision-making and stakeholder-centric design. It explores how distributed Ledger [1] systems can improve transparency, accountability, and cooperation among diverse actors by providing shared, verifiable, and accessible information. The analysis draws on real-world case studies from different geographic and institutional contexts to demonstrate how theoretical constructs translate into practical implementation. In doing so, the chapter highlights both the governance benefits and the socio-technical complexities associated with blockchain adoption.

The scope of this chapter encompasses theoretical foundations, applied use cases, and future prospects of blockchain-enabled water management systems [17]. It addresses the integration of blockchain with IoT, artificial intelligence, and geospatial systems, while critically examining challenges related to scalability, legal enforceability, data privacy, and institutional readiness. The chapter further discusses policy considerations and regulatory frameworks necessary for sustainable [3] adoption and concludes with actionable recommendations for water managers, policymakers, and technology practitioners. By coupling advanced technical capabilities with governance principles, the chapter outlines a pathway toward more equitable, transparent, and resilient water governance systems.

8.2 LITERATURE REVIEW

REF.	*AUTHOR(S)*	*YEAR*	*DOMAIN / FOCUS AREA*	*TECHNOLOGIES USED*	*KEY CONTRIBUTIONS*	*IDENTIFIED GAPS*
[1]	Asgari & Nemati	2022	Smart Water Systems	Blockchain, DLT	Systematic review of blockchain in water management	Limited implementation evidence
[2]	Chang *et al.*	2021	Agricultural Water Mgmt.	IoT, Blockchain	IoT–blockchain framework for irrigation optimization	Scalability and governance issues
[3]	Arasi *et al.*	2025	Urban Water Ecosystems	Blockchain, IoT, AI	Autonomous, data-driven water governance	High computational overhead
[4]	Nododile & Nyirenda	2025	Smart Metering	Blockchain, IPFS	Secure, scalable data storage architecture	Latency and off-chain dependency
[5]	Homaei *et al.*	2025	Water Infrastructure Security	AI, Blockchain, Digital Twins	Cyber-physical water security framework	Limited field deployment
[6]	Xia *et al.*	2022	Intelligent Water Mgmt.	Blockchain	Conceptual governance architecture	Lack of real-world validation
[7]	MDPI Review	2024	Water Resource Governance	DLT, Smart Contracts [14]	Transparency and efficiency in governance	Weak regulatory analysis
[8]	Singh *et al.*	2024	Resource Management	Blockchain, IoT	End-to-end water governance design	Legal enforceability unaddressed
[9]	Eslamian & Eslamian (Eds.)	2025	Environmental Systems	Blockchain	Cross-sector analysis of blockchain adoption	Mostly conceptual

REF.	*AUTHOR(S)*	*YEAR*	*DOMAIN / FOCUS AREA*	*TECHNOLOGIES USED*	*KEY CONTRIBUTIONS*	*IDENTIFIED GAPS*
[10]	JSSS Consortium	2025	Water Diplomacy	Blockchain, AI, IoT	Automation of water diplomacy processes	Governance complexity
[11]	Kumar *et al.*	2025	Water Informatics	IoT, Blockchain	Review of informatics-driven water systems	Limited trust mechanisms
[12]	Hameed *et al.*	2022	IoT Security	Blockchain	Decentralized authorization model	Not water-specific
[14]	AlZubi *et al.*	2023	Smart Infrastructure	IoT, Smart Contracts	Utility Automation using Smart Contracts	High deployment cost
[15]	Reyna *et al.*	2018	IoT–Blockchain Integration	Blockchain, IoT	Early identification of integration challenges	Outdated scalability assumptions
[16]	Xue *et al.*	2025	Water Resource Mgmt.	Blockchain, IoT	Survey of techniques and future directions	Limited policy focus
[17]	Atzori	2017	Decentralized Governance	Blockchain	Institutional implications of decentralization	No empirical validation
[18]	Szmigiera	2020	Public Governance	Blockchain	Transparency and accountability analysis	Sector-agnostic
[19]	Casino *et al.*	2019	Traceability Systems	Blockchain	Trust and traceability in supply chains	Water domain not explored
[20]	Hunter *et al.*	2023	Water Distribution	IoT	Smart metering efficiency gains	Data integrity not ensured
[21]	Kshetri & Voas	2021	Public Sector Blockchain	Blockchain	Adoption challenges and opportunities	Water governance not central
[22]	Cachin	2016	Blockchain Architecture	Hyperledger Fabric	Permissioned blockchain architecture	Governance-layer missing
[23]	NIST	2023	Standards	Blockchain	Security and architecture guidance	Not sector-specific
[24]	IEEE SA	2024	Smart Water Systems	IoT Standards	Reference architecture draft	Early-stage standard
[25]	W3C	2023	Digital Identity	Verifiable Credentials	Privacy-preserving data exchange	Adoption barriers

8.3 CHANGING PARADIGM OF WATER GOVERNANCE

The management of water resources is currently undergoing a paradigm shift as novel digital technologies are incorporated. Traditionally, water systems were selected based on efficiency, as well as through centralized and/or hierarchical institutions. Today's challenges, on the other hand, consist of the lack of accountability, inequitable water Distribution [20], and silos of data. Governments are trying to overcome these challenges through the adoption of "smart governance." Smart governance [17] is adopting the use of emerging technologies and real-time data management, such as the use of Blockchains, IoTs, and Artificial Intelligence in water resources.

8.3.1 Need for Transparency -and Trust

The governance of water resources systems in different countries is losing the confidence of the public due to lack of trust in systems [18]. These systems use poor data to allocate resources and do not track water quality. There are multiple conflicts that arise because default compliance lacks accountability due to lack of governance. The additional lack of trust becomes a severe obstacle for water management in times of scarcity and increased demand. Water governance is also data characterized by many stakeholders, including farmers, industries, utilities, regulatory bodies, and the public, which operate within a partitioned framework. Blockchain systems, as opposed to traditional databases, are more secure in that the information stored within databases is not susceptible to tampering. Furthermore, Blockchain technology [9] eliminates the need to put trust in institutions and instead put trust in technology since it offers proof of transfer and of water usage that is immutable and verifiable [25] record of trust of institutions.

8.4 BLOCKCHAIN FUNDAMENTALS

A blockchain is an innovative form of distributed Ledger [1] technology which changes dramatically the manner in which information is stored, verified, and disseminated by participants in a network. Fundamentally, a blockchain keeps track of time-stamped transactions in sequential order which are stored in blocks. Each of these blocks in turn contains a cryptographic hash of the preceding block, as well as a set of transaction records. These blocks are interconnected, forming a "chain" which is completely Verifiable [25] and tamper resistant. Blockchain technology [9] is not a database, as it cannot be controlled by a singular, dominating individual or organization. Rather, it is a collection of records which is simultaneously distributed and synced in real time to every individual "node" in the network. New transactions in the block "chain" require permission of all network participants through consensus mechanisms. The core principles of blockchain technology are presented in Figure 8.1, highlighting immutability, decentralization, transparency, and security.

Immutability means that once something has been recorded and verified, no one can change or delete it, and an audit trail is created that cannot be erased. Decentralization removes a single point of failure and control by distributing authority to all participants in the network, thereby reducing the likelihood of manipulation or collapse of the system. Transparency [18] allows all verified participants to access and inspect the transaction records, which facilitates trust and increases accountability. Cryptographic techniques such as hash functions and digital signatures help maintain security by assuring the integrity of the data and the participants. All these principles together foster an environment that is free of trust, which means that the parties can transact and share information without the need for intermediaries or central control, and do so in complete confidence.

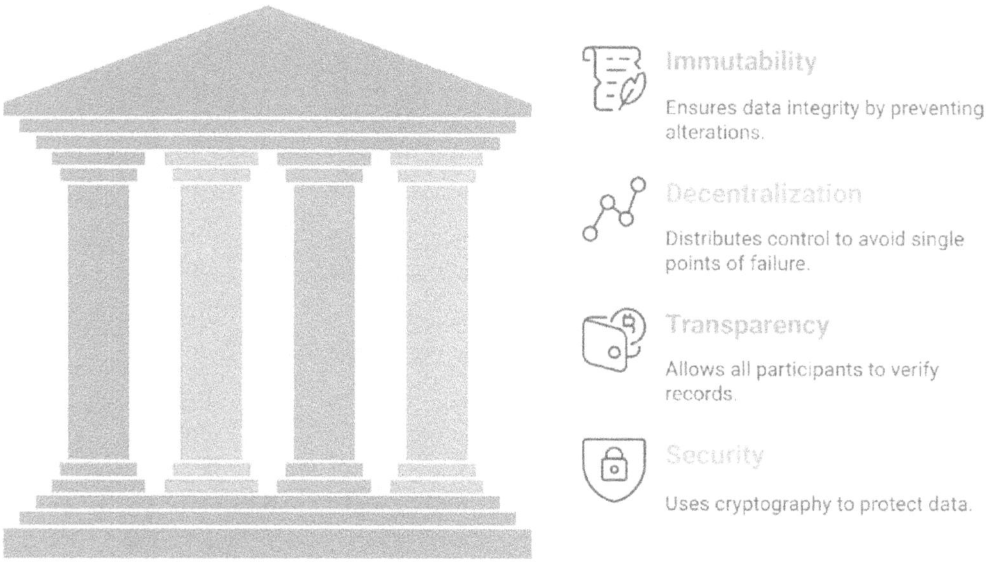

FIGURE 8.1 Four basic characteristics of blockchain technology.

8.4.1 Key Components

A blockchain's architecture [24] consists of four major components that make possible its decentralized functioning. Blocks and Transactions are basic data containers. Each block contains a set of transactions that are time stamped along with the metadata of the preceding block's hash and its own identifier. These blocks are sequentially linked, which forms an auditable chronological record. The added protocols define how network participants validate and agree on transactions awaiting addition to the blockchain. In Proof-of-Work, participants expend resources and time attempting to answer complicated puzzles. In Proof-of-Stake, validators are selected based on the network stake. Practical Byzantine Fault Tolerance (PBFT) also gives the ability to build consensus on the condition that some nodes act adversarially, which is useful in permissioned networks.

Elimination of intermediary and complex governance processes is possible through programmable self-executing codes also known as Smart Contracts [14]. They are stored in the blockchain and automatically enforce rules that are triggered upon the occurrence of certain conditions. In a distributed network, each node is connected to every other node, and each node keeps its own complete ledger. Due to the redundancy of copies, the system is reliable and resistant to single points of failure as each participant may independently verify the entire transaction history. The architecture of a blockchain network is outlined in Figure 8.2, demonstrating the interaction between transactions, consensus mechanisms, smart contracts, and ledger management.

8.4.2 Blockchain Types and Suitability

Blockchain networks are categorized into three distinct types based on access permissions and governance structures, each offering unique advantages for different use cases.

- **Public Blockchains:** Open participation (e.g., Ethereum).
- **Private Blockchains:** Restricted access for specific organizations.
- **Consortium Blockchains:** Shared governance among multiple agencies (ideal for water governance).

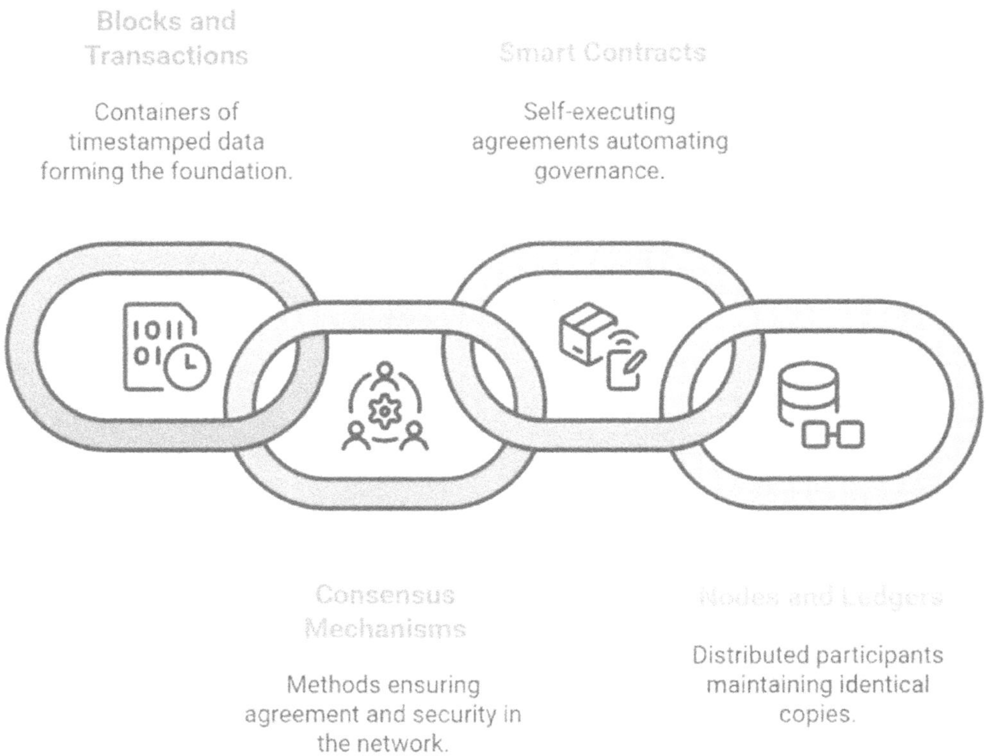

FIGURE 8.2 Core components of blockchain technology.

TABLE 8.1 Governance Models Comparison

MODEL	*CONTROL STRUCTURE*	*DATA TRANSPARENCY*	*SCALABILITY*	*EXAMPLE USE*
Centralized	Single authority	Low	High	Traditional water boards
Decentralized	Distributed nodes	High	Medium	Blockchain pilot networks
Distributed (Hybrid)	Consortium model	High	High	Smart Water Councils

Public blockchains accept every participant without restrictions and permit validation, transaction processing, and unrestricted access to the entire ledger. This model uses the Ethereum blockchain for maximum transparency [18] and decentralization and is probably the slowest and most private blockchain available. Private blockchains allow participation to certain organizations or controlled entities, thus controlling access to the data and the validation of transactions. These blockchains are faster and more private than the public blockchains, making them suitable for the internal operations of single-organization water utilities.

Consortium blockchains are strategically placed between the two extremes of public and private blockchains. Some remain fully public or controlled by a single entity. This setup is tailored to water governance cases involving multiple stakeholders, like municipal governance and water-regulating bodies, water utilities, and environmental NGOs. Controlled access is needed for collaborative governance. This model is suitable for water governance cases with multiple stakeholders because it is fully controlled by a predetermined group of organizations. It provides weaker controls, a balance between privacy and transparency, allowing the team to cover critical water data resources and jointly enforce governance without exposing sensitive data.

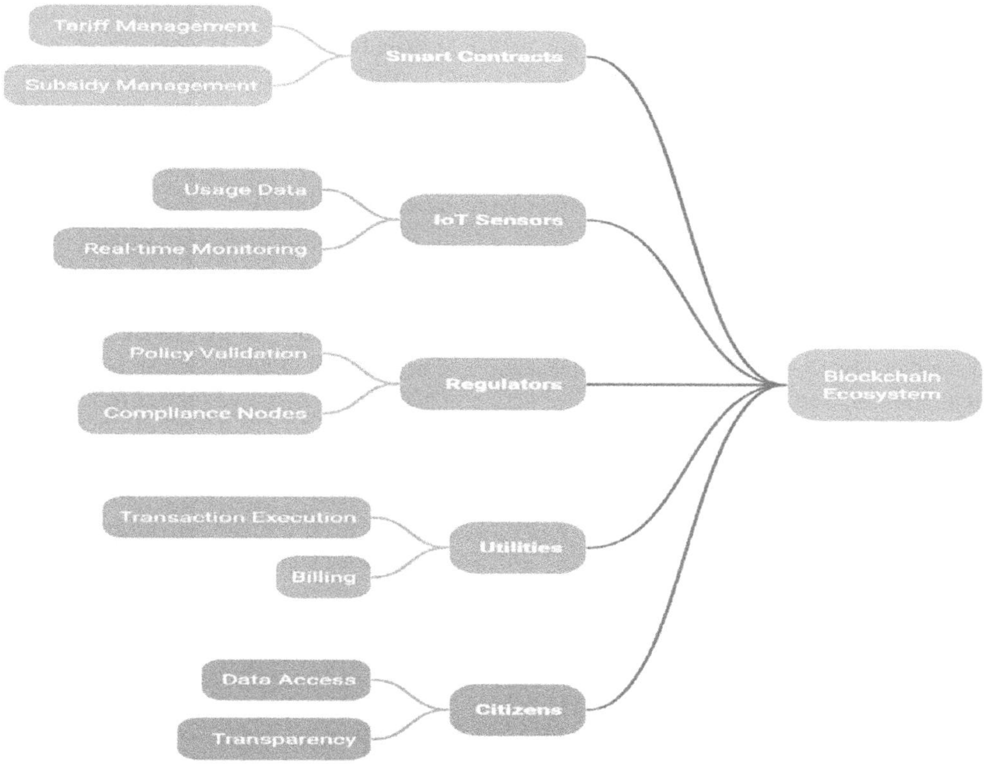

FIGURE 8.3 Core components of blockchain-based water governance systems.

8.4.3 Need of Blockchain for Water Governance

Blockchain tackles fundamental inefficiencies and deficits of trust in traditional water governance by creating an authoritative, trusted, and unified repository of all water transactions and usages, including monitoring, pricing, quality certification, and trading rights. Blockchain can cut administrative, delay, and corruption costs by removing brokers and verifiers as intermediaries. The attest proven in technology enables tracing of water rights ownership, allocation, and transfers across complex multi-jurisdictional frameworks. The Ledger [1] for every transaction from extraction permits consumption records and simplifies dispute resolution and accountability with a single source, auditable, time-stamped history. This accountability with a single source, auditable, time-stamped history enhances compliance and dispute resolution mechanisms. Moreover, seamless alignment of global frameworks of sustainability with United Nations Sustainable [3] Development Goals 6 ensures Water and Sanitation (WASH) for all, as well as equity and intergenerational [8] sustainability. Blockchain aids measurement, accountability, evidence-based policy formulation, and management of water and resources. Figure 8.3 illustrates the key stakeholders and functional components of a blockchain-based water governance system, highlighting the roles of citizens, utilities, regulators, IoT sensors, and smart contracts.

8.4.4 Applications of Blockchain in Water Governance

As Blockchain technology [9] can address various issues regarding water quality as well as improve the transparency [18] of billing systems, the prospects of enhancing water governance through Blockchain technology [9] appear optimistic. The application of Blockchain technology [9] is possible at various levels, both within the value chain and on the outer fringes of the value chain. The governance or operational level of the value chain systems on both urban and rural sides can be technologically optimized at the outer fringes of the value chain. The application of DLT (distributed Ledger [1] technology) has the potential for improving water governance at multiple levels and across different domains.

8.4.4.1 Water Rights and Ownership Tracking

Management of water quarantine at the contract level involves the unlocking of value decisions which stand in the center of value setting. Existing water records are maintained either in paper-based formats or in isolated digital systems with unrestricted access.. This congested set of systems leads to decision failures and, in cases, disputes resolvable through payment concerning make-better and take-better water resources in many cases. Water entitlements can be registered digitally through Blockchain technology [9], and conflicts over ownership through transaction and registration at any time can be resolved. Each allocation, transfer, and complete cycle of revocation concerning specific resources transacted over is digitally signed, stamped, verified, and timed.

The enduring record-keeping remains crucial for arid areas with intricate water use frameworks due to their claims, seasonal water Distribution [20], cross-generational [8] transfers, and other historical factors. Farmers, the business sector, and city governments can readily access the original Distribution [20] documents to confirm and trace the water rights and associated claims. Automated Smart Contracts [14] can manage the enforcement for usage caps, scarcity prioritization, and regulatory requirements. Furthermore, blockchain water rights records can improve water market with lower costs from removed middlemen due to transparency [18] and equitable pricing that meets true supply and demand.

8.4.4.2 Smart Contracts [14] for Water Trading

Automated contracts enable water users (such as farming and light industry) to trade water at much lower transaction costs by programmable agreements that self-activate under certain conditions. Trades automatically occur once set conditions such as limits, pricing, and drought variables are switched on, bypassing regulatory processes. Payments and detention are triggered automatically and administered seamlessly, resulting in equitable outcomes.

These self-executing contracts can include complex rule trading with priority levels, seasonality, and required environmental flows. For example, during times of drought, self-executing contracts can automatically reallocate water supplies, with advanced compensation determined and immediately transferred, from lower priority agricultural users to higher priority, critical city users and municipal users with advanced compensation determined and immediately transferred. The contracts can also integrate real-time data from IoT sensors to optimize pricing to facilitate conservation during low water reserve levels.

Moreover, Smart Contracts [14] improve microeconomic efficiency of the water market by facilitating micro-payments and permitting intervallic water release under conditions that traditional contracts would consider excessive. With the rules in place, a farmer with a water surplus can lease to his neighbor, virtuously devoid of any manual interaction under complex contract self-execution. The system guarantees conservation water control compliance with absolute control and full jurisdiction retention. The proposed system architecture shown in Figure 8.4 emphasizes the integration of IoT sensors, distributed ledgers, web-based applications, and stakeholder participation to achieve transparent and efficient water resource management.

The trust and immutability feature integral to a blockchain-based water governance system depend on underlying consensus and cryptographic architecture [24] which vary in implementation. The Telangana water grid system used Hyperledger [22] Fabric's Practical Byzantine Fault Tolerance (PBFT) consensus. where validator nodes (State Water Board servers, Tech Mahindra infrastructure, and district authority nodes) needed to reach two-thirds agreement before transactions are recorded, guaranteeing Byzantine fault tolerance where one-third of the nodes can act maliciously or fail, and the system remains functional.

Each water extraction record underwent a three-layer verification process: IoT sensors digitally signed the data using ECDSA cryptography, validator nodes that executed chain code (Hyperledger's [22] Smart

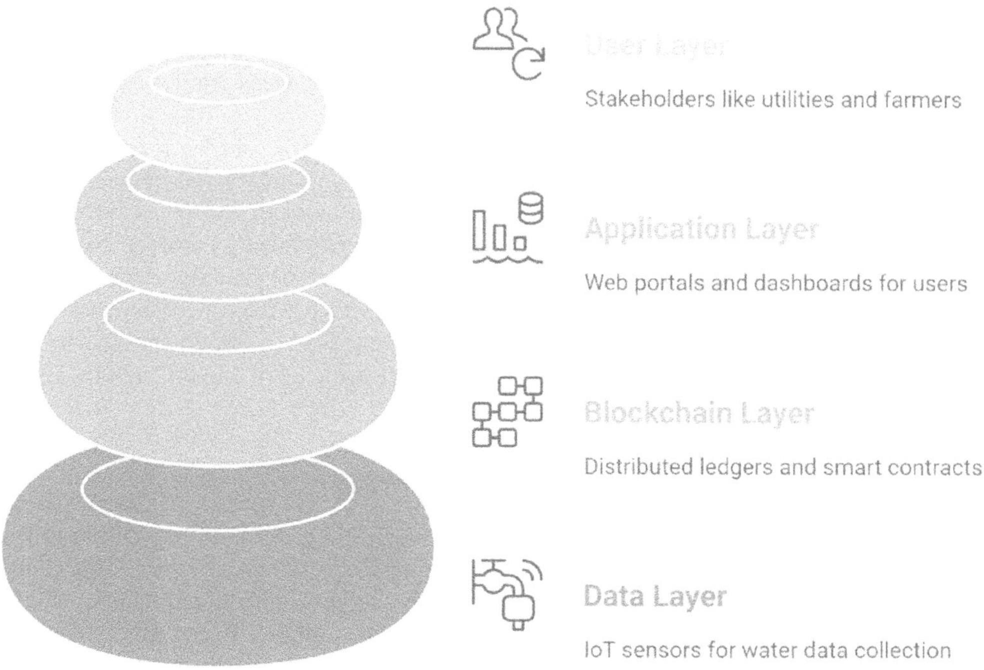

FIGURE 8.4 Proposed architecture of blockchain-enabled smart water governance system.

Contracts [14]) to check the signatures and business logic compliance, and consensus nodes that formed Merkle tree structures that other transactions are hashed to create single block identifiers. Temporal security layers in the BWSSB implementation in Karnataka were through time-stamped hash chains, where quality measurements are made, and each references the cryptographic hashes of previous measurements, thus making it nearly impossible to tamper with historical data without recalculating the entire chain history, which is a resource-expensive action considering the network's computational resources.

The deployments of Hyperledger [22], while purportedly blockchain-based, still require trust to be issued to a confined group of approved validation organizations, and thus fail to deliver the decentralization blockchain architecture [24] promises; as in the case of the Water Board server hack in February 2022, which documented a deliberate attempt to inject falsified data, it was self-evident that the PBFT security architecture [24] rests entirely on the assumption that a simple majority of validators will always act in good faith a trust assumption that is almost indistinguishable from the access control model of any conventional data base.

The rules of water governance are operationalized as workflows of Smart Contracts [14] which code compliance enforcement as automatic enforcement, without human oversight. Meanwhile, implementation complexities introduce vulnerabilities which are absent in systems without Automation [10] [10]. Maharashtra constructed Smart Contracts [14] in the programming language "Solidity." They included contracts with multi-stage execution logic. When farmers started the trades, contracts first asked Oracle nodes that were linked to WARMS databases about current water availability. They checked both parties' water entitlement token balances, conditional regulatory constraints, and transaction fee limits and performed the atomic swap that prevents execution failures where tokens are either transferred or not at all.

Access control layers included role-based permissions. Farmers, water authorities, and auditors are considered different classes of stakeholders who, through public and private cryptographic key pairs, have separate restricted read and write privileges. Farmers are restricted to trade initiation, as are authorities who empower multi-signer systems requiring three of five officials to lower control thresholds. Non-compliance automatic enforcement was exposed by the Murray-Darling's carryover calculation error. One defunct Smart Contract containing multimedia floating bugs was evaluated. All audits were exposed to over 89 controlled transactions that toggled access during relayed 'tokens' for environmental proofs. All were discoverable. Analysis of post-incident contract without formally including verification of mathematical proofs was equally lacking. Assets equated to millions were accessible for code fragmentation through behavior.

Singapore's NEWater system used ZK-SNARKs to enable industrial clients to confirm proprietary water quality specifications without revealing protected consumption data, thus addressing the issue with confidentiality and limited disclosure. They proved the quantum of water consumed without revealing key proprietary data, and without enabling complete disclosure, generated a cryptographic seal for the individual pieces of information. Advanced methods demonstrate that the primary water utilities lack the required skills for such sophisticated extension of cryptography, which is the core of the solution provided by Karnataka. The hybrid framework is designed to exclusively store cryptographic hashes on-chain, because with such restrained utility personnel support, the intricate logic of Smart Contracts [14] can't be sustained.

8.4.4.3 Decentralized Water Quality Monitoring

Real-time recording of water quality as it is measured through IoT gadgets, of parameters such as pH, turbidity, dissolved oxygen, and others, can be instantly uploaded to a blockchain net. This combats the potential of changing the data and helps the general public, the government, and health institutions to independently verify the validity of the data without relying on the utilities. Every sensor is attached to a key that records the times and dates of the sensor data, creating a permanent, unalterable record of water quality in the water Distribution [20] networks.

Predictive modeling and anomaly detection frameworks can leverage the rest of the record to determine the times and locations of potential public health crises, spotting the lack of infrastructure in these locations before the situations arise. AI can notify what parameters of water quality would constitute the minimal threshold which would result in a public emergency, and what emergency measures must be taken by the authorities. Advanced research can be facilitated since the data is publicly available, empowering society to position itself as the primary monitor of water quality. Deployed technological devices can safeguard the public through Automation [10] of systems to keep track of the water quality.

8.4.4.4 Transparent Tariff and Subsidy Management

Most governments offer funding for water irrigation and conservation; however, current systems usually direction systems often have multiple inefficiencies, dishonest practices, and high instances of funding. Blockchain increases accountability by ensuring that subsidies are provided only when steps of a Smart Contract are met, with clear specifics such as confirmed installation of drip irrigation systems and authentic water savings documentation. Decreased leakages and fraudulent claims are often substantially protected through Smart Contract protocols.

Technology eliminates ghost beneficiaries and duplicates and creates a path for audit bucket and Distribution [20] allocation. With the Smart Contract system, water tariff structures create programmable policies that automatically adjust to reflect the appropriate tier based on the predetermined consumption levels, which includes billing Automation [10] as well as significant payment reconciliation. Low-income households that qualify for subsidized rates are instantly verified and receive discounts without delays resulting from the longstanding bureaucracy.

Furthermore, through the unreachable Ledger [1] of Blockchain technology [9], water conservation incentive payments are automatically paid and monitored by IoT devices as consumption declines and settles below the target threshold. This form of transparency fosters public confidence in government programs. The ability to monitor the specific consumption of various user categories alongside the subsidies that were given allows for real-time assessment of the effectiveness of the subsidies. This in turn permits adjustments to be made for the water pricing strategies that policies are based upon.

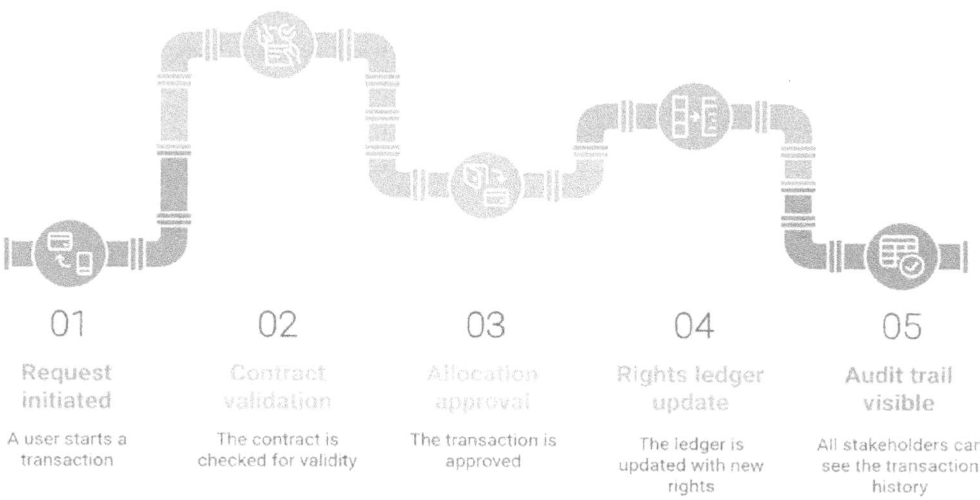

FIGURE 8.5 Overview of Smart Contract process for water governance.

The digital blockchain records permanent tokens, or "water credits," demonstrating a user's entitlement to a specific quantity of water, measured in either cubic meters or acre-feet. This encourages users to trade water tokens on digital platforms, which motivates efficient water use and conservation through market-based efforts. Figure 8.5 illustrates the workflow of a smart contract in water governance, from transaction initiation to audit trail generation.

8.4.4.5 Tokenization of Water Credits

Blockchain allows "water tokens" representing a unit of water entitlement, typically measured in cubic meters or acre-feet. Users can buy, sell, or exchange tokens on digital marketplaces, fostering efficient usage and conservation incentives through market-driven mechanisms.

Tradeable liquid assets can be water credits, tradeable and transferable across a range of industry sectors. Farmers employing water saving technologies directly earn tokens, which they can later sell to corporations needing tokens, thus incentivizing conservation and earning surplus tokens. Reserved tokens can be allocated for specific provisions, which can ensure ecomaps, thus ensuring that the ecosystems that do have tokens reserved for them are able to maintain minimum flows for the ecosystems that need them tokenized.

Water tokens can be fractionalized and used for micropayments, thus enabling smaller-scale farmers to enter water markets that were previously reserved for larger-scale users. Token transfer and management are done on the blockchain via Smart Contracts [14], which then apply automatic regulatory controls on token transfer to ensure compliance with environmental limits, regulatory priorities, and other parameters. The token markets, through their emerging pricing signals, provide participants with the

TABLE 8.2 Comparison of Blockchain Applications in Water Governance

USE CASE	*BLOCKCHAIN FUNCTION*	*BENEFITS*	*PILOT EXAMPLE*
Rights tracking	Immutable ledger	Prevents duplication	Australia Water Ledger
Subsidy management	Smart Contracts [14]	Eliminates fraud	India Pilot under Jal Shakti
Quality monitoring	IoT integration [15]	Tamper-proof data	Israel Smart irrigation
Water trading	Tokenization	Market-driven allocation	Spain Agriculture [2] pilots

information required to strategically align their water use with prevailing socio-economic conditions, thereby influencing positive changes in the overall pattern of water resource usage and socio-economic productivity Distribution [20].

8.5 CASE STUDIES OF BLOCKCHAIN IN WATER GOVERNANCE

Blockchain technology [9] is appearing as a revolutionary instrument in the field of governance of water, proposing transparent and decentralized methodologies for the administration of water rights, monitoring of quality, and Distribution [20] systems. Practical examples, like California's water trading systems and India's groundwater management systems, show how distributed Ledger [1] technology can increase accountability and equitable access while minimizing potential conflicts.

8.5.1 Water Ledger [1] Project in Australia

In Australia, there are water trading platforms implemented with Blockchain technology [9] to provide effectiveness and transparency in the allocation of water rights. This also deals with the sparse availability of water. The Water Ledger [1] Project also deals with this problem. It began in the Murray-Darling Basin, and it allows people to trade water rights directly with each other. This is done through "Smart" Contracts, which use Blockchain technology [9] to streamline conformance and compliance and "bridge" the gaps between different levels of trading. The platform digitizes water entitlements and automated allocations which allows the municipalities, environmental water holders, and other parties to trade anywhere and anytime through the decentralized trading system, which ensures privacy.

Impact: It is estimated that 70% of all admin delays are done away with by this implementation. It also eliminates transaction layers which speed transactions from weeks to mere seconds. With transaction transparency, stakeholders have more trust with one another since there are no "double-trading" issues. Pricing of water also had better control and better desirable levels achieved. The more the drought, the better the price. It gave a much better framework for determining water availability, control, and other parameters. Moreover, most of the disputes which were recorded are due to the trading data that was held by the control authorities. Prior to digital transformation, disputes among authorities were prevalent due to the complex and heterogeneous nature of jurisdictional record-keeping systems.

8.5.2 Smart Irrigation Monitoring in Israel

In Israel, advanced Agriculture [2] integrates IoT-based irrigation sensors with blockchain to confirm efficiency and verify water consumption. Smart sensors monitor and compute flow rates, soil moisture, and other crop parameters and send the information to the blockchain ledger. Through automated Smart Contracts [14], farmers are eligible to receive subsidies, and their on-chain water usage data gets recorded. This is done without manual paperwork or inspections, as the Smart Contracts [14] verify compliance with the best practice guidelines and conservation targets.

Impact: The system reduced overuse of water by 20%, given that farmers are provided with accurate, real-time, and automated feedback of their consumption habits. Better compliance reporting reduced disputes between farmers and water authorities, as both parties worked with the same verified data. Through accurate records, farmers that needed assistance were able to receive targeted support. Government bodies reduced 40% of their administrative costs with automatic subsidy verification, while farmers received faster disbursement of the subsidies. The integration produced irrigation datasets across various crops and climate zones.

8.5.3 India: Telangana Water Grid Management Pilot (2020–2022)

In partnership with Tech Mahindra and the State Water Board, the government of Telangana set out to implement a water Distribution [20] monitoring system that uses Blockchain technology [9] in the districts of Hyderabad, Ranga Reddy, and Medchal-Malkajgiri. This pilot monitored the water Distribution [20] to around 2.3 million residents through 47 pumping stations, receiving water from the Krishna River.

Implementation: To enable the creation of immutable records of water extraction, treatment, and Distribution [20] using Hyperledger [22] Fabric, the system was designed to enable the use of IoT sensors. IoT sensors located at pumping stations and capable of monitoring flow rates at 15-minute intervals would have their data hashed and later stored on the chain. Contracts of cryptocurrency-based payment systems, where payment is made during the completion of a certain contract, flagged differences during the sending of water to clients where the consumption was higher than the quota by 12% set in the contract.

Measurable Outcomes: With water monitoring sensors, the project was able to monitor and control the extraction of water and unauthorized water extraction that was not paid for during the 18 months of the project, decreasing water extraction without paying monitoring to 4.2 million liters. Further, the water theft was reduced by 24% in the controlled areas. Due to a reviewable system with accessible records, the resolution of a dispute was decreased from 45 to 8 days, reflecting administrative efficiency. However, the system was not without its problems, with 31% of IoT sensors losing connectivity in the monsoon having operational costs that exceeded estimations by 37% set at ₹8.2 crore ($1 million).

8.5.4 Maharashtra's Water Rights Trading Platform: Agricultural Water Rights Trading Platform (2021–2023)

The Maharashtra Water Resources Regulatory Authority, along with partnered blockchain water trading platforms, has pioneered a water trading facility in the drought-affected Ahmednagar and Solapur blocks that includes 1,847 registered farmers and spans 23,000 hectares.

Implementation: On a permissioned blockchain network built on Ethereum, the water entitlement is tokenized, and farmers are allowed to trade in allocations of water for irrigation. All farmers are allocated digital tokens that denote their seasonal water quota for a particular growing season (summer or winter). These tokens, which are allocated based on land holdings and existing crop patterns, are digitized and stored on the blockchain. Through automated Smart Contracts [14], the blockchain is deployed to facilitate the execution of trade activities.

Measurable Outcomes: Three hundred and twelve farmers can trade water rights that is equivalent to 18.6 million liters. Water rights were defeated during the trade through the economic value of 2.1 crores which crossed the abundant reserved water threshold of the bank. Farmers are still able to increase their crop yields by 11% more than the control groups. Despite the increase of 11% in crop yields among control populations, the transaction costs are still thought to be very high by farmers at the price of 450 rupees per trade. It should be noted that only 23% of registered farmers participated in the trade, which primarily consisted of educated landowners possessing a smartphone and internet. High schooling and a lack of smartphone and internet access were significant barriers for the farmers.

8.5.5 2022–Present: Monitoring Water Quality in the City of Bengaluru

BWSSB has partnered with the Bangalore-based research institute IISc and IBM India to implement a water quality monitoring system in Blockchain technology [9] to 68 blocks which provide water to 8.4 million people.

There are a total of 340 monitoring points that measure pH level, turbidity, chlorine concentration, and bacterial contamination of water in real time using water quality sensors. Data recording on the blockchain does not allow manipulation of individual sensors, and so sensors have to validate their data through a consensus before recording the information. Mobile applications allow the public to obtain real-time quality alerts about the water, and the historical data is stored in a blockchain as a non-removable record.

During the first 14 months, the system managed to assess 2.3 million records in total, and in the process, 89 contamination events that needed urgent attention were uncovered. There was a 41% drop in citizen complaints due to the accessibility of the data which was a gain in public trust. There was also a dramatic drop in response time to quality violations, going from 4.2 hours to 47 minutes. The major issue that came out of all of this was the fact that blockchain costs are extremely high. The costs were around 15 lakhs rupees gaining 2.8 TB of data. The proposed semi-centralized model achieved a 68% reduction in costs by storing only cryptographic hashes on the blockchain while retaining the remaining data in off-chain storage systems.

8.5.6 Blockchain for Transparency in Jal Jeevan Mission in India

Within India's Jal Jeevan Mission aimed at providing every household with piped water, subsidy disbursement for rural water supply schemes in Maharashtra and Telangana are monitored in real time on the blockchain. The implementation of Smart Contracts [14] on the blockchain allows for automated verification of pipeline installation, quality testing, and household connection milestones, and release payment to contractors and implementing agencies. This automated approach to payment processing aims to eliminate the costly problem of premature and fraudulent payment processing.

Impact: The blockchain implementation in rural water supply schemes dramatically reduced fraud by ghost schemes and free billing, with audits estimating financial irregularities at 35%. The implementation of mobile applications and blockchains enabled villagers to monitor and track project implementation and utilization of project funds, fostering citizen trust. Allocated resources to the projects enabled the better implementation of various departments, gram panchayat, and NGO collaboration. These schemes supplemented by real-time systems reduced project completion time by 25%. Coupled with the real-time monitoring provisions, the change in project completion also highlights the system's ability to provide valuable catchment for fund allocation. These funds, recorded immutably, provide support for corruption deterrence, while the seamless transfer of fund record allows for accountability on contractors and officials who fail to perform.

There must be comprehensive participation from all stakeholders as well as unified data formats from all entities for the successful implementation of any blockchain. Successful implementations show

TABLE 8.3 Comparative Summary of Blockchain Case Studies

COUNTRY	*PRIMARY USE*	*BLOCKCHAIN TYPE*	*OUTCOME*
Australia	Water rights trading	Consortium (Hyperledger [22])	Increased efficiency, transparency
Israel	Smart irrigation monitoring	Private (Ethereum)	Optimized resource usage
India	Subsidy management	Permissioned (Corda)	Fraud reduction, faster fund release

that engagement with community stakeholders as well as farmers, utilities, and regulators should be done as early as possible to alleviate concerns and build consensus around new workflows. Common data formats, measurement standards, and interoperability criteria should be set and agreed upon to prevent siloed systems and the information exchange barrier. Standalone implementations of blockchain tend to be mostly ineffective. Integration with the Internet of Things (IoT) and Artificial Intelligence (AI), however, is known to exponentiate blockchain's value. Case studies illustrate that with the addition of real-time sensors and data, and smart analytics, Blockchain technology [9] is able to create smart water systems with automated predictive management and decision-making.

Scaling and mainstream adoption of technology is hindered without legal recognition of digital records. Pilot projects faced challenges that stunted blockchain-based water contracts from being enforceable because of lacking formal legal status. Foreign jurisdictions that automated legislation to recognize digital signatures, Smart Contracts [14], and blockchain records to be legally binding for quicker adoption with fortified institutional confidence. Besides, clear frameworks to address the data privacy, liability, and disputes on the decentralized systems have proven crucial to large-scale deployment from single pilot projects across entire water basins or national systems.

8.6 INTEGRATION WITH OTHER TECHNOLOGIES

The disruptive potential of blockchain within water governance will not be realized in silos abstractedly, but alongside other emerging complementary technologies that enhance data collection, evaluation, and automated decision-making. Apart from Blockchain technology [9], the gathered IoT (Internet of Things) sensors, artificial intelligence, cloud technology, and geo-information systems technology are harnessed within a synergy where each technology augments the functionalities of the other. This provides integrated end-to-end intelligent [6] water management solutions that facilitate seamless real-time monitoring, immutable data keeping, predictive analytics, automated governance, and consolidated sophisticated open systems that address an array of contemporary water resource management [11] challenges.

8.6.1 IoT and Blockchain Convergence

The Internet of Things (IoT) is important for smart water governance and management systems because it enables real-time monitoring and collection of flow rates, pressures, turbidity, and water quality at different levels in a Distribution [20] network, in addition to providing collection systems for residuals (Slavik and Morgan, 2021). Real-time data streams from IoT sensors that are captured and directly written to a blockchain Ledger [1] become permanent and immutable with an accompanying timestamp, creating evidence of an audit trail for every record, and eliminating any potential for retrospective data access or manipulation.

Every sensor reading may be a certified transaction able to automatically execute a Smart Contract for maintenance notifications upon a drop in pressure indicating a leak, or for adjusting tariffs dynamically [12] during pre-specified peak periods. This degree of integration removes delays stemming from manual data replication, saves time through automated fraud verification, enhances regulatory compliance, and lowers the risk of compliance fraud through continuous monitoring. The integration permits the separate predictive maintenance from sensor-pattern predictive analytics stored in a block chain of the other participants in the value chain. All participants in the value chain from the utility to the end consumer can access the singular, trusted and verified data, enabling informed decision-making, resolution of disputes and streamlining the performance of the entire water infrastructure ecosystem. Figure 8.6 illustrates the integration of IoT and blockchain technologies within the proposed water governance framework, highlighting the interaction among field devices, edge gateways, blockchain nodes, and application services.

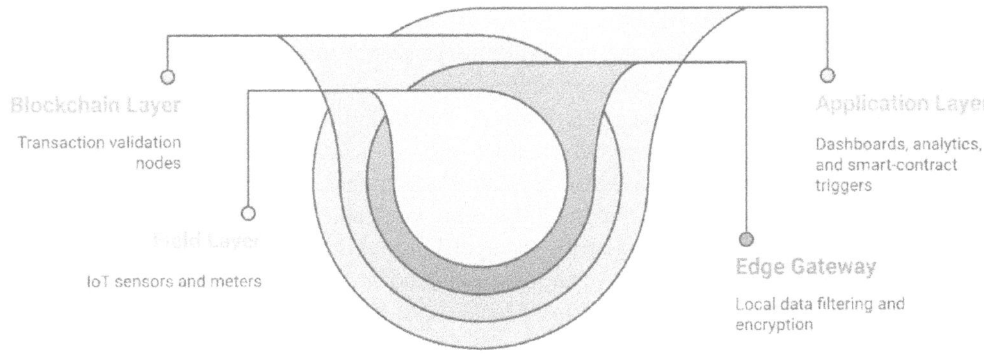

FIGURE 8.6 Interplay of IoT and Blockchain for water governance.

8.6.2 AI and Blockchain Synergy

AI assists in the decision-making process and, through the application of smart analytics, derives water actionable insights from the raw data. Blockchain technology [9] greatly simplifies these processes by securely storing and providing access to verified data sets. Models that predict future demand for water estimate based on weather forecasts, consumption patterns, and demographic trends. Those that identify leaks estimate based on patterns through prediction anomaly detection algorithms and redefine optimization of the Distribution [20] schedule to alleviate demand on energy and pressure within the network.

Blockchain technology [9] solves the problem of altered or inauthentic training data and datasets used in AI. Blockchain technology [9] gives an immutable Ledger [1] of undoubted data and allows AI to access unaltered and untainted historical data. This access is critical to unsupervised machine learning. The access also helps AI systems learn unbiased training data without the risk of tampering.

The combination of AI and Blockchain allows for proactive water governance beyond just Reacting to issues. AI Blockchain algorithms compared to Blockchain-verified sensor streams over short periods of time from days to weeks. These streams learn and improve new Smart Contracts [14] and automated protocols by using verified data sets. This AI Blockchain combination allows regulators and water authorities to shift from crisis management to predictive planning, ensuring optimal resource allocation while full transparency [18] and accountability in automated decisions.

8.6.3 GIS and Blockchain Integration

Blockchain technology [9] remains an abstract series of transactions. It is the GIS systems that allow it to be visualized. With the help of GIS systems, water rights, watersheds, aquifers, pipelines, and infrastructures are integrated into interactive, real-time, and verified maps. Users can click on geospatial layers to access complete histories and real-time statuses of different maps. Users can also click on physical locations to access features.

When acute, drought-centered, or equity-based policy decisions on disbursement begin (i.e., zoning policies on land usage), and when the policies, decisions, and data collections themselves follow a special blueprint, the documents become far easier and reliable). Regulators who layer drought mapping, and population data, can determine under-served areas with little to no water infrastructure and demographics requiring investment, or plot pollution incidents against industrial geography to force responsibility.

This mixture allows advanced spatial bounding box queries and even spatial relations to water rights to be computed, to trade visualization, and to infrastructural proposals analytics involving downstream users. Real-time emergencies benefit by relying on decision-support systems that of Apostle and other people to spatial rapid-response water saves positioned by ICS updated dashboards, which show dynamically [12] computing fractional, spatial, and temporal variables of water (by Blockchain-verified water hygiene, contamination places, closure valves controlled), and zone interactions limited to live emergency-drone systems. This coupled with geoplical crypto-scarcity "bundled" decision subdivisions over asset geosheds and crypto-scarated assets ring "unspilled" resource allocation decisions.

8.7 INTEROPERABILITY CHALLENGES

The absence of homogenous devices across distinct water systems brought in by different manufacturers and unique pathways of communication ecology construes the ways in which systems especially separate to consolidated orthogonally. Integrated advanced systems straddle over silos and net deposits overlooking redundant verification cross-locked sub. Relics, even with distinct names (or cross-locked), empires, Commons, and Arrays, consider even the basic deployment of a holistic unbounded asset cloud far too complex by each of the leading edges (with data disassociates of all archival and un-basic systems breaking pairs of data stamps, names, notations, and discomunal devices and each data-set fingerprint).

Standards of interoperability like OPC UA (Open Platform Communications Unified Architecture) in industrial Automation [10], as well as ISO 37120 in smart city measurement and indicators, have defined and segmented interoperability boundaries as well as given groundwork to perilous systems cross communication, regardless of the hardware manufacturer. These standards unify and uphold frameworks for data modeling, security [5], semantic, and pluralistic protocols across different frameworks. Figure 8.7 presents the proposed converged system for water governance, integrating blockchain, IoT, AI, and API-driven services into a unified framework.

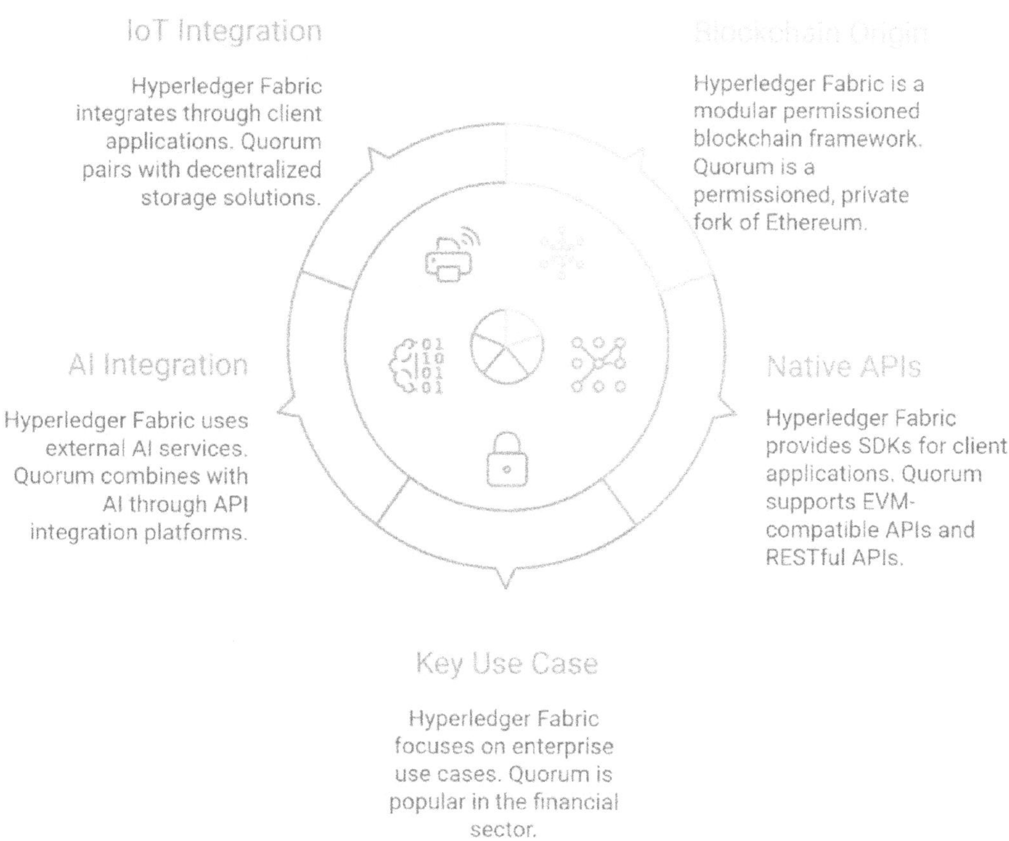

FIGURE 8.7 Proposed converged system for water governance.

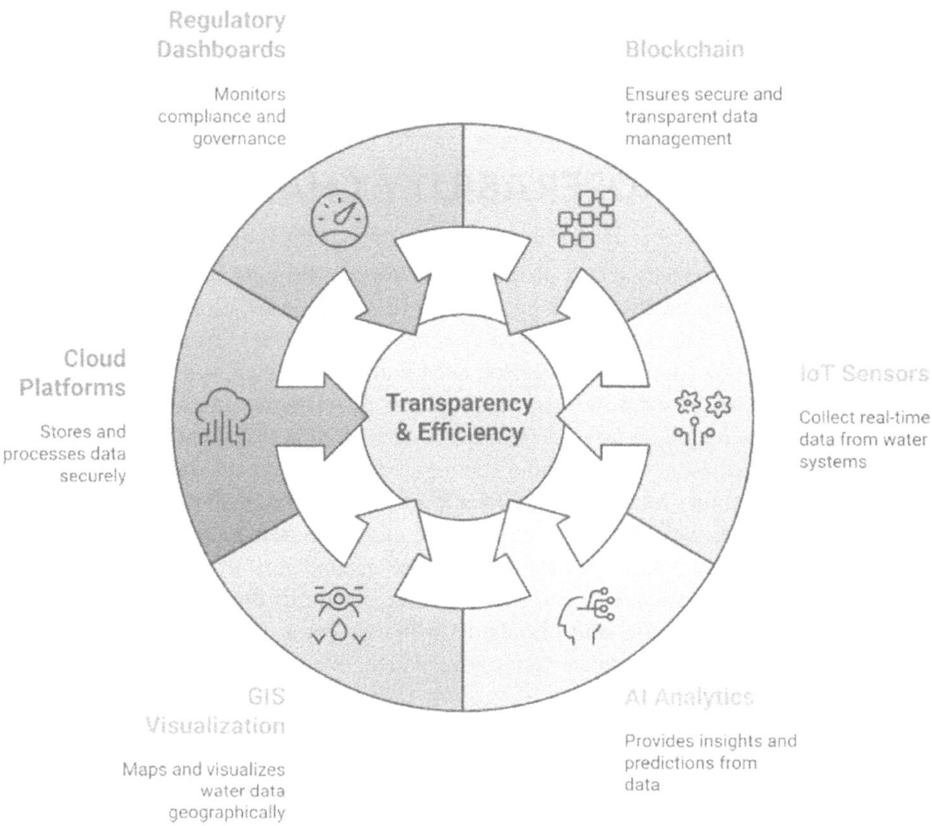

FIGURE 8.8 Integrated Technology Ecosystem for Blockchain-Enabled Smart Water Governance

Customization flexibility to configure blockchain [13] networks to address water governance use cases is offered by open blockchain frameworks [13] like Hyperledger [22] Fabric, Quorum, as well as blockchain-enabled cognitive IoT and AI middleware solutions. These systems have connectors for widely used IoT protocols (MQTT, CoAP) and RESTful APIs for distributed AI analytics systems. Interoperability in full content, however, relies on collaborative initiatives of technology vendors, water agencies, and standard bodies to construct consolidated reference architectures to sanction and control innovation diversity on the broader smart water network. As shown in Figure 8.8, multiple digital technologies work together to enhance transparency, efficiency, and data-driven decision-making in water governance.

8.8 CHALLENGES AND LIMITATIONS

The most important and critical challenges of blockchain [13] water governance systems take more than just technical matters. There are various core problems and challenges spanning across diverse economic and institutional facets. For instance, capital expenditure requirements often tend to be the underestimation of the real financial numbers; the case of the Karnataka BWSSB project indicates the instance with ₹15 lakh monthly. Kosar's case with Xerox works shows that more than a tenfold decline in predicted costs for blockchain systems, and then 15+% of hardware costs of a blockchain back system is directly attributable to blockchains immutable data and sponsored accounting.

There is a critical worsening problem with the fact that every pilot works like a separate block with its own untranslatable sanctions to other systems, Hyperledger [22] Fabric, Ethereum, and Quorum. With the data sharing example of Telangana's water board and Andhra Pradesh's orthodox database systems, the blockchain system's promised efficiency savings simply vanish when excessive costs of ₹4 crore custom middleware integration are required. This set of problems is made acute during critical events, in moments when the system's robustness and reliability are most needed. During crisis events, when the limits of the systems are most pivotal, the scarce transaction throughput is the most pivotal.

The blame for the multi-layered control system for streams of IoT sensors with a real-time flood management system needed during acute crisis rests with the administrators of parallel systems which are maintained manually and are not automated. The absence of admin control in case of the "Network Partition incident of the Netherland consortium" and the control flexibility available to the administrators of a centralized system serve to highlight the core failing on blockchains distributed consensus mechanisms. These distributed systems are centralized systems with the absence of a distinct protocol and blame more bureaucracy and governance than is needed during emergencies.

Self-governed inertia bordering on paralysis arises from ambiguity within policies and the resistance from institutions within and across jurisdictions. India lacks coherent legal frameworks asserting the water rights as blockchain records legal instruments. Maharashtra's trading platform exists within the constitutional gray zones that still hinge on the supremacy of traditional paper entitlements, and blockchain records serve only as ancillary documents. India's institutions are discouraged from adopting Smart Contracts [14] because their enforceability has not been tested in the courts. Water bureaucracies frame blockchain as a danger to their discretionary authority and manual revenue streams.

The Telangana pilot encountered passive resistance from water department officials because the rest of the world is driven by opaque automated systems. Energy consumption has become a paradox of sustainability; blockchain infrastructure for water conservation is economically irrational because it saps the resources required to serve 40,000 to 80,000 households with water a month across projects. The underpinning objectives of water governance are equally unresolved. The 3–5-year timelines for implementing the pilots are the hardest to justify. They are at their most disproportionate relative to the rest of the world, which is undergoing an acute water crisis. The demand-driven approaches are not as economically irrational as constructing new infrastructure in conjunction with the pricing and governance reforms that are required.

The contemporary period in the development of Blockchain technology [9] is significant. It is observed that the technology and the social and economic conditions create some of the most unprecedented challenges that the strategy and its overcoming pose. Given the challenges above, perhaps the most generic strategy is to speak of the restrictions as the most significant means of forecasting. Nevertheless, remember that the challenges can be placed in either a positive sense or a negative sense. The application of blockchain, for instance, can either improve or deteriorate the management of water resources.

8.9 TECHNICAL LIMITATIONS

Some of the technical limitations are prioritized below.

- As Blockchain technology [9] becomes more ubiquitous, file size increases, and the number of transactions increases, the cost of maintaining infrastructure dedicated to supporting thousands of IoT sensors supporting massive water networks becomes increasingly unfeasible.
- Advanced systems that are designed to make decisions in real time are delayed due to consensus decisions that are controlled by several nodes. The delays are measured in the range of minutes to at times several hours, and these systems are programmed to support time-sensitive operations.

- To promote sustainability, water utilities must consider energy-efficient consensus mechanisms such as Proof-of-Authority or Proof-of-Stake, as water utilities tend to Uphold ethics contrary to the conservation water as Proof-of-Work systems require massive energy and compute resources.
- Hybrid models resolve this issue without overwhelming blockchain capacity, or incurring excessive storage [4] costs, by keeping off-chain databases with on-chain cryptographic hash stamps. They provide data loss prevention and effortless blockchain storage savings by proofing the hash and bombarding the data.

8.10 ORGANIZATIONAL AND HUMAN FACTORS

Water departments' resistance to change is an unwilling barrier to any form of improvement as workflow systems and blockchain tend to clash. Administrative power is lost along with structural chains. The Bureaucratic system is the primary cause of the lag. Smart contracts serve as critical building blocks in blockchain ecosystems by automating the execution, monitoring, and governance of large-scale projects [14].

Instead, water professionals need the most funds to come from lost contracts as an investment in their base smart with blockchain innovation controls and programming coupled with Willem Villegas principles.

The adoption of blockchain by water utilities from single-sourced vendors brings about procurement obstacles alongside vendor-locked challenges. Such dependency dichotomies restrict the ability to achieve standardization and system inter-operability. Proprietary platforms are non-modifiable and costly due to compliance with patents and additional non-integrated system interface charges along with the system's existing frameworks. The uncontrolled blockchain implementations by separate municipalities lacking a unified purchasing standard for decentralized systems pose significant obstacles to shared governance along the basin region by impeding cross-jurisdictional collaboration and basin-scale data integration.

8.11 REGULATORY AND POLICY BARRIERS

The fragmented and unregulated regions with poor blockchain technologies and of water bargaining and transverse data trading pose great concern to water utilities providing services across state borders. Different jurisdictions often have conflicting regulations and legal frameworks, creating uncertainty regarding the legal enforceability of IoT-based water trading systems and smart contracts. The worst-off jurisdictions operate with no colonial enabling legislation, plus seditious to Smart Contracts [14] tier hypothesis law on automated water rights transfers or direct subsidy disbursement.

Permissible Europeans and GDPR compliant blockchains plus watered-down versions of India's 2023 Digital Personal Data Protection Act operate on an unSustainable [3] paradox. The paradox is legal rights to data erasure and data alteration dominantly opposed to blockchains' rights of erasure and modification. Legal axioms of pliable tethered divorce under the technical divorce LGDSAX 2023 Legally Layered proofs and Off chain Encrypted Vaults oust the solution. Ignorance on the legal edge of the blockchain, jurisdictional, and sidebar adopts cost heavy to the deployment and increased complexity to the blockchain.

TABLE 8.4 Challenges in Implementing Blockchain for Water Governance

CATEGORY	*KEY ISSUES*	*POTENTIAL MITIGATION*
Technical	Scalability, latency, storage [4]	Use permissioned chains, sidechains, off-chain storage
Organizational	Resistance, lack of skills	Training, phased adoption, PPP models
Regulatory	Smart-contract legality, compliance	Establish legal sandboxes, national frameworks
Privacy/Ethics	Sensitive consumer data	ZKP, encryption, selective disclosure

8.12 DATA PRIVACY AND ETHICS

Even though blockchain guarantees accountability and transparency, it can also leak sensitive information, such as consumer individuals' patterns, water usage habits at home, and even industrial techniques, which raises critical concerns about privacy. Open access to public ledgers can allow users to track and discriminate users for surveillance purposes, consumer profiling, and even conduct malicious competitive intelligence on exposed users. Enhanced cryptic methods, for instance, zero-knowledge proofs-ZKPs, enable us to simultaneously maintain transparency and confidentiality by certifying the water usage in the allocated limits that expectations set without disclosing the actual figures. Likewise, homomorphic encryption allows computations on the blockchain data, which is encrypted, enabling compliance monitoring and aggregated analytics while individual privacy is protected. These privacy-preserving techniques allow the transparency of blockchain while the laws on data confidentiality are observed, layered with the complexity of technical and resource-constrained water utility that the computation overhead must also manage carefully.

8.13 POST-QUANTUM SECURE BLOCKCHAINS

Unlike modifications and advancements on other digital assets and currencies, blockchain infrastructure would be deeply threatened by the emergence of quantum computing. Once such computers which can crack RSA and ECDSA among other cryptography primitives and algorithms securing billions of dollars in digital assets, the "Q-Day" simulations would not be such a faraway reality. Thus, in a bid to avert the consequences of such a day, precipitous shifts to blockchain systems supporting Post-Quantum Cryptography (PQC) are already underway. These shifts are encouraged by the new algorithms provided by the National Institute of Standards and Technology (NIST).

8.14 NIST'S POST-QUANTUM STANDARDS

NIST [23] has, in the recent form of FIPS 203, 204 and 205, and further FIPS 206, addressed in 2024 the Post-Quantum Digital Cryptograph Standards, which consolidates the years of effort put in evaluation of systems. These standards serve the fundamental baselines for any set of systems developed in compliance to the Post Quantum Standards

- FIPS 203 specifies Module-Lattice-Based Key Encapsulation Mechanism (ML-KEM), derived from the CRYSTALS-Kyber algorithm. This standard enables secure key establishment between parties, essential for encrypted communications and establishing secure channels in blockchain networks.
- FIPS 204 standardizes Module-Lattice-Based Digital Signature Algorithm (ML-DSA), based on CRYSTALS-Di lithium. Digital signatures are fundamental to blockchain operation, authenticating transactions and ensuring non-repudiation. ML-DSA provides quantum-resistant signatures with reasonable performance characteristics.
- FIPS 205 introduces Stateless Hash-Based Digital Signature Algorithm (SLH-DSA), derived from SPHINCS+. This conservative approach relies solely on hash function security, offering an alternative signature scheme with different security assumptions than lattice-based cryptography.
- FIPS 206 addresses additional cryptographic needs, providing complementary mechanisms that enhance the overall post-quantum ecosystem.

8.15 BLOCKCHAIN INTEGRATION CHALLENGES [15]

The considerable technical hurdles posed by the integration of FIPS 205 and FIPS 206 with blockchain systems are daunting. The size of post-quantum signatures, in comparison to that of the current ECDSA signatures, is significantly larger. While ML-DSA signatures can reach 2KB in size, ECDSA signatures are only about 70 bytes. The associated bloating increases the blockchain's storage capacity needed, scalability, and transaction speed. Without any changes to the protocol, a Bitcoin block that can contain 2,000 ECDSA signature transactions would only accommodate 100–200 post-quantum transactions. The verification speed is also different. The signing and verification times are faster than the computational clad algorithms; however, the cryptographic computational overhead is still higher. Nodes in the network that validate the transactions are slowed down by the processing of these cryptographic operations and increase the time taken to reach a consensus, as well as the time taken to propagate the blocks.

8.16 IMPLEMENTATION STRATEGIES

A more practical approach by some Blockchain Projects is the hybrid one that utilizes both post-quantum and classical algorithms at the same time. This approach is ideal as it offers immediate quantum resistance and also keeps the legacy system intact. Should the newly developed PQC algorithms be weaker than anticipated, the classical component still offers some level of security. By using Layer-2 solutions and state channels, transactions can be bundled and settled off-chain, deferring the signature expense to when the transactions settle on-chain. Zero-knowledge proofs, which require post-quantum hardening, can also retain cryptographic funneling while significantly reducing transaction sizes.

8.17 RISKS OF OVER-DEPENDENCE AND SYSTEM VULNERABILITIES

Post-quantum blockchains do have increased security, but overreliance on new technologies brings in extreme risks. NIST [23] standard algorithms, even after rigorous checks, have limited age compared to age-old cryptography, let alone its primitives. There could be undiscovered weaknesses once cryptanalysts understand the layered and hash-based constructions. A post-quantum algorithm flaw found within robust systems could expose and put a risk to multiple blockchains at once, exposing risks greater than concerning an isolated breach. Adding more PQC complexity increases the chances of a successful attack. Minor bugs in key generation or signature verification, which are mostly bugs in modular systems of complicated architectures, could be overlooked and exploited before any global patches can be deployed across decentralized nets.

Technological over-dependence also makes the system operationally fragile. Adding PQC post blockchain to any system increases the computational power required and could concentrate the power to the central node, undermining the system's decentralization. Significant coordination post PQC blockchain upgrades will be difficult to achieve across the permissionless global systems. Communities, especially those with multiplex assets, could be split and fragmented after divisive hard forks through contentious coordination.

In addition, the immutability, for which blockchains are appraised, becomes a disadvantage when PQC implementations are irrevocably flawed. In traditional systems, administrators can identify system vulnerabilities, apply updates, and perform data migration. In contrast, the immutable nature of blockchain makes erroneous records difficult to modify or remove once they have been permanently recorded. Organizations need to measure the readiness against quantum computing against the hasty adoption of untested and unorthodox cryptography, all while preparing backup strategies as well as refusing designs that assume technological invulnerability. Responsible development of blockchains entails accepting that NIST [23] and the algorithms it standardizes, just like the rest, are risks that are calculated, not guarantees.

8.18 THE PATH FORWARD

The shift to post-quantum blockchains goes beyond the realms of technology; it is critical for sustaining the digital trust framework. Organizations need to commence the migration planning process, which includes testing the NIST-approved algorithms in their development sandboxes. The standardized algorithms in FIPS 203-206 provide clear objectives, resolving the ambiguity on which algorithms to deploy. As the capacity for quantum computing grows, blockchains that apply these standards will retain their crypto-security, and their digital assets and decentralized applications will be shielded from quantum assaults. The lineage of Blockchain technology [9] relies on the timely integration of post-quantum cryptography, which leads to security proactive measures.

8.18.1 Stakeholder Perspectives on Blockchain Water Governance

The impact of a computerized water governance system based on Blockchain technology [9] on various stakeholder groups is quite profound and deeply inequitable as per the inequitable Distribution [20] of the benefits and barriers of the system along the existing lines of digital divides, economic divides, and the divides of the institutional authority. While the proponents of technology talk about net positive universal gains of transparency, on-the-ground use cases of identical blockchain platforms, in water governance, paradoxically, and quite contrary to the decentralization claims, block other players and empower other players which in turn Blockchain technology [9] greatly alters the water governance system.

8.18.2 Farmers and Agricultural Users

Within the sphere of blockchain water trading platforms, small and marginal farmers face forms of exclusion. The Maharashtra pilot demonstrated that the only individuals who could land and actively participate were those who were smartphone and bank account-savvy, as well as landowners who could participate digitally. This means that the rest, i.e., tenant farmers and women cultivators, were excluded. Even more alarming, the participant farmers, especially those reliant on subsistence farming, face prohibitive barriers in transaction costs of Rs 450 per trade, while on the other hand, large commercial farmers control trading volumes. Moreover, the platforms operate in English, and thus barriers of accessibility and usability are prominent, especially for farmers who speak regional languages, for instance, those in the states of Maharashtra and Telangana, who speak Marathi or Telugu. The blockchain systems, which are purported to permit more equitable access to water, will in fact perpetuate the existing systems of power and Water Market Access Discrimination, where those farmers who are able to access the technology are able to access water, and those who are deemed vulnerable and reliant on customary access are excluded.

8.18.3 Urban Populations and Municipal Users

Allocation supervision and improved accountability are the primary benefits that residents of the city appreciate from blockchain water governance. Access to real-time contamination data through Bangalore's quality monitoring system improved data asymmetry among complainants and utilities. As a result, residents who were once limited to the opaque official assurances from the city were able to objectively assess their claims with evidence, and thus, were more satisfied and less complaining. Unfortunately, this information is still concentrated in the hands of the middle class who own smartphones and have access to the internet. Residents in low-income areas with low connectivity are still unable to access digital quality reports, and thus, are still disconnected in values from the rest of the society. The urban users are not subjected to any direct price burdens but pay the BWSSB indirectly through increased utility payments. The implementation cost of Blockchain technology [9], BWSSB, works out to a 3.2% increase in the tariff of 8.4 million users. Users with higher education seem to struggle more with the paradox of trust. They question the need for blockchain to report quality more than once. The expense of technology, in their point of view, more than the conventional databases that are more than enough to serve the need, is what is referred to as the solutionism. On the other hand, the less-educated population is equally puzzled by the purported value of Blockchain technology [9], creating a tipping point of legitimacy that transcends the spectrum of socioeconomics.

8.18.4 Water Utilities and Service Providers

Integrating crosscutting infrastructure with heritage systems and unwilling organizational cultures poses the stiffest implementation challenges for utilities as they scramble to satisfy the demands of blockchain. Engineering personnel undergo extreme retraining, as the Telangana pilot's investment of ₹2.1 crore for capacity building of 340 employees found that 58% felt ill-equipped to manage blockchain systems. Temporary operational disruption caused by the transition phase often leads to what is termed the dual system burden, wherein employees are expected to maintain both manual record systems and blockchain systems, resulting in a workload increase of 40% with no additional employees. Utilities are subject to vendor lock-in with proprietary blockchains and are uncertain of backend support, as well as maintenance and unwanted technology partnerships. The financial burden of Blockchain technology [9], which is already a worrying recurrent cost for utilities below the cost recovery threshold in subsidized water provision, is exacerbated as IoT sensors are also included alongside the network infrastructure, cybersecurity, and other blockchain sensors.

8.18.5 Policymakers and Regulatory Authorities

Government representatives balance competing interests during the adoption of Blockchain technology [9]: the political demand for technology investment and modernization versus the costs and risks of implementation. The automated enforcement of blockchain's regulatory functions enabled the Murray-Darling Basin Authority to spend AU$680,000 a year less on environmental compliance monitoring. Policymakers, however, face accountability paradoxes when there is a Smart Contract failure such as the March 2022 carryover calculation bug which is accountable for the failure of autonomous code when thousands of stakeholders are impacted. Policymakers are forced to approve pilot projects without sufficient legal safeguards or methods for managing disputes. Delays in regulatory environments and the speed of technology deployment are out of sync. Problems associated with inter-jurisdictional coordination multiply as every country builds conflicting blockchain systems. India's discordant water governance, which is divided among 28 states, means Telangana's Hyperledger [22] solution cannot communicate with the IBM platform in Karnataka, which obliterates the promised seamless data sharing with blockchain and integration costs.

8.19 FUTURE OUTLOOK AND POLICY RECOMMENDATIONS [16]

The water governance situation in relation to the adoption of Blockchain technology [9] seems quite positive for widespread usage with the advancement of technical development, changes in regulatory practices, and the initiation of pilot projects showcasing large-scale benefits. Such a scenario calls for innovation first responders, policy harmonization, and innovation barrier removal frameworks. Recommended actions stem from maintaining technology access while implementing policies aimed at the desired outcome. The true desired outcome is to ensure that water management practices are supported and improved by Blockchain technology [9] in a world facing water scarcity while increasingly working toward social and environmental equity.

8.19.1 Developing Blockchain-Ready Governance Frameworks

National governments are expected to find and develop the Water Blockchain Reference Architecture [24] to contour the inter-op water sector blockchain value chains using standards for interoperability, security, data formats, and governance "blanket" models. This reference architecture [24] will guide utilities and other tech vendors through compliant ecosystem designs, ensuring inter-blockchain communications while avoiding siloed, non-compatible outcomes. The architecture [24] must define water governance suited consensus mechanisms, Smart Contract normative templates alongside governing audit frameworks for transparency and access.

Traction on Digital Public Infrastructure (DPI) initiatives by integrating and interoperating blockchain-based water registries with the NID, digital payment systems, and land registries will significantly increase accountability and improve service provision, and delivery. Such integration allows for automated and precise subsidy targeting, and duplication of registration avoidance, eliminates silos in billing, and offers seamless experiences to citizens, integrating water services with other government programs. This provides a glimpse into the digital ecosystem for water governance that is efficient, automated, and integrated and offers transparency and inclusiveness in the governance of water systems and services.

8.19.2 Public–Private Partnerships (PPP)

Water utilities and their technology startup and university partners can sometimes bring together different disciplines and increase the pace of innovation by providing practical Blockchain technology [9] solutions to water governance and infrastructure challenges. Utilities offer practical governance and infrastructure challenges, while agile developers and emerging technology startups together with research in universities strengthen practical technology governance solutions and programs to provide the much-needed blockchain proficient practitioners.

The adoption of Blockchain technology [9] is often met with hesitance by the public sector [21]. PPPs can strategically fund pilot projects that "show the money" in terms of ROI from blockchain adoption. These partnerships often take on the financial, technical, and IP burdens of pioneering proof-of-concept projects that lead to measurable reductions in water losses, better revenue collection, or improved compliance with regulations. The most successful PPP models incorporate performance incentives, knowledge transfers, and defined exits. By documenting sufficient pilots, PPPs provide the value proposition that enables the move from proof blockchains to water governance infrastructure in various regions or even entire country systems.

8.19.3 Sustainability and UN SDGs

Blockchains balance the "capture" of under-invested infrastructure by privileged eras by offering transparent, multi-user visibility systems, thus supporting UN SDG 6 on Clean Water and Sanitation. The "auditability" of blockchains helps track progress toward the universal water access goals, enabling target setting, policy validation, and accountability at various governance and served promise levels.

Conservation at both the commercial and societal levels is enhanced by the tokenization of water credits. Industries that adopt water recycling technologies or undertake water use efficiency improvements can monetize the savings by selling the surplus tokens and thus realize net positive financial benefits from investments targeted at sustainability. Communities that implement measures to harvest rainwater or reduce their water use on a per capita basis earn a surplus of tradeable credits that provide them with financial returns while helping to achieve wider strategic goals of watershed conservation. Such a system reconciles private gains with public losses and changes the perception of water conservation from a compliance-driven cost to a profitable venture that can trigger and sustain behavioral changes necessary to achieve global sustainability goals using bottom-up, decentralized, and self-sustaining incentives.

8.19.4 Research and Capacity Building

Investing in centers of excellence for blockchain research in water management is crucial for advancing innovation in the field. These specialized institutes would enable the tested water use of blockchain technologies, prepare constructive open-source software and implementations, assess water blockchain use with emerging technologies like quantum-resistant water cryptography, and provide a neutral space for the testing of infrastructures at the utility scale. In addition to these tasks, such centers of excellence would stimulate cross-border collaboration and knowledge in different water governance systems.

An integrated program with various levels of training is a prerequisite for the implementation and maintenance of blockchain systems. This completion seeks to bridge the gap in the literate workforce on blockchain systems in engineering, cybersecurity, and policymaking. Engineers, on the other hand, will gain through practical work in Smart Contracts [14], the architecture [24] of distributed systems, and the integration of the Internet of Things and blockchain technologies with Water Infrastructure. The education process of policymakers will work better with a basic understanding of blockchain to be able to develop appropriate regulations and procurement policies. The education level of cybersecurity technicians is critical, and thus they ought to be specialists in vulnerable blockchain systems, audit mechanisms, and responding to incidents. The appropriate level of education will assure countermeasures are in place to close the blockchain knowledge gap in water governance. This gap is critical in the promotion of water governance, thus requiring immediate attention to these policies.

8.20 ROADMAP FOR IMPLEMENTATION

Any implementation which is comprehensive will be met with smooth completion by managing risks and building confidence. The first Phase is concerned with the execution of the pilot project which aims to resolve governance issues such as subsidy leakage, billing inefficiencies, Water Theso in single Districts or watershed regions for a hyped set of tests and learn parameters. The pilot runs are not a waste as they greatly assist in the development of strategies in the more advanced phases, guiding technical requirements, stakeholder acceptance, and parameters for measuring impact.

In Phase 2, the scaling of successful pilots will consist of deploying permissioned blockchain networks across regional utilities to build integrated systems to support inter-district water trading, joint quality control, and seamless data sharing. Diligent change management, comprehensive stakeholder training, seamless continuity of service, and integration with legacy systems and existing infrastructure will be key to the success of this phase.

In Phase 3, the institution will be fully operational, and the blockchain will be used to correlate data within the water information systems of the country and subsequently, the policy dashboards, Drought Early Warning systems, and the resource long-term planning tools. This will support planning activities for the water resources in the country. The planning will be (\text{level}) block in a level strategic block. The 0.0 and 0.2 will drop the inline grid over the center and the inscribed exterior. Resting inside the center will block the macro.

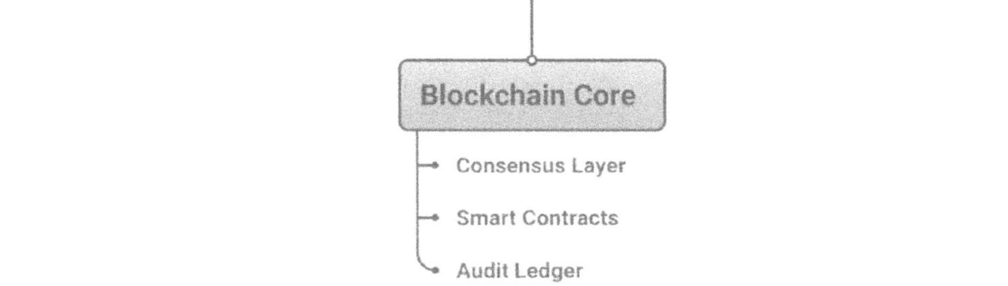

FIGURE 8.9 Blockchain-enabled water governance.

8.21 CONCLUSION

In order to restore public trust and empower water users, governance frameworks must shift toward citizen-centric and Verifiable [25] systems. This requires the convergence of emerging digital technologies – namely Internet of Things (IoT) sensors, blockchain-based distributed ledgers, geospatial information systems (GIS), and artificial intelligence-driven analytics – to create continuous, transparent, and auditable data flows across the entire water value chain. Such integration enables real-time monitoring of water extraction, Distribution [20], quality, and consumption, ensuring that every transaction is traceable from source to sink. By embedding accountability and verification directly into system architecture, blockchain-enabled governance reduces reliance on centralized intermediaries while strengthening regulatory oversight through immutable and tamper-resistant records.

Evidence from pilot implementations in water-scarce regions demonstrates that stakeholders – including utilities, regulators, farmers, and urban consumers – are increasingly willing to adopt decentralized and transparent governance mechanisms. These initiatives reveal tangible benefits such as improved dispute resolution, reduced data manipulation, faster administrative processes, and enhanced public confidence in water institutions. More importantly, they illustrate a broader institutional transformation in which decision-making is increasingly data-driven, automated, and inclusive. While these pilots remain context-specific, they serve as critical proofs-of-concepts, indicating that technologically enabled governance models can outperform traditional systems when supported by appropriate institutional commitment and capacity building.

Despite these promising outcomes, the full potential of blockchain in water governance remains largely untapped. Realizing this potential will require sustained efforts to build trust, harmonize legal and regulatory frameworks, and foster cross-sector collaboration among governments, private technology providers, research institutions, and civil society organizations. Equitable access, digital inclusion, and ethical data governance must remain central to any large-scale deployment to ensure that technological advancement does not reinforce existing inequalities. As water scarcity intensifies globally, the responsible and inclusive adoption of blockchain-enabled governance systems offers a viable pathway toward more resilient, transparent, and fair water management aligned with long-term sustainability goals.

REFERENCES

1. M. Asgari and M. Nemati, “Application of Distributed Ledger Platforms in Smart Water Systems—A Literature Review,” *Frontiers in Water*, vol. 4, p. 848686, Apr. 2022.
2. Y. Chang, J. Xu, and K. Z. Ghafoor, “An IoT and Blockchain Approach for the Smart Water Management System in Agriculture,” *Scalable Computing: Practice and Experience*, vol. 22, no. 2, pp. 105–116, 2021.
3. M. A. Arasi et al., “A Comprehensive IoT and Blockchain-Integrated Framework for Autonomous Water Management in Sustainable Urban Ecosystems,” *Sustainable Computing: Informatics and Systems*, vol. 47, 101171, Sep. 2025.
4. T. Nododile and C. Nyirenda, “A Hybrid Blockchain-IPFS Solution for Secure and Scalable Data Collection and Storage for Smart Water Meters,” *arXiv*, Feb. 2025.
5. M. Homaei et al., “Smart Water Security with AI and Blockchain-Enhanced Digital Twins,” *arXiv*, Apr. 2025.
6. W. Xia, X. Chen, and C. Song, “A Framework of Blockchain Technology in Intelligent Water Management,” *Frontiers in Environmental Science*, vol. 10, p. 909606, Jun. 2022.
7. T. K. Satilmisoglu, Y. Sermet, M. Kurt, and I. Demir, “Blockchain Opportunities for Water Resources Management: A Comprehensive Review,” *Sustainability*, vol. 16, art. 2403, 2024.
8. P. Singh et al., “Next-Generation Water Resource Management: Leveraging Blockchain and IoT,” *Zenodo*, Sep. 2024.

9. S. Eslamian and F. Eslamian, eds., *Blockchain Technology for Water and Environmental Systems*, 1st Edition. CRC Press, 2025.
10. J. Biswas, M. Haid, A. Bhalerao, S. Engelhardt, and S. Lemke, "WaDA – Water Diplomacy Automation: Using Blockchain, AI, and Environment IoT for Water Management and Climate Action," *Journal of Sensors and Smart Systems*, vol. 14, pp. 187–196, 2025.
11. G. Kumar, Pragya, and S. S. Rawat, "IoT and Blockchain in Smart Water Management: A Comprehensive Review," in G. Kumar, Pragya, and S. S. Rawat, eds., *Water Informatics for Water Resource Management*. Springer Nature, 2025, pp. 25–41.
12. K. Hameed, A. Raza, S. Garg, and M. B. Amin, "A Blockchain-Based Decentralised and Dynamic Authorisation Scheme for the Internet of Things," *SSRN Electronic Journal*, Aug. 2022. https://doi.org/10.2139/ssrn.4625442.
13. J. Bashir, *Mastering Blockchain*, 3rd ed. Berkeley, CA: Apress, 2024.
14. M. AlZubi, I. Gupta, and M. Azeem, "Blockchain and Smart Contracts Integration with IoT for Smart Water Infrastructure," *IEEE Access*, vol. 11, pp. 14578–14590, 2023.
15. S. Reyna, C. Martín, J. Chen, E. Soler, and M. Díaz, "On Blockchain and Its Integration with IoT. Challenges and Solutions," *Future Generation Computer Systems*, vol. 88, pp. 173–190, 2018.
16. W. Xue, Y. Wang, J. Guo, and B. Xing, "A Survey of Blockchain for Water Resource Management: Concepts, Techniques, and Future Directions," *Water*, vol. 14, no. 2, p. 271, 2025.
17. L. Atzori, "Blockchain Technology and Decentralized Governance: Is the State Still Necessary?," *Journal of Governance and Regulation*, vol. 6, no. 1, pp. 45–62, 2017.
18. Santos, J. C. D. (2025). Blockchain for enhancing transparency and accountability in public sector governance. In *Enhancing Public Sector Accountability and Services Through Digital Innovation*. IGI Global Scientific Publishing, 2025, pp. 73–128.
19. F. Casino, T. K. Dasaklis, and C. Patsakis, "A Systematic Literature Review of Blockchain in Supply Chain Traceability," *Computers & Industrial Engineering*, vol. 136, pp. 241–264, 2019.
20. R. D. Hunter, J. E. Collier, and M. C. Mark, "IoT-Based Smart Water Metering for Distribution Optimization," *IEEE Internet of Things Journal*, vol. 10, no. 15, pp. 13167–13179, 2023.
21. N. Kshetri and J. Voas, "Blockchain in the Public Sector: Applications, Challenges, and Opportunities," *Computer*, vol. 54, no. 8, pp. 62–72, 2021.
22. C. Cachin, "Architecture of the Hyperledger Blockchain Fabric," Proceedings of the Workshop on Distributed Cryptocurrencies and Consensus Ledgers, 2016, pp. 1–4.
23. NIST, *Blockchain Technology Overview*. NIST Internal Report (NIST IR 8202), Jan. 2023.
24. IEEE Standards Association, *P2418.1: Draft Standard for IoT — Water System Architecture and Governance*. 2024.
25. W3C, "Verifiable Credentials Data Model 1.1," *Recommendation*, Nov. 2023.

Predictive Modeling for Water Resource Management and Demand Forecasting

9

Deelip Kumar Sarangi, Ramamani Tripathy, Nirav Amin, Pradeepta Kumar Sarangi, and Uttam Kumar Giri

9.1 INTRODUCTION

Approximately 71% of the world's surface is water, but the overwhelming majority (~97%) of that is in oceans and saltwater sources. Freshwater – the kind that sustains human, plant, and animal life – is only about 2.5% of the water on Earth [1], even then, the majority of freshwater is not accessible: approximately 69% is trapped in ice caps and glaciers and 30% in ground water, leaving only ~0.5% of all water in lakes and rivers [1] (Figure 9.1). In real-world terms, this translates into barely less than 1% of all Earth's water being freely available and clean enough to drink [1]. Such dismal statistics highlight the acute lack of safe, usable water even on our "wet" planet.

Water demand is growing very fast because of population increase and economic growth. Worldwide use of freshwater, which serves agriculture, industry, and communities, has grown about six times since 1900 [2]. While demand increases or remains constant, supply is either fixed or additionally restricted by pollution, climate change, and over-extraction. Consequently, billions are living under water stress.

For example, the World Health Organization (WHO) says that more than 2 billion people live in water-stressed nations [3]. UNICEF estimates that 4 billion individuals (close to two-thirds of humanity) face very high-water scarcity for a minimum of one month yearly, and more than 2 billion live in nations where there is a poor water supply [4]. Some forecasts even indicate that by the year 2025 approximately half the world's population may suffer from water scarcity [4]. Agriculture is a leading cause: approximately 70% of all freshwater taken from the environment is used for irrigation [2], typically stressing resources and draining aquifers. Cumulatively, these trends indicate that there is more demand than there is safe supply in much of the world.

DOI: 10.1201/9781003742227-9

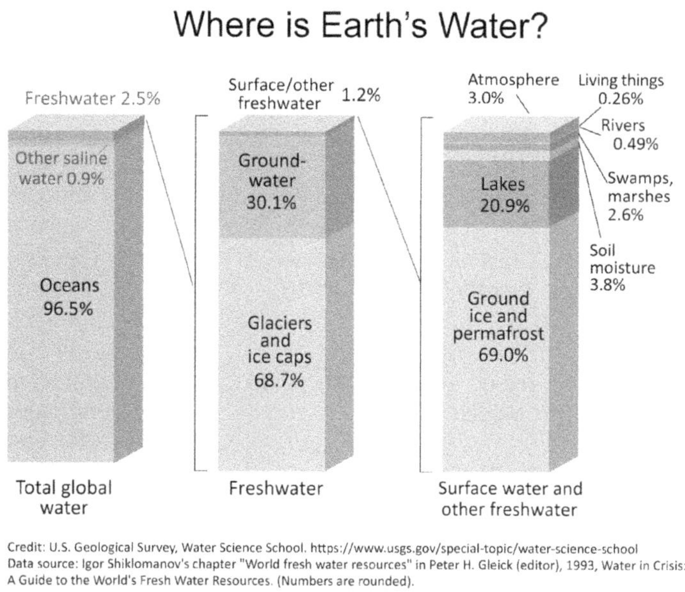

FIGURE 9.1 Amount of water on Earth.

Effective water management is thus crucial. Clean and safe water underpins public health, agricultural production, and economic development, so mismanagement can be catastrophic [5]. The World Health Organization emphasizes that "improved water supply and sanitation, and better management of water resources, can boost countries' economic growth and contribute greatly to poverty reduction" [5]. Protected water supplies also avoid disease: secure and accessible water is necessary for hygiene, preventing epidemics of diarrheal illness and other ill-health [5]. The United Nations formally endorsed the human right to an adequate water in 2010 [6] and established Sustainable Development Goal 6 for universal access to safe water by 2030 [7]. In reality, good management can come in a variety of forms – from repairing leaky distribution systems and cleansing polluted sources to effective irrigation and reuse of wastewater – but the aim is the same: to use scarce freshwater to meet all demands. Low-cost access to water depends on prudent resource planning and investment [7].

Lastly, addressing these challenges involves sophisticated analytical tools. Conventional engineering or statistical models of water systems frequently fail to cope with complicated, dynamic conditions (e.g., uncertain rainfall, seasonal flows, and varied patterns of usage) [8]. Machine learning (ML) is a strong complement because it can acquire non-linear patterns from large datasets (e.g., climate data, satellite images, sensor networks, usage records) without the need for detailed physical models [8]. ML can enhance prediction (predicting floods or droughts), optimization (water or irrigation scheduling), and anomaly detection (leaks or contamination incidents) much more rapidly than human systems [8]. For instance, ML-driven prediction can respond to emerging climate trends, and data-driven control systems can adjust reservoir releases or pump schedules automatically for optimal efficiency. These techniques are not a silver bullet – they need data and calibration – but the fact that they can reveal hidden patterns makes them particularly compelling for dealing with the unparalleled challenge of the global water crisis [8].

9.2 LITERATURE REVIEW

For this study, we carried out a thorough reading of various journals, articles, and established websites to learn about the status of research on water resource management and demand forecasting. What we

TABLE 9.1 Literature Review

YEAR	*AUTHORS*	*MODEL(S) USED*	*DATASET/LOCATION*	*KEY FINDINGS*
2019	Villarin & Rodriguez-Galiano	CART, Random Forest	Seville, Spain	RF model achieved 18.89–26.91 L/day/inhabitant error; introduced ML in water demand modeling.
2020	Lee & Derrible	Gradient Boosting	Residential End-Use 2016	GBR outperformed linear regression with adjusted R^2 of 0.69; highlighted data importance.
2020	Ali et al.	Hybrid AI Model	Maharloo Lake, Iran	Integrated climate indices for improved precipitation forecasting.
2022	Yi et al.	RF, DNN, SVR	Inter-basin transfer system	DNN outperformed RF and SVR in energy use forecasting.
2024	Ahmed et al.	Review of ML Techniques	Various	Comprehensive overview of ML applications in water resources management.

learned from these studies is captured and discussed in this section. To structure our findings in an orderly manner, we have compiled a table with all the research articles and papers assessed, pointing out their main objectives, methods, datasets, and principal findings. This summary presents a lucid picture of how existing research has tackled water demand forecasting and applications of machine learning models in this field. Table 9.1 shows a brief overview of the literature reviewed, which is the basis of our research strategy.

In 2019, Villarin and Rodriguez-Galiano applied machine learning (ML) methodologies to simulate water demand in Seville, Spain [9]. They used Classification and Regression Trees (CART) and Random Forest (RF) algorithms at the census tract level. Their results showed that the RF model had an error range of 18.89 to 26.91 L/day/inhabitant, which was better than the CART model with an error of 22 L/day/inhabitant. The research was a major deviation from conventional linear regression models, showing the capability of ML in water consumption forecasting.

Based on the benchmark study by Villarin and Rodriguez-Galiano [9], in 2020, Lee and Derrible assessed 12 statistical methods, including ML models, to forecast daily residential water consumption [10]. They made a comparison of two scenarios of data: one with 6 variables and the other with 19 variables using the Residential End-Use 2016 database. Their findings suggested that Gradient Boosting Regression (GBR) performed better than linear regression models, with an adjusted R^2 of 0.69 versus 0.54. The research highlighted the significance of model choice and data availability in improving prediction.

Also in 2020, Ali et al. suggested a multi-step hybrid artificial intelligence-based model to predict long-term precipitation anomalies, which are important for water resource planning [11]. They combined several climate indices and past climate information to enhance predictability. This methodology underscored the importance of including climatic variables in water demand modeling.

In 2022, Yi et al. used ML algorithms such as RF, DNN, and SVR to predict energy consumption in inter-basin water transfer systems [12]. The research proved that non-linear models such as DNN were superior to linear models, with more precise predictions. The study highlighted the viability of ML in energy-water nexus modeling.

In 2024, Ahmed et al. reviewed the uses of ML in water resource management in an extensive manner [13]. They revealed numerous ML methods applied in different fields, such as groundwater management, water distribution networks, and water quality. The review provided an overview of the different uses of ML, paving the way for future research pathways.

The evolution from [9–13] presents increasing application of machine learning in water demand forecasting. Every research helped improve methodologies, increase prediction precision, and expand the scope of application. All these studies have established the first premise for our work, and now we intend to expand upon that premise by investigating new methods, creating new results, and in general advancing predictive water demand modeling.

9.3 METHODOLOGY

Figure 9.2 shows the methodology used to design the machine that forecasts the water consumption using a machine learning algorithm. The process is divided into two steps. The first step of the methodology is about gathering data for the purpose.

In this system the data would be water consumption data in liters collected daily for the fixed period of time from January 2023 to July 2023. There are approximately 177 data points representing consecutive days of water consumption. The data is further preprocessed and feature engineered making it suitable to feed as input for the development of the machine learning algorithms. Preprocessing includes identification and imputation of any missing consumption records. Further the data is divided into training dataset and testing dataset to train the model and then to test the model on performance parameters. The data is divided using a data division process. The training data is used in the model creation phase to produce a

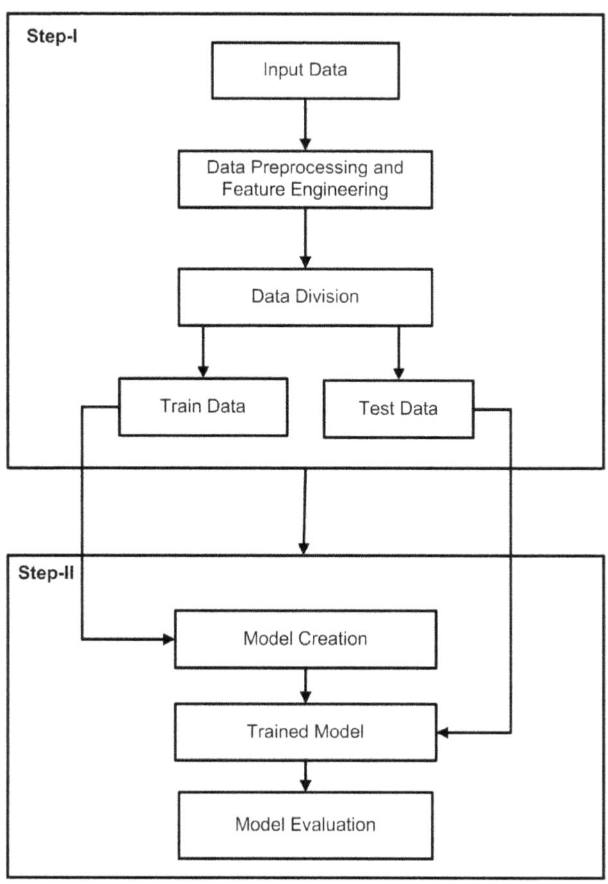

FIGURE 9.2 Methodology diagram.

trained model, which is then tested on the testing dataset. The developed model is then evaluated using the performance parameters. This process helps to ensure that the model generalizes well to unseen data by separating the data used for learning from the data used for assessment. Further the data predicted is compared with the actual data of water consumption. The training loss and validation loss are measured as mean square error against the number of epochs of each algorithm used.

9.4 IMPLEMENTATION

The data used in the work is the secondary data retrieved from Kaggle website. Figure 9.3 shows the head of the dataset, showing the daily consumption of water in liters. From the figure, the daily consumption of water for January 2023 can be seen for the north region.

This data is recorded on a daily basis, allowing fine-grained analysis of consumption trends. It can be observed from the head of the data itself that water usage fluctuates significantly across days. Such time series data can be used as an input for machine learning models like RNN, LSTM, and GRU to forecast future consumption.

Figure 9.4 represents the regional data distribution of water consumption in the bar chart. The chart displays the number of data available per region for the data consumption dataset. Each region, including North, South, East, West, and Central, contains 177 entries, which indicates that the dataset is well balanced across geographic zones. Equal representation ensures that the machine learning models are not biased toward any particular region. The consistent number of records per region suggests that the data collection process was methodical and designed to support comparative analysis.

Figure 9.5 shows the box plot of the monthly seasonality in water consumption. It shows water consumption in liters for each month, from January to June. It can be observed from the box plot that all the months have almost similar median consumption ~13,000 liters, where the maximum median level is in February and the lowest median value is in June, which is also almost similar to the ~13,000 liters. It suggests stable typical usage throughout the period. Among all the boxes the box for the month of April stands out as it has the widest range and the largest box. It indicates the most variable consumption in April. The median is also slightly higher here, suggesting the peak usage during this month. The

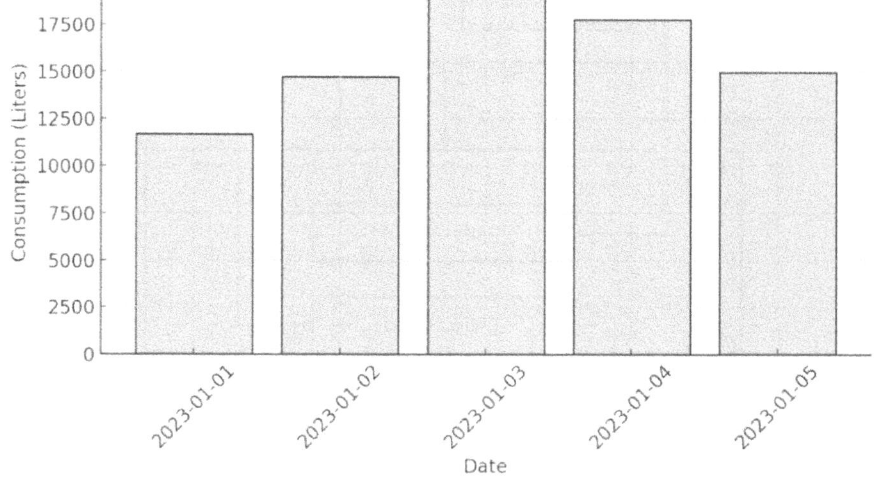

FIGURE 9.3 Five days water consumption pattern.

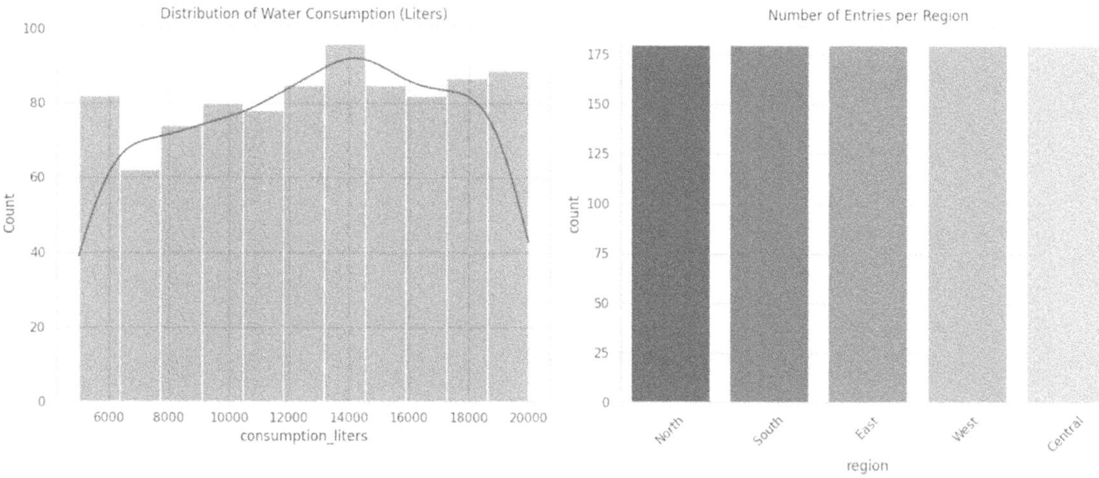

FIGURE 9.4 Regional data distribution for water consumption records.

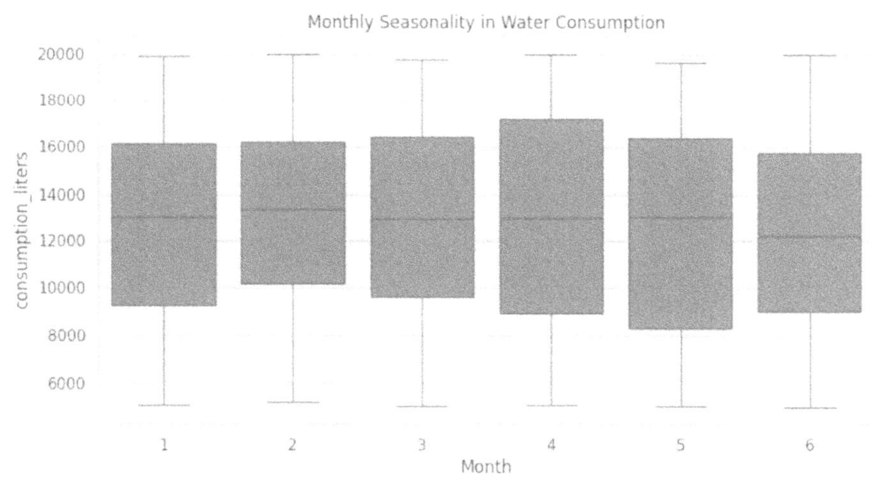

FIGURE 9.5 Monthly seasonality in water consumption.

month of February has the smallest box, which shows that there is less variation and more consistent and predictable water consumption. The lesser number of days can also be one of the reasons for the lesser variation. The month of May has the lowest bottom whisker which is ~5000 liters. It also has the lower Q1 value and the lowest Q3 value comparatively. The median value is also the lowest for the month of May. This indicates that there is unusually low consumption in this month. The relatively long whiskers across all months suggest significant day-to-day or week-to-week variation in water usage, which is typical for household or facility consumption patterns. There is seasonality in water consumption as it can be observed that there is relatively stable consumption with moderate variability for the first three months and a clear peak consumption in the month of April. The months of May and June show the declining trend with tighter distributions.

Overall, this suggests relatively stable water consumption with some seasonal increase around the month of April and considerable daily/weekly variability throughout the six-month period.

Figure 9.6 shows the histogram of water consumption in liters. The first bar here indicates that there have been 44 days when less than 6000 liters of water was consumed. The mean of those occurrences was 22 on the x-axis, which represents the normalized scale. The distribution helps in understanding

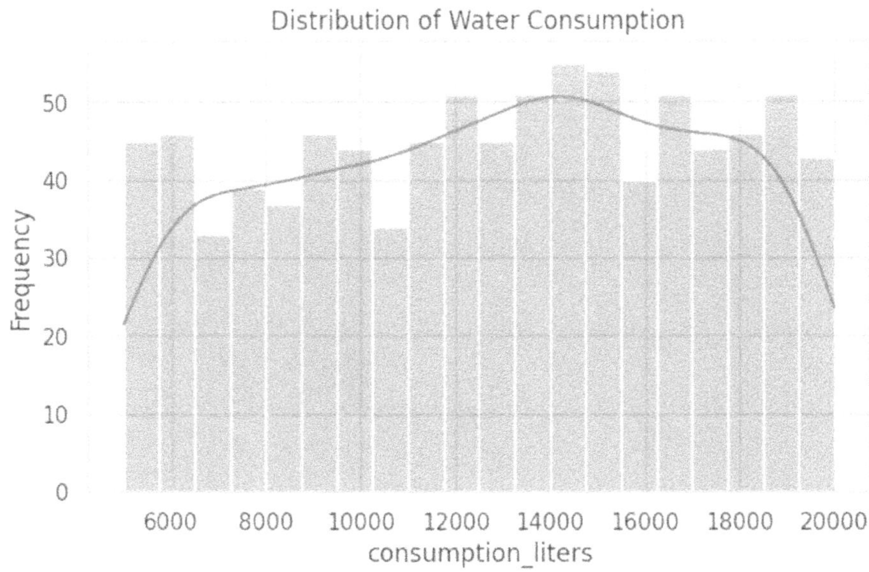

FIGURE 9.6 Distribution of water consumption.

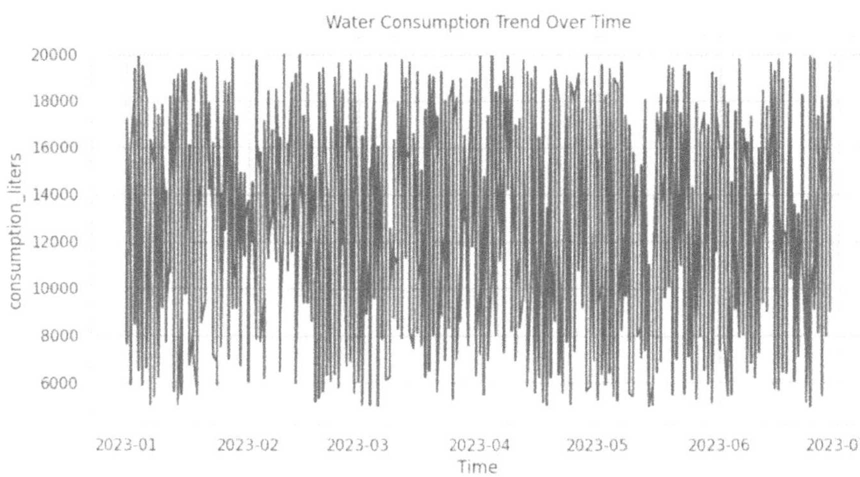

FIGURE 9.7 Water consumption trend over time.

the common consumption ranges, outliers, and skewness in the data. Such visualizations are crucial for understanding the underlying structure of the dataset before applying machine learning models. This study helps in feature scaling and normalization decisions. It also helps in identifying the model which would perform better. It supports in anomaly detection by highlighting rare consumption patterns.

Figure 9.7 presents raw daily water consumption data. It displays the time series from January 2023 to June 2023. The data shows that there is inherent volatility with consumption value ranging from ~5000 to ~20,000 liters daily, presenting a fourfold variation in daily consumption. The high variability shows that the data has a highly irregular pattern which can be influenced by many unpredictable factors. Visually it can be seen that there is a possible cluster of weekly or biweekly high consumption days which is followed by lower consumption days. These can be operational patterns, behavioral cycles, or weather patterns.

Figure 9.8 shows the daily water consumption patterns overlaid with a 6-month rolling mean for January 2023 to July 2023. The raw data shows that there is significant variability, with values fluctuating between ~5000 and ~20,000 liters. This volatility shows the inherent unpredictability associated with

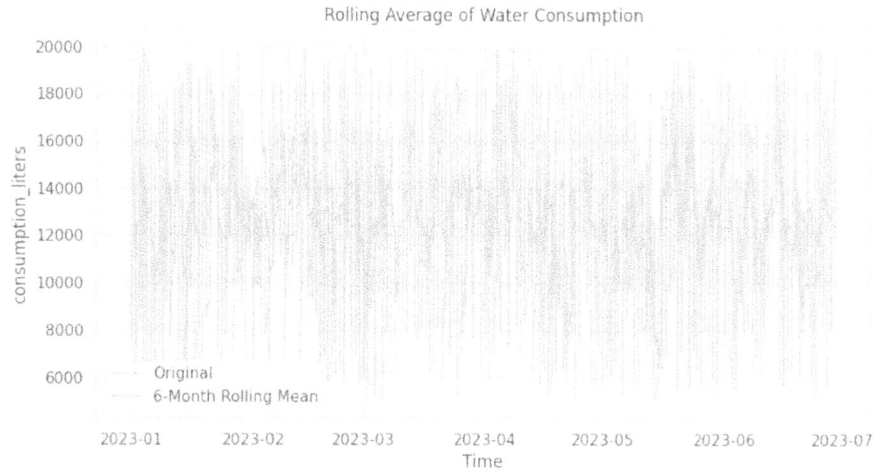

FIGURE 9.8 Rolling average of water consumption.

daily water consumption patterns. The consumption may be affected by weather conditions of the day, even the day of the week, and sporadic events planned. The six-month rolling mean is smoother, which helps in identifying the underlying trend in water consumption. The rolling average has a relatively stable trajectory. A slight upward trend can be observed from March to April with a gradual decline in the following period. This also aligns with the observation made from the box chart. There is significant variation in the mean value and the actual daily consumption which shows that it is challenging to forecast the water consumption.

(a) Implementation of RNN Model

Figure 9.9 shows the performance of the machine built with the Recurrent Neural Network (RNN) model in forecasting water demand. The model is forecasting water demand daily for 177 days.

The blue line represents the actual observed water consumption values, while the red line represents the prediction made by the model. It can be observed here from the graph that the model has captured the trend; however, its prediction has been in the range of ~10,000 liters to ~14,000 liters daily. It can be seen that the model has failed to capture the amplitude of fluctuation. The prediction has a slight lag; this

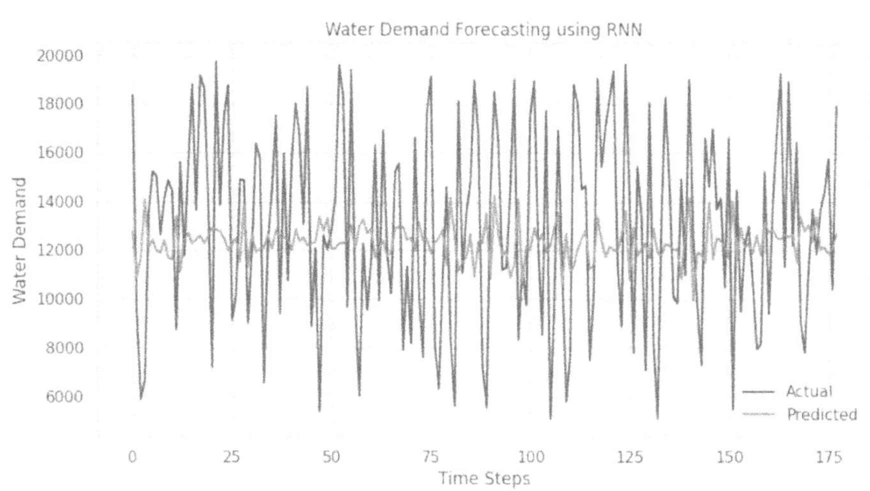

FIGURE 9.9 Water demand forecasting using RNN algorithm.

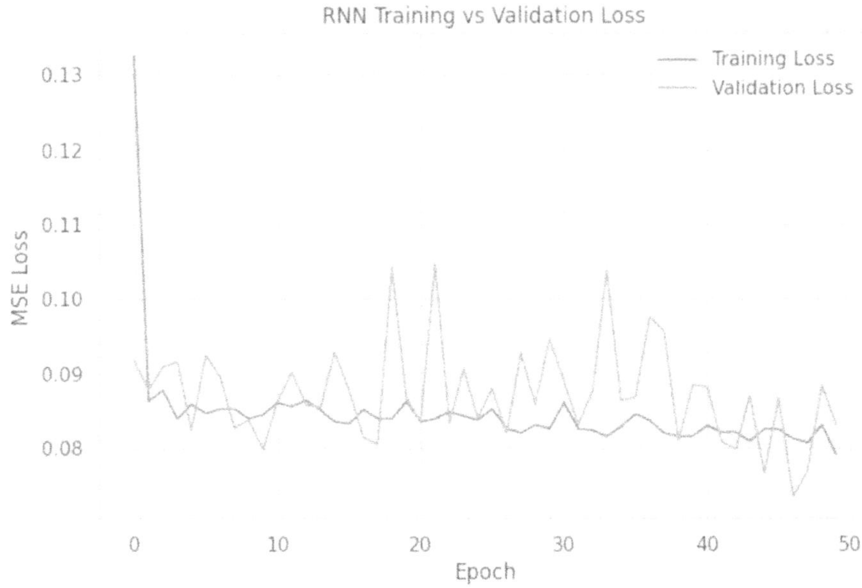

FIGURE 9.10 Training loss vs validation loss of the RNN machine.

delayed prediction is a characteristic of the RNN architectures. The RNN-based model has consistently made predictions in the small range that is of ~10,000 to ~14,000 liters, and the prediction has clustered around the mean level of ~12,000 to ~13,000 liters. The consistent failure in predicting the volatility can be problematic for operational planning, where accurate prediction of demand is critical.

Figure 9.10 shows the learning curve for the RNN model. It shows the training loss and the validation loss in MSE against the epochs of the machine built. The model shows the rapid learning in the first 3 epochs. The training loss drops to ~0.084 from ~0.133. This quick and steep decline shows that the machine has learned the basic pattern quickly. However, after the initial 3 epochs the training loss reduces to only ~0.082 in next 24 epochs and with 49 epochs the training loss reduces to ~0.079. It can be observed here that the training loss keeps getting reduced gradually but it is very slow after the first 3 epochs. The validation loss follows a more erratic pattern compared to the training loss. The validation loss here is not gradually falling as in the training loss. It is fluctuating in the range of ~0.074 to ~0.105 throughout the training process. The oscillation is significant between 15 and 40 epochs. It has notable spikes when epochs are 20, 22, and 35.

The validation loss fluctuates, which shows that the model has not reached a stable optimum level. The training curves reveal that the RNN model learns effectively from the training data, but it struggles to consistently generalize to unseen data which is evident even by the validation loss volatility.

(b) Implementation of LSTM Model

Figure 9.11 shows the performance of the machine built with the Long Short-Term Memory (LSTM) model in forecasting water demand. The model has made predictions for approximately 90 days.

The blue line represents the actual water consumption data, while the orange line displays the prediction made by the model. The prediction made by the model is fluctuating in the narrow band of ~12,000 to ~13,000, which is narrower than the prediction band of RNN model. The model captures the general consumption trend but has a very limited ability to predict the extreme variability of the actual data. The predicted consumption is in the range of ~12,000 to ~13,000 liters, while the actual consumption is in the range of 5000 to 20,000 liters, which shows that the model adopts a conservative forecasting strategy that prioritizes mean accuracy over capturing the complete range of possible outcomes. This limits the

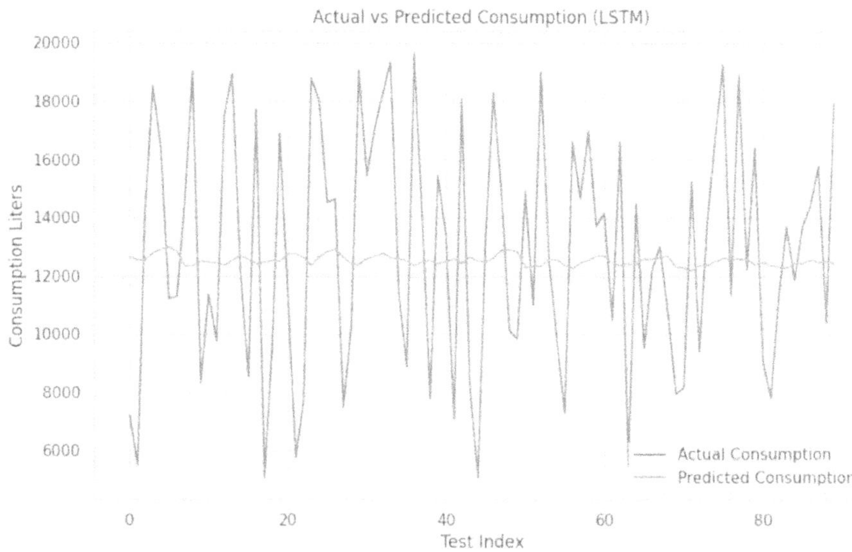

FIGURE 9.11 Water demand forecasting using LSTM algorithm.

FIGURE 9.12 Training loss of the LSTM model.

applicability of the model for capacity planning and resource allocation, where accurate prediction of the peak demand is essential.

Figure 9.12 shows the training dynamics of the LSTM model. It shows the training loss in Mean Squared Error (MSE) against the epochs of the machine built with the LSTM algorithm. The training loss here begins with the MSE of ~0.08365 which is notably lower than the RNN's initial loss which was ~0.133. The loss is gradually decreasing and forming lower high and lower low as the epochs are increasing. However, the lowest MSE it achieves is ~0.0822 which is not much lower than the initial value of ~0.08365 which shows a faint effect of increased epochs as compared to the effect observed in the machine built with the RNN algorithm.

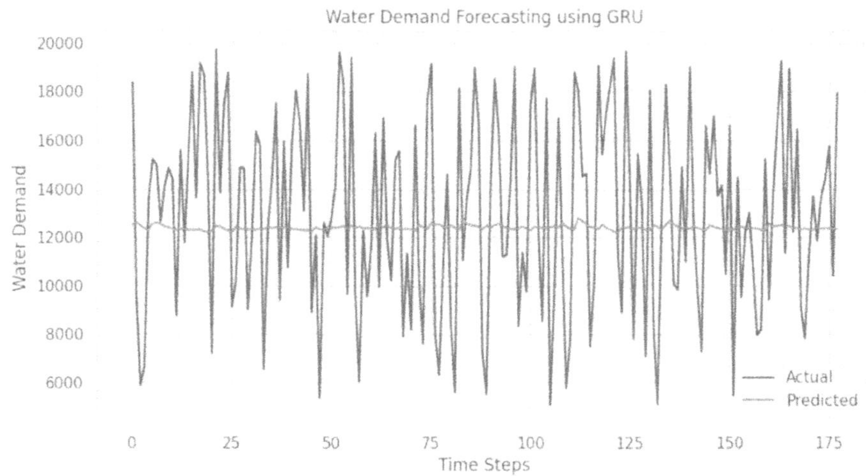

FIGURE 9.13 Water demand forecasting using GRU algorithm.

(c) Implementation of GRU Model

Figure 9.13 shows the performance of the machine built with the Gated Recurrent Unit (GRU) model in forecasting water demand. It predicts the water consumption for 177 days.

The GRU model predicts the water demand with similar characteristics as it was predicted with the RNN model and the LSTM model. The GRU prediction model has maintained the prediction in the narrowest range of ~12,000 to ~12,500 liters. The model fails to represent the substantial day-to-day variability. Like the RNN model and the LSTM model, the GRU model underestimates the low and high of water consumption. The predicted value loosely tracks the broader trends in the actual consumption. For example, from 100 to 130 days, the actual consumption value is slightly higher which is also followed by the predicted consumption. It shows that the model has learned to identify the seasonal pattern. The water consumption time series data contains substantial stochastic components that are inherently difficult to predict using historical patterns alone.

Figure 9.14 shows the learning curve for the GRU model. It shows the training loss and the validation loss in MSE against the epochs of the machine built. The GRU model shows the steepest learning curve initially. It has the training loss reduced from ~0.120 to ~0.086 in the first 4 epochs, which shows that the model has efficiently captured the primary pattern in the early training. However, by 30 epochs the MSE reduces to ~0.084 only, which is a very minute change. The MSE oscillates in a small range in its entire span and has not given any lower training loss value, which is lower than ~0.084 approximately.

The validation loss shows significant erratic behavior compared to the training loss. It drops from ~0.0845 to ~0.082 in the first few epochs. There is a persistent oscillation in the range of ~0.079 to ~0.0995 throughout the training. There are substantial spikes at 11 (~0.099), 38 (~0.099), and 45 (~0.0995) epochs. The amplitude of this fluctuation is substantial relative to the loss magnitude. The frequent spike shows the sensitivity of the model toward specific patterns or outliers in the validation set. It shows inconsistent generalization performance.

9.5 CONCLUSION

While seeing the training loss and validation loss chart of the model, it is observed that with the first model, the RNN model, the validation loss oscillates in a range and does not stabilize. It can be concluded

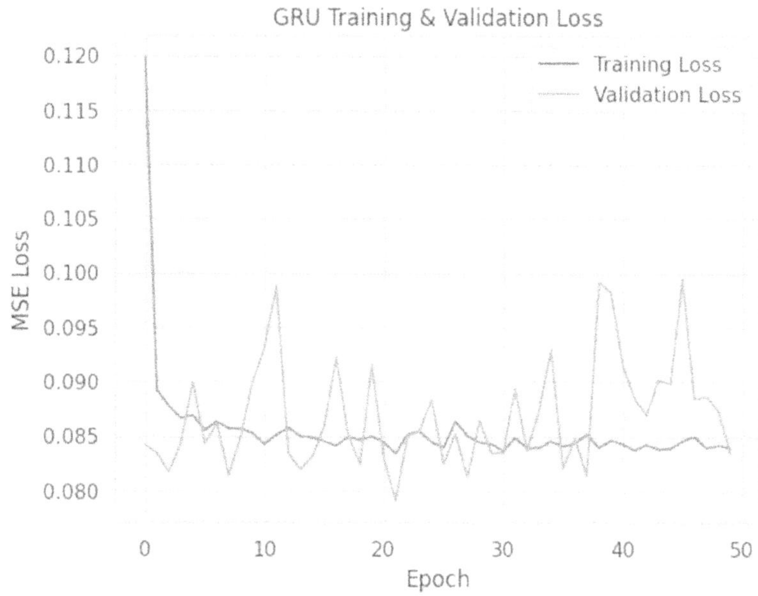

FIGURE 9.14 Training loss vs validation loss of the GRU machine.

from this that the machine struggles to consistently generalize to unseen data. Hence, there is a need for another sophisticated model. However, seeing the validation loss and training loss chart of the LSTM model, it can be seen that the LSTM model has more erratic behavior than the RNN model. The training loss is observed to be lowest at ~0.079 with 49 epochs in the RNN model, while it is lowest at ~0.0826 with 48 epochs in the LSTM model. The GRU model has also shown erratic behavior, exhibiting the validation loss against the epochs. There is a persistent oscillation in the range of ~0.079 to ~0.0995 throughout the chart. Hence, it can be concluded that all the machines have nearly comparable performance; however, the RNN model has shown slightly better performance.

The similarity between the machine built with GRU, LSTM, or RNN model prediction shows that the selection of architecture may not be the solution to optimum prediction, but the problem lies with the inherent uncertainty of the data, and the insufficient input features to explain consumption variability. There is possibly a need for probabilistic forecasting approaches rather than point prediction.

All the architectures converge to similar means following strategies, indicating that improving forecast accuracy for the datasets may require alternative approaches such as ensemble methods, hybrid modeling, probabilistic forecasting, and incorporation of variables, i.e., weather, day of week, or holidays.

REFERENCES

1. U.S. Geological Survey. (2019). *The Distribution of Water on, in, and Above the Earth.* U.S. Department of the Interior. https://www.usgs.gov/media/images/distribution-water-and-above-earth
2. Ritchie, H., & Roser, M. (2024). *Water Use and Stress. Our World in Data.* Retrieved October 12, 2025, from https://ourworldindata.org/water-use-stress#:~:text=A%20growing%20global%20population%20and,plateauing%2C%20or%20at%20least%20slowing
3. World Health Organization. (2023, September 13). *Drinking-water (Key Facts).* Retrieved October 12, 2025, from https://www.who.int/news-room/fact-sheets/detail/drinking-water#:~:text=,and%20numerous%20neglected%20tropical%20diseases

4. United Nations Children's Fund (UNICEF). (n.d.). *Water Scarcity (Key Facts).* Retrieved October 12, 2025, from https://www.unicef.org/wash/water-scarcity#:~:text=,be%20living%20in%20areas%20of
5. World Health Organization. (2023, September 13). *Drinking-water (Key Facts).* https://www.who.int/news-room/fact-sheets/detail/drinking-water
6. United Nations General Assembly. (2010). *Resolution A/RES/64/292: The Human Right to Water and Sanitation.* https://www.un.org/en/ga/search/view_doc.asp?symbol=A/RES/64/292
7. United Nations. (n.d.). *Sustainable Development Goal 6: Clean Water and Sanitation.* https://sdgs.un.org/goals/goal6
8. Ahmad, T., & Ali, S. (2021). Machine Learning Applications in Water Resource Management: A Review. Water Resources Management, 35(14), 4591–4612. https://doi.org/10.1007/s11269-021-02890-1
9. Villarin, M. C., & Rodriguez-Galiano, V. F. (2019). Machine Learning for Modeling Water Demand. *Journal of Water Resources Planning and Management*, 145(5), 04019017. https://doi.org/10.1061/(ASCE)WR.1943-5452.0001067
10. Lee, D., & Derrible, S. (2020). Predicting Residential Water Demand with Machine-Based Statistical Learning. *Journal of Water Resources Planning and Management*, 146(1), 04019019. https://doi.org/10.1061/(ASCE)WR.1943-5452.0001119
11. Ali, M., Deo, R. C., Xiang, Y., Li, Y., & Yaseen, Z. M. (2020). Forecasting Long-Term Precipitation for Water Resource Management: A New Multi-Step Data-Intelligent Modelling Approach. *Hydrological Sciences Journal*, 65(16), 2693–2708. https://doi.org/10.1080/02626667.2020.1808219
12. Yi, S., Kondolf, G. M., Sandoval-Solis, S., & Dale, L. (2022). Application of Machine Learning-Based Energy Use Forecasting for Inter-Basin Water Transfer Project. *Water Resources Management*, 36(14), 5675–5694. https://doi.org/10.1007/s11269-022-03326-7
13. Ahmed, A. A., Oyedele, L., Sayed, S., Abdoulhalik, A., & Moutari, S. (2024). Applications of Machine Learning to Water Resources Management: A Review of Present Status and Future Opportunities. *Journal of Cleaner Production*, 441, 107151. https://doi.org/10.1016/j.jclepro.2022.107151

10 Machine Learning in Weather Forecasting and Water Management

Chetan Sharma and Sakshi

10.1 INTRODUCTION

The sphere of weather forecasting and water management has evolved greatly and impressively over the ages. The evolution of the study of cloud formations and natural pointers has been massive since the beginning of the study itself and has seen the incorporation of high-quality satellite imagery and computer simulations and models, leading to a process of constant innovation [1]. This chapter examines the practical role of machine learning (ML) in the development of meteorology and hydrology and its transformational nature in the respective disciplines. Previously weather prediction was mostly based on the knowledge and experience of the weather experts who visually observed patterns in the sky, wind directions, and air pressure variations. It is a habit of farmers, sailors, and early-day meteorologists to monitor nature keenly and subsequently predict which weather changes are going to occur in the near future. On the other hand, traditional water management was mainly the process of responding to emergency problems like flood prevention and irrigation that often relied on past trends and local experience.

The change came with the rise of technological advances. The twentieth century brought in radar, weather satellites, and automated sensors, thus causing a major revolution in the field of weather forecasting and water management. These tools provided a plethora of previously unimaginable data, allowing for more accurate and more timely predictions. However, the large amount and complexity of this information posed some new challenges: how to manage, question, and analyze this information effectively and accurately? Machine learning is a subdivision of artificial intelligence, which allows systems to learn and progress through experience without necessary programming [1]. ML algorithms have been vital to meteorology to analyze complex meteorological data sets, discover patterns, and make predictions with increased accuracy like never before. Data gathered using satellites and ground sensors is analyzed using ML methods, enabling forecasters to make more precise predictions and give advanced warnings of meteorological events like hurricanes and heatwaves.

DOI: 10.1201/9781003742227-10

ML has transformed hydrology practices in the management of water. The algorithms of machine learning can analyze previous information about rainfall, river levels, and consumption of water to project the availability of water in the future. This enables effective management of water resources to be proactive. Such forecasts are crucial in agricultural planning, water resource management in the city, and predicting extreme events like floods and droughts [2].

Moreover, machine learning (ML) is taking a central role in climate change studies, and it helps researchers to understand and predict its impact on weather trends and water resources. Machine learning models can determine trends and give predictions on how the climatic conditions are going to be in the future through examination of large volumes of data. This fact can further be utilized towards the formulation of policies intended to mitigate the effects of climate change and adaptation to its effects. Moving deeper into the twenty-first century, the integration of machine learning into the weather prediction and water management industry marks a paradigm shift in how things are done [2]. It is not necessary to focus on how much more data to get, but rather to use data in a smarter way. With machine learning, it becomes possible to make use of the capabilities of big and complex data and convert it into useful and valuable information. This research is an important step towards the improvement of our understanding and ability to interact with our natural environment, therefore, allowing the making of more informed decisions in the areas of weather forecasting and water resource management. These issues will be widely discussed in this chapter, as they will prove the necessity of machine learning in the sphere of meteorology and hydrology. It will also point out how machine learning is constantly changing the way we understand and manage the ever-changing dynamics of weather and water.

10.2 PRINCIPLES OF MACHINE LEARNING

One of the examples of technical improvement is the use of machine learning (ML) in weather forecasting and water management. To be able to comprehend the importance of machine learning in different fields of work, it is necessary to know the fundamental concepts of machine learning [3]. In its essence, machine learning involves training determine programs to foresee, or to choose, based on information analysis. This methodology has a number of basic principles and methodologies.

- **Supervised Learning:** This is a learning process that employs labeled data to train models to make predictions relying on past cases. One such example is that a model might learn to predict the rainfall by using past weather records alongside the current rainfall measurements [4].
- **Unsupervised Learning:** Unlike supervised learning, unsupervised learning involves the training of models based on data not pre-labeled. The models learn to identify the patterns and correlations in the data automatically, which plays a major role in identifying unusual weather patterns or new variables in climate modeling [4].
- **Reinforcement Learning:** It involves training models based on a reward and punishment mechanism. This environmental application may involve improvement of efficiency and effectiveness of water distribution networks by encouraging conservation [4].
- **Deep Learning:** This is a sub-field of machine learning that implies the application of multilayered neural networks, commonly referred to as deep networks, in data interpretation and analysis. This particularly comes in handy when dealing with complex data, such as that which has been captured by satellite photography, to predict weather conditions or assess water resources.
- **Decision Trees and Random Forests:** These are powerful methods that are applied in both classification and regression exercises. These tools could be used in the sphere of hydrology to predict the likelihood of floods or to estimate the impact of various factors on the quality of water.

- **Support Vector Machines (SVM):** SVMs are very effective when it comes to solving classification and regression problems. The first solution may be to classify land territories with respect to drought or flood-prone areas.

Natural Language Processing (NLP): It has commonly been associated with its use in the analysis of language. Nevertheless, it can also be used in the analysis of textual data, including weather reports, to determine important patterns and make predictions.

Clustering and Dimensionality Reduction: The techniques simplify the complex environmental data and help to find out the major trends and patterns that are relevant to managing weather conditions and water resources.

Time Series Analysis: It is an important technique in the meteorological and hydrological fields to study data over a time frame so that one can predict future events, including rainfall patterns or river-water levels.

All these strategies have a clear and useful contribution to the understanding and successful approach to the environmental challenges posed by weather and water resources. They are more accurate in the forecasts, optimize the resources, and are more prepared for extreme situations [5].

10.3 WEATHER FORECASTING WITH THE HELP OF MACHINE LEARNING

Machine learning (ML) applied to weather forecasting is an indication of a significant leap in our ability to predict weather trends with greater precision. In this part of the chapter, the authors explore how the concept of machine learning can be applied in the development of predictive weather patterns with convincing case examples that demonstrate the success in this area [6].

10.3.1 Weather Patterns Predictive Models

ML models have revolutionized weather forecasting and are trained on large volumes of data collected using a variety of data sources such as satellites, weather stations, and ocean buoys. Neural networks, decision trees, and ensemble techniques are some methods used in these models to forecast the meteorological conditions. Neural networks are also good at addressing the nonlinear nature of weather data which makes them suitable for the prediction of complex weather phenomena like hurricanes and thunderstorms.

The ability of ML models to develop and improve with time is a major strength compared to traditional methods. As more data become increasingly available, these models can be retrained to be more accurate. Moreover, the ML models can incorporate other types of data, e.g., historical weather data, real-time measurements of sensors, and even social media trends, to generate comprehensive and comprehensive weather predictions [6].

10.4 CASE STUDIES: WEATHER PREDICTION SUCCESS

- **The European Centre for Medium-Range Weather Forecasts (ECMWF):** It is a machine learning method that is employed to improve the accuracy of weather forecasts at a medium range. Their strategy, incorporating machine learning into the traditional numerical weather prediction models, has proved to increase the capacity to forecast extreme weather features in the next several days by a large margin.

- **National Oceanic and Atmospheric Administration (NOAA):** Predicting the path of hurricanes through machine learning (ML) has been game-changing at NOAA. The machine learning models have helped to improve the accuracy of predicting the path of the hurricane through the analysis of past hurricane data, and in that way, provide the chance to make the evacuation plan and resource assignment timely.
- **Deep Thunder by IBM:** The project demonstrates what machine learning can do in highly localized weather forecasting. Deep Thunder applied computerized projections and machine intelligence to create weather forecasts of high accuracy and fine-tuned information. These predictions came in handy as far as urban planning and emergency response were concerned.
- **The AI Flood Forecasting of Google:** With the help of machine learning (ML), Google and the Indian government predicted floods on the Ganges and Brahmaputra rivers. This approach provided accurate information on floods which proved crucial in the disaster evacuation and management process.

10.5 IMPROVING WEATHER FORECASTING ACCURACY

Machine learning (ML) has been essential to enhance the accuracy of weather forecasts. The main way these results are achieved is through the use of modern algorithms and sophisticated methods of data processing, as well as including satellite and ground data. These elements are critically discussed in this part of the text.

10.5.1 State-of-the-Art Algorithms and Data Processing Methods

The ability of the ML algorithms to process and analyze complex weather data is based on their advancedness. Some of the basic algorithms and strategies are displayed in Figure 10.1.

1. **Convolutional Neural Networks (CNNs):** CNNs are widely used in image recognition functionality and demonstrate an extraordinary level of performance regarding the ability to analyze satellite images and radar data to identify meteorological phenomena such as night clouds, weather systems, and the formation of cyclones.
2. **Recurrent Neural Networks (RNNs):** RNNs are effective at processing sequential data and can therefore be applied to time-series data in weather forecasting. They can predict the future weather with the help of past weather data.

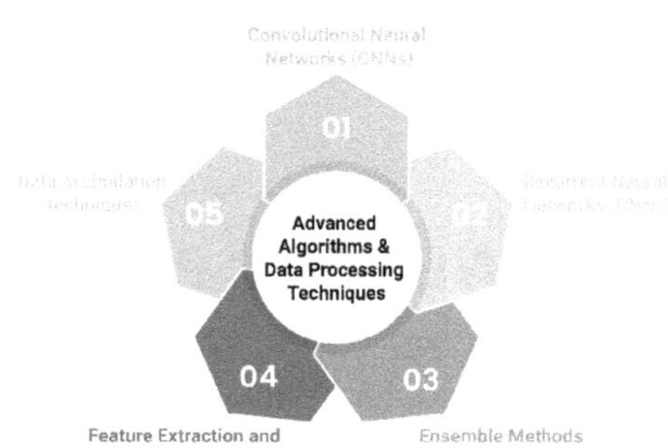

FIGURE 10.1 State-of-the-art data processing and algorithms.

3. **Ensemble Methods:** These are methods that join a number of machine learning models to enhance the precision and reliability of the forecasts. Ensemble techniques are highly beneficial when weather forecasting because they incorporate information from multiple datasets and models, resulting in more robust and reliable redictions.
4. **Feature Extraction and Dimensionality Reduction:** Methods such as Principal Component Analysis (PCA) are applied to reduce complex meteorological data in order to locate important elements that determine the weather patterns.
5. **Methods of Data Assimilation:** It is instrumental in using observational data as an input for weather models. Techniques like the Kalman filtering and variational techniques can be used in order to enhance the model with the help of real-time data and, therefore, enhance the accuracy of predictions.

10.5.2 Combining Ground Data and Satellite to Make Better Predictions

The combination of ground and satellite information is important in supplementing the accuracy of the forecast. The satellite information gives an all-around and broad view of weather patterns, including cloud coverage, temperature, and other atmospheric factors. Ground data, on the other hand, has the particular and detailed data on the weather, including temperature, humidity, and wind speed, that are received at weather stations of a particular location [7].

1. **Data Fusion:** This is the procedure to combine both satellite and ground data to create an all-around and more accurate representation of the meteorological conditions. Machine learning algorithms are able to easily combine these heterogeneous data sources, using their own merits.
2. **Real-Time Processing of Data:** With machine learning (ML), real-time information about ground stations and satellites can be processed and assessed rapidly which can lead to accurate and timely weather predictions. It is particularly essential in the prediction and management of extreme weather conditions.
3. **Predictive Analytics:** Predictive analytics may use both satellite and ground data to ensure that the ability to predict meteorological phenomena is much more accurate. The ML models can predict the course and intensity of storms correctly, analyzing the changes in atmospheric states registered by satellites with the help of the data obtained in the ground stations.

   In such a manner, the precision of weather predictions has been significantly enhanced with the use of the most advanced machine learning algorithms and techniques, along with the consideration of the combination of satellite and ground data. The development is not only useful in daily weather forecasting but also essential in planning and controlling extreme weather conditions, which ultimately leads to safety and preparedness. Due to the constant advancement of technology, there is a tremendous possibility of other advancements in the area of meteorology which offer exciting vistas into the future.

10.6 WATER MANAGEMENT BASED ON MACHINE LEARNING

Machine learning (ML) in water management has led to innovative approaches to the major problems in this area. In this part of the chapter, the author will examine the real-life applications of machine learning to the distribution and saving of water resources, and how it can be used to forecast floods and droughts [8].

10.6.1 Water Resource Allocation and Conservation

1. **Predictive Analysis of Water Demand:** It is possible to train machine learning to analyze trends in water usage by the various sectors of the economy, such as agriculture, industry, and domestic water usage, to predict future water needs. The use and availability of water resources should be sustainable enough and must receive the highest priority in terms of optimal resource allocation of water resources [9].
2. **Water Quality Monitoring:** The analysis of collected data through machine learning is now applied to sensors installed in rivers, lakes, and reservoirs, which allows monitoring water quality in real-time. This involves the detection of pollutants or changes in the PH levels that are essential for human consumption, as well as the health of the ecosystems.
3. **Irrigation Management:** ML models can be used to increase irrigation schedules and optimize water allocation in agriculture. By analyzing the information related to the soil moisture, weather forecasts, and types of crops, these models can suggest highly effective irrigation plans and, therefore, reduce water wastage and improve harvests [9].
4. **Detection of Leaks in a Distribution System:** Machine learning algorithms can determine trends that indicate leakages in water distribution systems. By analyzing flow rate and pressure data, these systems can effectively pinpoint the precise locations or points of leakages, hence making it easier to repair them within a short time and to reduce water wastage.

10.6.2 New Solutions to Flood and Drought Forecasting

1. **Flood Forecasting Techniques:** Machine learning techniques are being developed in order to ensure more accurate predictions of floods. These models make use of the river gauges, rainfall trends, and topographical maps data to forecast the occurrence and severity of flooding. This makes evacuations quick and the catastrophe preparedness more effective [10].
2. **Drought Prediction and Management:** Machine Learning (ML) is very instrumental in predicting drought. By analyzing the past weather conditions, the degree of moisture in the soil, and satellite images, the ML models can forecast drought events. This contributes to coming up with early response and mitigation measures.
3. **Climate Change Impact Assessment:** The implementation of machine learning technologies is being used to model the impact of climate change on water supply. This includes the prediction of changes in the patterns of precipitation, melting of the snow, and river flows, which are essential for the formation of water management policies in the long term [10].
4. **Integrated Water Systems Modeling:** The advanced machine learning strategy is underway to create compound frameworks that consider a number of elements, including land use, population growth, and climate change. The models provide a wide understanding of sustainable management of water resources.

10.7 ONLINE WATER QUALITY SURVEILLANCE

Implementing machine learning (ML) in water quality monitoring is an important development in the environmental management industry at present. ML has made it possible to implement and use data-driven techniques to assess and maintain water quality, therefore establishing new opportunities. This part of the chapter focuses on the use of machine learning in water quality analysis and provides real-life examples to illustrate its impact [11].

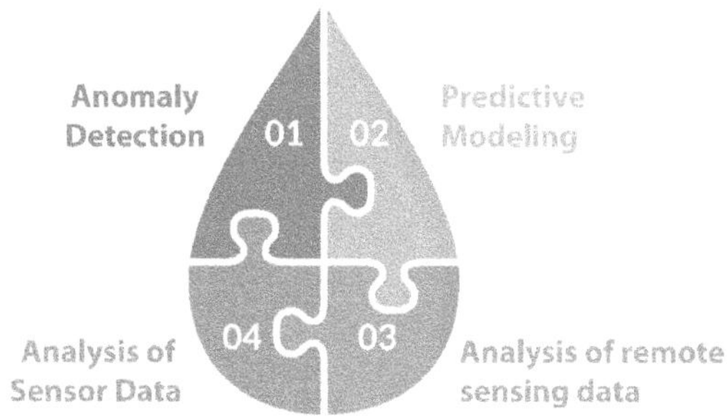

FIGURE 10.2 Water quality analysis by machine learning.

10.7.1 Machine Learning Used to Analyze Water Quality

The water quality can be monitored using the techniques of machine learning in several critical ways which are depicted in **Figure 10.2.**

1. **Predictive Modeling:** Machine learning-based algorithms can be used to predict possible cases of contamination based on trends and patterns of previous data on water quality. This preventive approach allows taking quick measures before the worsening of water quality.
2. **Remote Sensing Data Analysis:** With the help of satellite and aerial imaging, which is analyzed through the use of machine learning systems, it is feasible to observe water bodies on a large scale. Machine learning algorithms can recognize any changes in the color of the water, its degree of turbidity, and the existence of overgrown algae, which are signs of a number of issues connected with the quality of the water [12].
3. **Sensor Data Analysis:** Complex sensors that are used in water systems generate enormous amounts of data. ML models are also efficient in processing this data to provide real-time information on water quality indicators, including pH, dissolved oxygen, and pollutant concentration.
4. **Anomaly Detection:** Machine learning algorithms can point out anomalies in the data on water quality, and these anomalies signify potential cases of downstream pollution or some kind of system malfunction. Early detection is necessary so that it can reduce any dangers to the health of the people [12].

10.7.2 Worldwide Impact and Real-Life Examples

1. **The Chesapeake Bay Program (USA):** It is a program that monitors and improves the water quality of the Chesapeake Bay based on machine learning. ML models are based on several data sources, including river flow, land use, and pollution loads, to provide meaningful information that can be used to develop conservation and restoration efforts. The project has registered massive improvements in water transparency and a reduction in nutrient contamination.
2. **Smart Water Management System Amsterdam (the Netherlands):** Amsterdam smart water management system uses machine learning to analyze data resulting from a network of sensors located across the water system of the city. This approach has led to better detection and remediation of the sources of pollution, therefore improving the overall health of the aquatic life of the city.
3. **Lake George Initiative (USA):** This project utilizes ML to monitor the water condition in Lake George. These models study a number of features like the nutrient levels, invasive species, and the water circulation with the help of the information gathered by over 500 sensors. This initiative has been imperative in preserving the environment of the lake and has become a model of conservation of fresh water.

4. **Ganga River Pollution Monitoring (India):** The Indian government has used the machine learning method to monitor and improve the water quality of the Ganges River. The machine learning models use data provided by multiple sources to identify locations with high levels of pollution and assess the effectiveness of the remediation process. As a result, water quality has been improved in various parts of the river.

10.8 CHALLENGES AND LIMITATIONS

Weather forecasting and water management machine learning (ML) have numerous challenges and constraints with regard to their usefulness and reliability.

1. **Data Quality and Heterogeneity:** Data quality and heterogeneity are significant problems. Good ML models require large and high-quality data. But the presence of data discrepancies, empty values, and errors may significantly bias the model outputs, leading to erroneous predictions. Satellite images, sensor data, and history are data sources that make training and deploying models more complicated.
2. **Computational Needs:** It is extremely intensive to work with large volumes of data and execute complex ML models. This may be an issue in real-time applications or resource-constrained applications that require quick data processing.
3. **Model Interpretability and Trust:** Interpretability and complexity normally do not go hand in hand. Complex models can be more precise, but they are less obvious, and thus predictions are difficult to interpret and believe in. This is needed in situations of high stakes such as catastrophe management.
4. **Generalization and Overfitting:** ML models that were trained through a particular dataset might not be useful when predicting different regions or scenarios in the future, particularly climate change. The model can be less generalizable to other data sets by overfitting to particular data sets.
5. **Ethical and Environmental Issues:** The ethical problems are data privacy and ML models' biases. Of importance are environmental concerns, particularly the carbon footprint of the large-scale ML operations.

10.9 SOLVING DATA QUALITY AND AVAILABILITY PROBLEMS

1. **Data Quality:** The quality of training and analysis data has a significant impact on the ML model. The incomplete, biased, or inaccurate information may result in poor predictions and judgments. The bad satellite images or sensor data used as weather forecasting models may lead to inaccurate predictions.
2. **Data Availability:** Complete and varied data is needed by ML. Nonetheless, data accessibility may be challenging, particularly in developing countries that have low monitoring systems. Such locations do not have historical weather and water statistics, and hence the ML models do not work.
3. **Data Integration:** It is not an easy task to combine data from numerous sources that have a wide variety of formats and standards. Sensors, satellites, and stations provide weather

and water data that have to be preprocessed to be used in ML models, and it needs computer resources and experience.

4. **Temporal and Spatial Resolution:** Weather predictions should be accurate and based on the spatial and temporal resolution. The limitations of ML-based projections are the data granularity that is hard to obtain.

10.10 FUTURE DIRECTIONS

10.10.1 Machine Learning and the Environment: New Trends

Machine learning (ML) has recently become a massive change in the discipline of environmental science. The future of machine learning in this domain is due to many nascent developments that are offering new techniques and solutions to complex environmental problems. A few of the new trends are presented in Figure 10.3.

1. **Deep Learning in Climate Modeling:** Deep learning models are increasingly applied in climate modeling to enhance the models. Such neural network-based models are capable of analyzing large volumes of various climatic features, which lead to more accurate and detailed climate projections.
2. **IoT and ML Integration:** The Internet of Things (IoT) is changing how data is collected in the environmental research area. Incorporated with IoT devices, ML provides real-time and high-resolution data, allowing more dynamic and more precise monitoring of the environment and decision-making.
3. **In Satellite Imagery:** Machine learning algorithms are being enhanced to identify features in satellite images, like the change in land cover, deforestation, or glacier motion. This effect simplifies fast and accurate monitoring of the environmental changes on a global scale.
4. **ML Algorithms That Use Less Energy:** With the growing concerns about the environmental impacts of computers, there is an existing trend of developing lower-energy-consuming ML algorithms. These algorithms aim at making the calculations of ML as less harmful to the

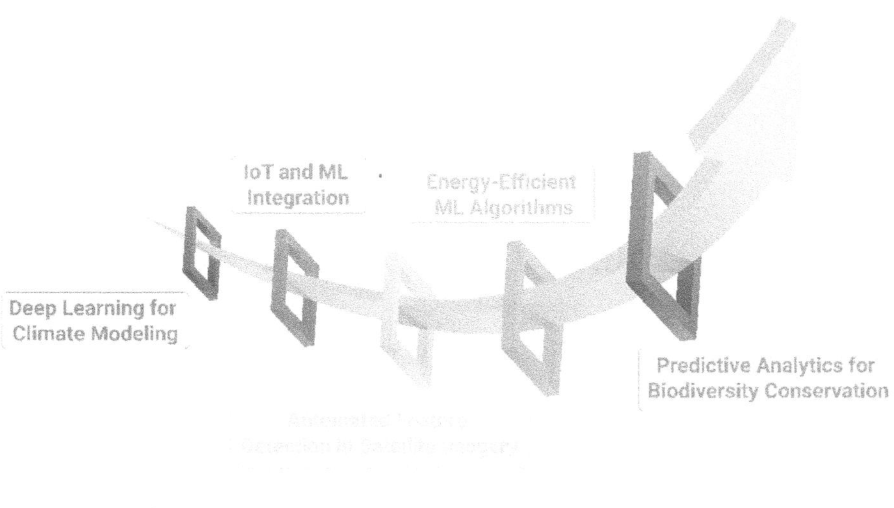

FIGURE 10.3 Machine learning in environmental science.

environment as possible by lowering carbon emissions, therefore facilitating the development of a balance between technical advances and environmental sustainability.

5. **Predictive Analytics of Biodiversity Conservation:** Machine learning (ML) will play a key role in biodiversity conservation because it will allow building prediction tools that will be able to assess the risk posed to various species and ecosystems. The models help in creating active conservation programs.

   All these developments indicate that there are positive and beneficial integrations in the application of machine learning to the area of environmental research and that the primary focus should be on creating viable, sustainable, and efficient solutions to environmental emergent issues.

10.11 THE FUTURE BREAKTHROUGHS IN WEATHER AND WATER MANAGEMENT

The intersection of machine learning (ML) and weather prediction and water management is about to experience significant improvements that can change both fields altogether. One of the opportunities that can be exploited is the development of hyper-localized weather forecasting models. Through high-resolution data and advanced machine learning methods, these models can provide accurate weather forecast at a micro-scale, e.g., individually to a local neighborhood or to specific agricultural fields [13]. The high level of detail will go a long way in enhancing local planning and response strategies, particularly in areas that are prone to extreme weather events. ML will radically transform our understanding of water management, i.e., in regard to water resource conservation and distribution. Future developments would be the development of intelligent water distribution systems where dynamism is established, in which real-time demand and supply data can be used to allocate water resources, which allows the efficiency to be increased significantly, and the wastage to be reduced. Besides, it is expected that machine learning-based water quality prediction models will be developed to an even higher level so that timely detection of contaminants and response to incidents of contamination can be implemented. Another interesting development is the integration of machine learning (ML) with the Internet of Things (IoT) technology in environmental monitoring [14]. It is possible that with the introduction of smart sensors to the water bodies and weather stations, which would transmit data to machine learning models, we would have constant and real-time monitoring and control of the environmental conditions. The effect of this will be more proactive and efficient decision-making processes.

Combined efforts of machine learning, big data analytics, and high-tech sensor technologies resulted in the massive progress of weather prediction and water management. These changes are the emergence of a new era, which is more accurate, efficient, and sustainable.

10.12 CONCLUSION

The machine learning (ML) application in the world of weather prediction and water management has demonstrated new earth-shattering discoveries, which means that the next level of forecasts and resource management will be much more accurate, efficient, and flexible. The important implications have included the possibility of machine learning to automatically work with big data, with numerous information sources, and apply advanced algorithms to obtain correct forecasts. Machine learning (ML) has played a major role in enhancing the precision of weather event prediction in weather forecasting. Moreover, ML has revolutionized the allocation of resources as well as quality control in water management.

Further evolution of machine learning in such areas in the future is aimed at attaining even greater integration and complexity. The expected improvements will focus on enhancing the quality of information processing in real-time, expanding the accuracy of predictions, and increasing access to machine learning technologies at different locations and among different groups of people. Data quality and the ethical consequences of the use of technology are the topics that must be subject to ongoing investigation. Also, it is imperative to ensure that there is a balance between the environmental impact of infrastructures used in processing data and the benefits generated.

With machine learning (ML), meteorology, and hydrology teaming up as we travel into the future, the ability to predict even more will be possible not only with better prediction models but also with a deeper understanding of the world around us. This allows the potential to take more informed choices and carry out environmentally conscious actions, which is a significant milestone in the existing relationship of human to the natural world.

Environmental sciences are in the middle of a phenomenal transformation, which has been enhanced by the current technological development. To be more specific, the so-called machine learning (ML) takes the lead of this change as it changes how we understand and interact with weather and water systems of the world we live in. This development includes not just the advancements in computer size or data collection methods but a total shift in the way we look at issues related to the environment.

As technology has improved, the machine learning models are also becoming sophisticated, and they can process large and complex data sets, which would have been unaffordable to us only a short time ago. With this technological advancement, it is possible to produce more advanced and accurate models of weather conditions and circumstances concerning water management. One such case is the integration of real-time data processing and predictive analytics which has significantly enhanced the precision of predictions, leading to an enhancement in preparedness for natural catastrophes.

In such a manner, the technology is enabling a higher level of collaboration and integration in most areas of the field of environmental sciences. The combination of machine learning with other fields like remote sensing, climate science, and hydrology enables the development of innovative solutions, including the development of large-scale climate models that can predict long-term environmental impacts.

The current development of technology in environmental sciences is not only an academic or scientific endeavor but also a crucial step toward a stronger and more sustainable intercourse with our environment. The more we advance, the larger the ability of technology to address the complex challenges faced by environmental growth, and this promises new possibilities to overcome the new challenges.

REFERENCES

1. A. Y. Sun and B. R. Scanlon, "How can big data and machine learning benefit environment and water management: a survey of methods, applications, and future directions," *Environ. Res. Lett.*, vol. 14, no. 7, p. 73001, 2019.
2. I. Gad and D. Hosahalli, "A comparative study of prediction and classification models on NCDC weather data," *Int. J. Comput. Appl.*, vol. 44, no. 5, pp. 414–425, 2022.
3. M. Drogkoula, K. Kokkinos, and N. Samaras, "A comprehensive survey of machine learning methodologies with emphasis in water resources management," *Appl. Sci.*, vol. 13, no. 22, p. 12147, 2023.
4. E. F. Morales and H. J. Escalante, "A brief introduction to supervised, unsupervised, and reinforcement learning," in *Biosignal processing and classification using computational learning and intelligence*, Elsevier, 2022, pp. 111–129.
5. K. Sharifani and M. Amini, "Machine learning and deep learning: A review of methods and applications," *World Inf. Technol. Eng. J.*, vol. 10, no. 07, pp. 3897–3904, 2023.

6. D. Markovics and M. J. Mayer, "Comparison of machine learning methods for photovoltaic power forecasting based on numerical weather prediction," *Renew. Sustain. Energy Rev.*, vol. 161, p. 112364, 2022.
7. J. Qin, H. Jiang, N. Lu, L. Yao, and C. Zhou, "Enhancing solar PV output forecast by integrating ground and satellite observations with deep learning," *Renew. Sustain. Energy Rev.*, vol. 167, p. 112680, 2022.
8. F. Ghobadi and D. Kang, "Application of machine learning in water resources management: a systematic literature review," *Water*, vol. 15, no. 4, p. 620, 2023.
9. H. Kavand, S. Ziaee, and M. Mardani Najafabadi, "The impact of water conservation policies on the reallocation of agricultural water-land resources," *Front. Water*, vol. 5, p. 1138869, 2023.
10. N. Nandgude, T. P. Singh, S. Nandgude, and M. Tiwari, "Drought prediction: a comprehensive review of different drought prediction models and adopted technologies," *Sustainability*, vol. 15, no. 15, p. 11684, 2023.
11. Y. Zhao, M. Chen, J. He, and Y. Ma, "Monitoring water quality parameters using multi-source data-driven machine learning models," *Eng. Appl. Comput. Fluid Mech.*, vol. 19, no. 1, p. 2509658, 2025.
12. M. Zhu *et al.*, "A review of the application of machine learning in water quality evaluation," *Eco-Environ. Heal.*, vol. 1, no. 2, pp. 107–116, 2022.
13. K. A. Adkins *et al.*, "Hyper-local weather predictions with the enhanced general urban area microclimate predictions tool," *Drones*, vol. 7, no. 7, p. 428, 2023.
14. T. Miller *et al.*, "Integrating artificial intelligence agents with the internet of things for enhanced environmental monitoring: applications in water quality and climate data," *Electronics*, vol. 14, no. 4, p. 696, 2025.

Climate-Resilient Strategies for Sustainable Water Management

11

Indu Bala

11.1 INTRODUCTION TO CLIMATE-RESILIENT WATER MANAGEMENT

Climate-Resilient Water Management System refers to the development of applications or strategies that will enable water, both natural or artificial, to withstand, adapt, and recover from the impacts of climate change. With the increase in global temperature, the weather patterns are continuously becoming unpredictable, and therefore, the water resources are also facing unprecedented stress. Ensuring reliable water availability for households, agriculture, industries, and ecosystems requires a shift from traditional management practices to advanced and scientifically proven approaches. In this chapter, we have introduced the context, challenges, and motivation for the strengthening of climate resilience in water resource planning and governance [1].

11.1.1 Global Water Challenges in a Changing Climate

The water systems all over the world are facing severe pressure due to climate-induced changes. In many countries, they are experiencing reduced rainfall, declining river flow, and a shrinking glacier, whereas in other countries, there are intense rainfall events that are leading to floods. The population growth, the expanded agriculture and industrialization, and urbanization have increased the demand for clean and reliable water supplies to many fields. In many developing countries, the inadequate water infrastructure, poor governance, and limited financial resources are making this situation worse. The mismatch between the increasing demand and the fluctuating supply has created the water scarcity problem all over the world, and it has become a most pressing global concern of the twenty-first century. Climate change has intensified these issues and is making the water management systems more complex and unpredictable [2].

DOI: 10.1201/9781003742227-11

11.1.2 Impacts of Climate Variability on Water Resources

Climate change is directly making an impact on the quality and quantity of water availability. The irregular rain patterns have influenced the underground water level, surface runoff, and the reservoir levels. The extended drought conditions have reduced the steam flow and affected the soil moisture level, leading to agricultural field stress by reducing the crop yield. On the other hand, the extreme rainfall in certain areas has caused floods, erosion, sedimentation, and water infrastructure damages. The rise in temperature has further increased the evaporation rate, which leads to a reduction in freshwater availability [3]. In addition to that, the sea level is continuously rising and has resulted in the salinity intrusion in coastal, aquifer, and river systems. All these changes have disrupted the domestic water supply, energy generation, agricultural productivity, and ecological balance. Therefore, there is a dire need to adopt strategies to overcome the water scarcity problem.

11.1.3 Need for Climate Resilience in Water Systems

The conventional water management systems have been designed for stable climate conditions and therefore are no longer sufficient to deal with this situation. The water systems must be flexible, adaptive, and robust enough to handle all sorts of uncertainties. Climate-resilient water systems must focus on long-term sustainability by integrating risk assessment, early warning systems, and nature-based solutions that need technical innovation and participatory governance. All these systems must ensure that communities and ecosystems can cope with or recover from climate impact faster by reducing vulnerabilities and enhancing water security [4]. Building resilient solutions also supports the sustainable development goals, particularly those related to clean water and sanitation, food security, climate action, and ecosystem conservation.

11.1.4 Scope and Structure of the Chapter

This chapter examines climate resilience strategies for sustainable water management in urban and rural areas under increasing climate variability. It integrates ecological, technological, institutional, and socio-economic approaches to address climate-induced water challenges. Section 1 introduces the concept of climate-resilient water management and outlines global challenges and the impacts of climate variability. Section 2 discusses the main climate change drivers affecting water resources. Section 3 presents the core principles of sustainable and resilient water management, including integrated water resources management, the water-energy-food-ecosystem nexus, equity, governance, and adaptive capacity building. Section 4 focuses on nature-based solutions, while Section 5 highlights technological interventions and innovations. Section 6 emphasizes community-based and policy-driven adaptation strategies, and Section 7 addresses climate-resilient planning for urban and rural areas. Finally, Section 8 concludes the chapter with future research directions and policy implications.

11.2 CLIMATE CHANGE DRIVERS AFFECTING WATER RESOURCES

The change in climate has altered the Earth's physical and ecological systems by directly influencing the availability, distribution, and quality of fresh water. Since water is the most sensitive component of the climate system, small changes in temperature or precipitation can significantly influence the hydrological

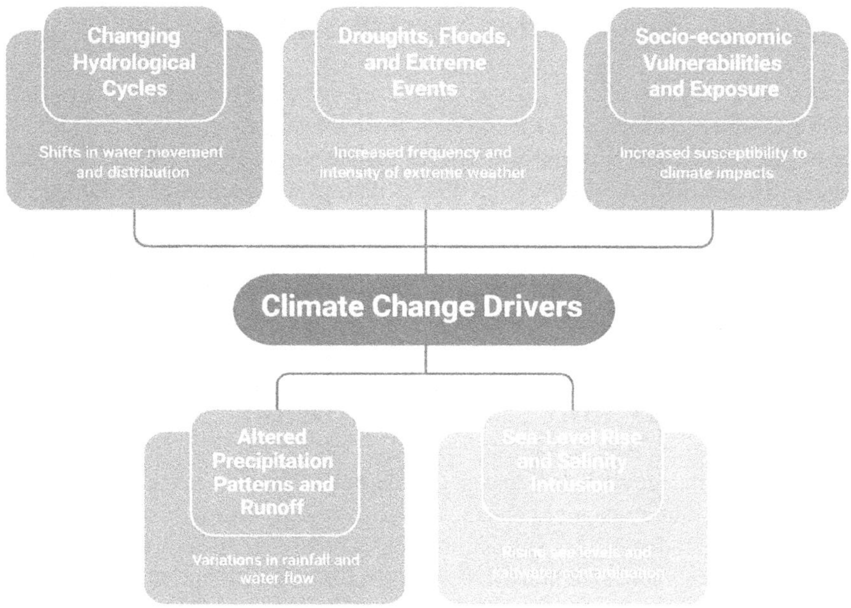

FIGURE 11.1 Drivers responsible for climate change.

shift. In this section, we have explained the major climate factors that are affecting the water resources globally and have also discussed how the changes are affecting both the natural ecosystems and human societies as shown in Figure 11.1.

11.2.1 Changing Hydrological Cycles

The hydrological cycles that govern the movement of water through evaporation, condensation, and precipitation are significantly disrupted by the rising global temperature. The warmer air holds more moisture that leads to intensified evaporation from the land at a water surface. This process contributes to the change in humidity level, cloud formation, and strong intensity. Snow melt patterns are also shifting these days. The mountain snow packs are melting early in the year, therefore, reducing the water availability during the dry season [5]. The change in soil moisture has further affected the groundwater recharge and agriculture productivity. All these disruptions challenge the water managers who rely on a hydrological pattern to plan the storage, irrigation, and supply system.

11.2.2 Altered Precipitation Patterns and Runoff

The change in climate patterns is causing the precipitation to become more unevenly distributed across some regions and seasons. Some of the areas are experiencing long dry spells, whereas the others have witnessed heavy rainfall within short periods. These changes in the rainfall pattern have influenced the timing and the volume of surface runoff that lead to the unpredictable reservoir levels and river flows. In many of the continents, the monsoon and the seasonal rainfall system have become severely erratic that they complicate the agricultural planning and the water allocation [6]. Reduced runoff in dry and semi-dry areas has contributed to the decline in water recharge, shrinking wetlands, and lowering groundwater levels. On the contrary, there are certain areas where excess runoff during intense storms has increased the risk of flooding, landsliding, and infrastructure damage.

11.2.3 Droughts, Floods, and Extreme Events

The frequency and intensity of extreme weather events are increasing due to climate change. The prolonged drought led to water shortage and reduced crop yield, forest fires, and depletion of both surface and groundwater resources. The drought is also weakening the ecosystems and making them more vulnerable to erosion and degradation. On the contrary, the floods disrupt the community's damaged water supply and contaminate the drinking water, causing economic losses [7]. A flash flood in an urban area also overwhelms the drainage systems, whereas a rewind flood destroys the agricultural land and settlements. These climate-induced extreme events are straining the resilience of existing water infrastructure and highlighting the need for adaptive planning and risk-based management.

11.2.4 Sea-Level Rise and Salinity Intrusion

The rise in sea level due to the melting glaciers and the thermal expansion of the ocean has posed a serious concern to the coastal freshwater resources. As the seawater moves inland, it infiltrates the water surface, water bodies, and underground aquifers. The salinity intrusion due to this has reduced the availability of fresh water for drinking, irrigation, and industrial use. The coastal rivers and bays are also facing a serious concern in ecological balance as the salt water affects the vegetation, soil quality, and fish population [8]. The low-lying island nations and the delta regions are more vulnerable because even a slight increase in sea level can lead to widespread contamination of groundwater. Protecting these regions from this situation requires innovative solutions such as aquifer recharge, salinity barriers, and substantial coastal zone management.

11.2.5 Socio-economic Vulnerabilities and Exposure

The climate-driven challenges have affected the vulnerable community to a different extent, especially in low-income and developing regions. The limited financial resources, inadequate water infrastructure, and dependence on climate-sensitive sectors like agriculture have increased the exposure to the water scarcity issue. The rapid urbanization has added additional pressure as the expanded cities require more water for domestic and industrial use. The rural population is generally facing reduced crop yields, land-use issues, and migration pressures due to long drought conditions. The social inequality is also getting deeper as the marginalized groups such as women and small farmers are bearing the greatest burden of water scarcity and climate impact [9]. Addressing all these socio-economic issues requires inclusion policies, community engagement, and targeted investments in adaptation measures.

11.3 PRINCIPLES OF SUSTAINABLE AND RESILIENT WATER MANAGEMENT

Sustainable and resilient water management requires a holistic approach that will balance human needs, ecological integration, and long-term resource security. With the intense climate change and water-related stresses, conventional approaches are no longer sufficient. They need integrated and adaptive frameworks to overcome the situation in such a way that they can withstand the climatic uncertainties while supporting the socio-economic developments [10]. In this section, we have outlined the key principles that could help in developing a sustainable solution for climate-resilient water management, as shown in Figure 11.2.

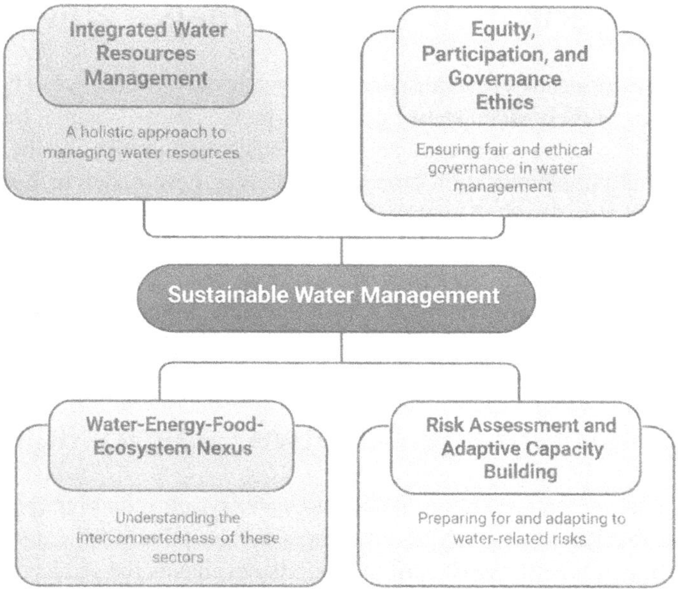

FIGURE 11.2 Sustainable water management.

11.3.1 Integrated Water Resources Management (IWRM)

Integrated Water Resources Management (IWRM) is an accepted framework that promotes the coordinated development and management of resources like water, land, etc. The basic notion behind IWRM is to maintain the socio-economic and environmental balance while ensuring that the water resources are utilized efficiently and effectively. The key element includes the cross-sectional coordination to encourage cooperation among the agricultural, industrial, and domestic sectors with the environmental agencies. It also includes the involvement of communities, policymakers, and the private sector in decision-making [11]. This framework is promoting the water conservation vision and efficiency rather than increasing the water supply. Moreover, it also helps maintain the ecological health of rivers, wetlands, and aquifers. IWRM facilitates building resilience while enabling flexible planning and reducing conflict among water users to promote long-term sustainability while addressing climate variability.

11.3.2 Water-Energy-Food-Ecosystem Nexus

The Water-Energy-Food-Ecosystem nexus understands that all these sectors are interconnected, and therefore, the action in one area has a repercussion on the other. For example, the agriculture sector requires water and energy, energy production requires water, and the ecosystem provides essential services such as water purification and flood control. Therefore, the main factors that are considered by this nexus while addressing these issues are to minimize the wastage of all these resources while improving the efficiency of all these sectors. This nexus also promotes practices like wastewater reuse for irrigation and the use of renewable energy for water pumping. The nexus understands the role of a natural system in supporting water availability and its quality. It also ensures that the national and local bodies reflect the closed sector dependency. Applying all these nexus approaches facilitates resilience to climate changes and enables the integration of solutions that optimize resources, reducing vulnerabilities for sustainable development [12].

11.3.3 Equity, Participation, and Governance Ethics

Effective water management needs fair, transparent, and inclusive governance. The equity ensures that all individuals, regardless of their social status, gender, or geographical location, have access to safe and drinkable water. This participation involves engaging the local communities and the stakeholders in planning, decision-making, and implementation processes. The core governance includes clear communication and access to information for all. This also monitors the overall system and evaluates the usage of water and its management decisions. To ensure inclusiveness, the marginalized groups are also included in decision-making while prioritizing the essential human and ecological needs. Good governance can reduce the conflict and could improve compliance with the regulations while strengthening resilience by fostering shared responsibility in water conservation and water management [13].

11.3.4 Risk Assessment and Adaptive Capacity Building

Climate changes introduce a high level of uncertainty to water systems, making risk assessment a critical component of resilient management. The risk assessment involves the identification of potential threats such as drought, floods, and infrastructure failures and also evaluates the vulnerability and exposure of the affected population and ecosystem. The Adaptive Capacity Building Program focuses on the development of the skills, knowledge, and policies to respond effectively to the changes in a climate [14]. It includes the early warning systems and the climate forecasting to support proactive planning. The scenario analysis is done to evaluate the future climate conditions and a well-informed infrastructure design. These capacity building programs are beneficial for water managers, farmers, and communities, and it also helps in designing the flexible policies that can be modified based on new information or changes in the climate conditions. Altogether, the risk management and adaptive capacity-building programs ensure that the water systems can adjust and respond to climate changes and may recover from all disruptions while creating the foundation for the long-term resilience program.

11.4 NATURE-BASED SOLUTIONS FOR CLIMATE-RESILIENT WATER SYSTEMS

Nature-based solutions exploit the power of natural processing and ecosystems to address the water-related issues caused by the climate changes. Unlike the conventional gray infrastructures that provide a single-purpose solution, NBS offers multiple benefits such as enhanced water quality, improved biodiversity, and less risk of floods. These approaches work with nature to make them environmentally sustainable, cost-effective, and adaptive. In these sections, various NBS approaches have been explored that can contribute to the resilient water system [15].

11.4.1 Watershed Restoration and Reforestation

The watershed restoration and reforestation also play an important role in regulating the water flow and thereby improving the groundwater recharge and reducing soil erosion. The healthy watershed acts as a natural sponge that absorbs the rainfall and releases it slowly into the stream and rivers, which helps in stabilizing the water supply throughout the year. The forest and the vegetation cover also play an important role in strengthening the water system and ecosystem health. It enhances the soil permeability, and vegetation improves the infiltration and supports the effective water ground recharge. The presence of a plant root also helps in stabilizing the soil and thereby controls the erosion process and reduces sedimentation in the rivers [16]. The forested watershed also facilitates climate regulation by moderating the local temperature and moisture conditions. In addition to that, the reforestation and vegetation growth help in promoting biodiversity by providing a suitable habitat for a wide range of plant and animal species.

11.4.2 Wetland Conservation and Urban Blue-Green Infrastructure

Wetlands are the most important component of the ecosystem on our earth as they provide essential services such as water purification, flood control, and habitat protection. Therefore, their conservation is playing an important role in climate-resilient water management.

11.4.2.1 Wetland Conservation

Wetlands play an important role in climate-resilient water management by storing excess water during the period of heavy rainfall and gradually releasing it during dry seasons. By doing so, they act as a natural buffer against climate variability. Moreover, they significantly reduce the risk of a flood by absorbing and slowing down floodwater and also support groundwater recharge through infiltration [17]. Further, wetlands filter pollutants and sediments that improve the water quality and contribute to carbon sequestration by capturing and storing carbon in vegetation and soil. Therefore, protecting the existing wetlands and restoring the degraded ones facilitate preserving habitats while supporting biodiversity and maintaining ecological balance.

11.4.2.2 Urban Blue-Green Infrastructure

With the expansion of the urban area, the pressure on water resources and drainage intensifies. It makes sustainable water management extremely important for the survival of an ecosystem. The blue-green infrastructure integrates the natural elements, like blue water bodies and green vegetation, into the urban planning to manage stormwater in a sustainable manner. These systems such as bioswales and rain gardens, green roofs, and permeable pavements, lakes, canals, urban ponds, as well as riparian buffers facilitate the regularization of water flow and reduce surface runoff. All these approaches collectively reduce the risk of urban flooding and thereby improve cooling effects within a city, improving air quality for urban resilience [18].

11.4.3 Managed Aquifer Recharge (MAR) and Groundwater Protection

Groundwater is an important resource in areas that experience erratic rainfall and have an increased water demand. Managed Aquifer Recharge (MAR) is an effective strategy that refills aquifers using surface water, treated wastewater, or stormwater to improve the groundwater availability. The techniques are very common to facilitate this process and to deal with the situation of water security during the drought period. It improves the water quality through natural infiltration within soil and rock layers and also prevents aquifer depletion and salinity intrusion in coastal areas [19]. However, it needs recharge efforts, effective groundwater protection, regulation of extraction, and water contamination prevention due to surface activities to ensure the long-term sustainability, as shown in Figure 11.3.

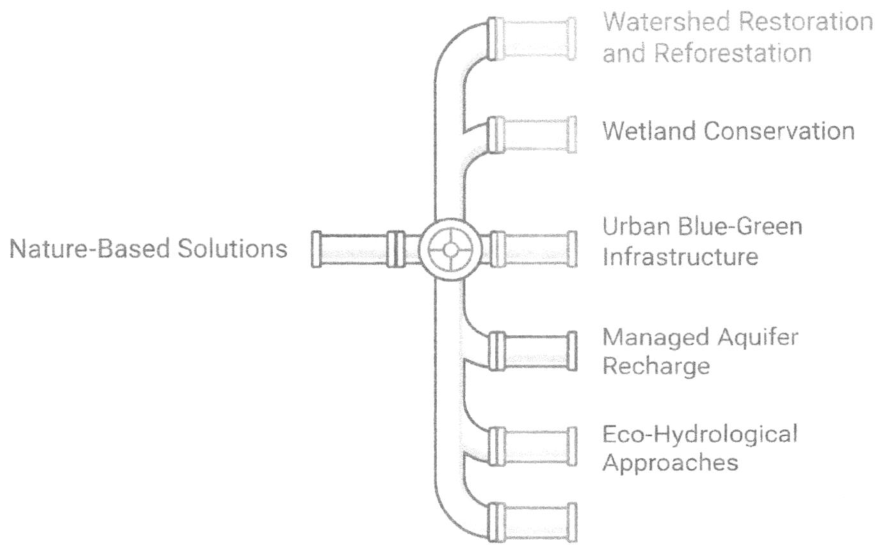

FIGURE 11.3 Nature-based solution.

11.4.4 Eco-hydrological Approaches to Flood and Drought Mitigation

It integrates the ecological principles with the hydrological techniques to improve sustainable water management. It regulates the water flow through ecosystem-based approaches and offers an effective solution to reduce the impact of floods and droughts. Floodplain restoration processes help rivers reconnect with natural floodplains to reduce floods and to improve groundwater recharge. River corridor rehabilitation restores the natural flow pattern while improving the sediment process to improve water quality. The soil moisture conservation strategies improve the soil water retention capabilities that could effectively handle drought conditions. In addition to that, ecosystems like forests could be beneficial as natural drought buffers that could store the water and release it gradually during dry periods. All these approaches can significantly improve resilience to extreme events [20]. Table 11.1 gives a comparison of all nature-based solutions for Climate-Resilient Water Systems.

11.5 TECHNOLOGICAL INTERVENTIONS AND INNOVATIONS

With the advancement in technologies, the way resources are managed has changed drastically. With the intervention of AI and ML, the quality of the water can be continuously monitored and the water can be dispersed efficiently. These technologies help in making fast decisions that could be beneficial for the development of the various water sources. When all these technologies are combined with the conventional techniques, the water resilience and the sustainability improve further. In this section, various technologies and technical interventions are discussed to support an effective reservoir water management.

TABLE 11.1 Comparison of Nature-Based Solutions for Climate-Resilient Water Systems

APPROACH	*BASIC NOTION*	*KEY FUNCTIONS*	*BENEFITS*	*CLIMATE RESILIENCE CONTRIBUTION*
Watershed Restoration & Reforestation	Restoration of forests and natural watershed processes	• Regulates water flow • Enhances infiltration • Stabilizes soil	• Improved groundwater recharge • Reduced soil erosion • Biodiversity enhancement • Micro-climate regulation	• Reduces flood peaks • Sustains base flow during dry periods • Moderates temperature
Wetland Conservation	Protection and restoration of natural wetlands	• Water storage • Filtration • Flood buffering	• Flood mitigation, improved water quality, groundwater recharge, carbon sequestration, habitat protection	• Buffers climate variability • reduces flood and drought impacts
Urban Blue-Green Infrastructure	Integration of water bodies and vegetation into urban design	• Stormwater management • Runoff reduction • Urban cooling	• Reduced urban flooding, enhanced air quality • Heat-island mitigation • Aesthetic and social benefits	• Enhances urban resilience to extreme rainfall & heat stress
Managed Aquifer Recharge (MAR) & Groundwater Protection	Artificial recharge of aquifers using natural infiltration	• Aquifer replenishment • Water quality improvement	• Increased groundwater availability • Drought security • Prevention of salinity intrusion	• Strengthens drought resilience • Long-term water security
Eco-hydrological Flood & Drought Mitigation	Integration of ecological processes with hydrology	• Flow regulation • Soil moisture retention • Floodplain reconnection	• Reduced flood intensity • Enhanced groundwater recharge • Improved water quality	• Mitigates extreme events like floods and droughts through natural buffering

11.5.1 Smart Water Technologies

Advanced technologies like the Internet of Things (IoT) and geographical information systems (GIS) can collect and analyze data in real time. It improves the monitoring, distribution, and management of water resources. The IoT devices use various types of sensors which can govern the quality as well as management of the water distribution system. It also analyzes flow rate and pressure within the pipeline network. They also enable real-time leakage detection in the water distribution system while monitoring water quality parameters. This real-time information helps predict and generate early warnings about floods, droughts, etc. Geographical information system (GIS) is another technology that facilitates the spatial mapping and analysis of water resources. It monitors land-use and land-cover changes to facilitate the identification of climate vulnerability zones [12]. It also plays an important role in planning and optimizing water resources. When combined with IoT, it leads to the development of a smart water system to improve readiness for climate-driven changes.

11.5.2 Water Recycling, Reuse, and Advanced Treatment Systems

Since freshwater is increasingly depleting, the recycling and reuse of water is the only solution for a sustainable water supply. These days, many advanced technologies are available to treat wastewater and its safe reuse for applications like agricultural irrigation, industrial cooling, urban landscaping, and groundwater recharge. They include membrane bioreactors, ultrafiltration, reverse osmosis (RO) processes, and decentralized wastewater treatment systems. These technologies help in reducing environmental pollution and improving water availability. It ultimately reduces the pressure on natural freshwater. The greywater recycling also offers a sustainable solution at the household and community levels by treating wastewater and reusing it for toilet flushing, gardening, and cleaning [16].

11.5.3 Climate-Smart Irrigation and Agricultural Water Efficiency

Drip irrigation technology delivers water directly to the plant roots and therefore reduces water requirements. The sprinkler irrigation system is another technique that effectively and efficiently distributes water compared to the conventional techniques. In addition, micro-irrigation systems help optimize water use on small farms to support the water requirements for the crops precisely.

By integrating sensory data, satellite data, and automated irrigation scheduling, farmers are now able to monitor soil moisture levels and soil conditions. Using this data they can optimize water requirements according to crop requirements. Breeding and biotechnology have further facilitated the development of crop varieties that need less water to yield more. Climate-smart irrigation practices combined with improved crop genetics improve food security while conserving water resources [5].

11.5.4 Desalination and Alternative Water Supply Technologies

Since the climate is changing and freshwater availability is reducing, many alternative water sources such as desalinated water and atmospheric moisture harvesting are now becoming extremely important.

11.5.4.1 Desalination Technologies

Desalination technologies convert seawater or brackish water into freshwater by using technologies like reverse osmosis (RO), multi-stage flash distillation, and electrodialysis. Recent advancements in renewable-powered desalination and energy recovery systems have improved the sustainability and cost-effectiveness of these processes.

11.5.4.2 Atmospheric Water Harvesting

New technologies need to be developed to extract the moisture from the air with the help of a condenser and provide nets by offering a novel solution for arid and remote areas.

11.5.4.3 Rainwater Harvesting and Storage Innovations

The new rainwater harvesting techniques utilize small water collection systems to stop filtration. Nowadays, with the help of technologies, the tanks can automatically distribute the water and can efficiently reuse the water. These approaches could be beneficial in storing and regulating the rainwater effectively while ensuring their quality for potable and non-potable usage. Various advanced technologies could reduce the dependence on the climate to improve resilience against drought conditions [19].

11.6 COMMUNITY-BASED AND POLICY-DRIVEN ADAPTATION STRATEGIES

To develop a climate-resilient water system, the integration of technology and ecological restoration processes is much needed. To do so, strong community participation is anticipated while supporting the new policies that could effectively cover the whole water framework. In this scenario, the local communities can play an important role in conserving water and managing water utilization in areas where water challenges are related to cultural practices. At the same time, the policymakers need to intervene and offer economic incentives to the people to ensure long-term sustainability. In this section, various such community-based policy-driven approaches are analyzed that could improve the water resilience process against climate changes [7].

11.6.1 Community-Led Water Conservation Models

The change in climate has affected the global water system, the hydrological cycles. As a result, the frequency of droughts, floods, and extreme weather events has been increased significantly. All these disruptions have posed serious concern regarding water security, public health, agriculture, and ecosystem stability. This chapter provides a comprehensive review on the strategies designed to enhance the climate resilience of water management systems for urban as well as rural areas [9].

From past few years, we all have witnessed the climate changes that have affected the global water system and the hydrological cycles. As a result, the frequency of the drought floods and the extreme weather events has been increased significantly. All these climate disruptions have posed a serious concern on the availability of our drinkable water, public health, agriculture, and ecosystem stability. In this chapter, the comprehensive review is provided on the strategies that could be designed to improve the climate resilience of water management system for urban and rural development.

The technology helps in making the fast decisions that could be beneficial for the development of the various water sources. When all these technologies are combined with the conventional techniques, the water resilience and the sustainability improves further. In this section, various technologies and technical interventions are discussed to support an effective reservoir water management [12–16].

Nowadays, with the help of technologies, the tanks can automatically distribute the water and could efficiently reuse the water. These approaches could be beneficial in storing and regulating the rainwater effectively while ensuring their quality for the potable and non-potable usage. Various advanced technologies could reduce the dependence on the climate to improve the resilience against the drought condition.

To do so, the strong community participation is anticipated while supporting the new policies that could effectively cover the whole water framework. In this scenario, the local communities can play an important role in conserving the water and to manage the water utilization in the areas where the water challenges are related with the cultural practices. At the same time, the policymakers need to intervene and to offer the economic incentives to the people to ensure the long-term sustainability. In this section, various such community-based policy-driven approaches are analyzed that could improve the water resilience process against the climate changes [15].

Community play an important role in sustaining water management, especially in rural and semi-urban areas. The community-based model exploits the local knowledge and could share the responsibility to protect the water resources. They can collaborate to restore watersheds by using the practices like soil conservation and the construction of recharge structures. The village-level water budgeting also help these communities to analyze and estimate the water requirement annually while allocating the resources for the domestic, agriculture, and other use. All these policies facilitate the farmers and the residents to manage the irrigation requirements while ensuring the equitable water distribution. It not only enhances their accountability, but also promote the equitable access by embedding water stewardship at the grassroots level.

11.6.2 Regulatory Frameworks and Water Governance Reforms

To deal with the climate-induced uncertainties, an effective governance framework is also needed to manage the water resources. They not only help in the regulation of extraction, protecting the ecosystem, but also ensure the equitable distribution of the water to promote sustainability. Various water-related laws and policies need to be revised and implemented at the national or state level to ensure the usage priority, quality standards, and conservation mandates. For example, the groundwater regulation needs, borewell licensing, and the restriction in the over-exploited zones. Similarly, the pollution control standards need to regulate the industrial effluence and urban wastewater discharge to preserve the water quality. Decentralized governance like panchayat municipalities could also manage the water resources effectively at the grassroots levels. These governance reforms will not only support transparent decision-making but also reduce the water conflicts among users [17].

11.6.3 Capacity Building, Education, and Stakeholder Engagement

The climate resilience process needs the awareness of an individual institution and communities to develop their initiatives to sustain water. Various training programs could be arranged to make people understand the basis of adopting the practices to conserve water. Various public awareness campaigns could be beneficial in educating the communities and people regarding water conservation, pollution, and other climate-related issues. Various government and non-government organizations will have to work closely to support the inclusive and shared decision-making processes. The literacy among users about the effective and efficient water utilization and climate-related changes in the school curriculum or university curriculum needs to be incorporated. Further, various knowledge exchange programs must be organized to promote learning from the successful stories [17].

11.7 URBAN AND RURAL CLIMATE-RESILIENT WATER PLANNING

To develop a climate-resilient water system, comprehensive planning is needed to deal with the challenges faced by urban and rural areas. With the population growth, the cities are most affected, and the rural areas are also experiencing the depletion of groundwater. Therefore, effective planning is much needed that would integrate the technology, innovation, and community participation to protect the whole water ecosystem.

11.7.1 Climate-Sensitive Urban Water Design (WSUD)

To deal with this situation, water-sensitive urban designs need to be adopted that will integrate the whole urban water cycle. It typically includes stormwater, drinking water, groundwater, and wastewater treatment for urban planning to enhance sustainability and climate resilience. These clever designs effectively reduce the surface runoff and promote the natural infiltration process to protect the water bodies from pollution. It needs a decentralized water treatment and reuse system that prioritizes the protection of natural water resources. By incorporating the water into the urban designs and aesthetics to recreate space while delivering the ecological benefits, these designs have the potential to transform conventional cities into smart cities where water is conserved while sharing the urban resources to attain sustainability [10].

11.7.2 Stormwater Management and Flood Risk Reduction

Due to the severe climate changes, the rainfall intensity is also increasing. Therefore, effective stormwater management systems are needed to reduce the risk of floods. The new strategies need to be developed with the emphasis on retaining and filtering the stormwater. Sustainable drainage solutions like rain gardens and retention ponds are used to slow down the runoff and for the temporary storage of the excessive rainwater. With the help of a real-time flood forecasting system using AI and ML techniques, early warning and proactive responses are possible. By incorporating these strategies, exposure to the floods and the damage due to the floods can be controlled in advance. Collectively, all these measures mitigate the risk of flooding and infrastructure damage while improving the water quality to support ecological health.

11.7.3 Resilient Rural Water Supply Systems

The rural areas are very much dependent on the groundwater and rainwater systems. Therefore, they are severely affected due to climate changes. To meet the demand for water, the solutions must be developed with an emphasis on drought certification, storage, and community involvement. The typical solutions involve rainwater harvesting and the proper utilization of the groundwater and surface water. There is a need to incorporate water storage structures like farm ponds, water tanks, etc. to store the monsoon water for future use during dry periods. The community-based models are also important for sustainable water distribution. Moreover, climate-smart agriculture practices need to be adopted to utilize the water efficiently.

11.8 CONCLUSION AND FUTURE SCOPE

Climate-resilient water management is a dire need due to severe climate change. To tackle this situation, emerging strategies need to be adopted to empower communities with dynamic policy frameworks. A dedicated research and cross-sector collaboration is needed to deal with the water deficit problem being faced by people all over the world to design innovative solutions. Therefore, future research should focus on the understanding of climate-water interactions to develop sustainable hybrid water management systems that integrate green infrastructure with conventional systems. To do so, multidisciplinary research and collaborations are needed that include environmental science, engineering, social sciences, and public policy.

Furthermore, with the advanced technologies like AI and ML, water monitoring systems continuously track water quality, detect leakages, and monitor reservoir storage while predicting system failures. These approaches can help in precisely allocating water to optimize irrigation scheduling. Furthermore, big data analytics could be useful for predicting risk by identifying drought conditions. By integrating these technologies with appropriate policy frameworks, systems can significantly enhance adaptive capacity and support effective water management. The future of water security is based on collective action, interdisciplinary collaborative research, and global commitment to safeguard this most important natural resource.

REFERENCES

1. Srivastav, Arun Lal, Rajni Dhyani, Manish Ranjan, Sughosh Madhav, and Mika Sillanpää. "Climate-resilient strategies for sustainable management of water resources and agriculture." *Environmental Science and Pollution Research* 28, no. 31 (2021): 41576–41595.
2. Chouhan, Shambhu, Smita Kumari, Rajneesh Kumar, and Pravin L. Chaudhary. "Climate resilient water management for sustainable agriculture." *International Journal of Environment and Climate Change* 13, no. 7 (2023): 411–426.
3. Ahmed, Mukhtar, Muhammad Abid, Naeem Abbas Malik, Sajid Ali, Adnan Zahid, Amjad Malik, and Mumtaz Cheema. "Climate-resilient strategies for sustainable management of water resources and agriculture." In *Climate Resilient and Sustainable Agriculture: Volume 2: Social and Transformative Strategies*, pp. 1–25. Cham: Springer Nature Switzerland, 2025.
4. Dolman, Nanco. "Integration of water management and urban design for climate resilient cities." In *Climate Resilient Urban Areas: Governance, Design and Development in Coastal Delta Cities*, pp. 21–43. Cham: Springer International Publishing, 2020.
5. Sikka, Alok K., Mohammad Faiz Alam, and Vidya Mandave. "Agricultural water management practices to improve the climate resilience of irrigated agriculture in India." *Irrigation and Drainage* 71 (2022): 7–26.
6. Javadinejad, Safieh, Kaveh Ostad-Ali-Askari, Vijay P. Singh, and Mohammad Shayannejad. "Reliable, resilient, and sustainable water management in different water use sectors." *Water Conservation Science and Engineering* 4, no. 2 (2019): 133–148.
7. Sikka, Alok K., Adlul Islam, and K. V. Rao. "Climate-smart land and water management for sustainable agriculture." *Irrigation and Drainage* 67, no. 1 (2018): 72–81.
8. Abdulameer, Layth, Mahmoud Saleh Al-Khafaji, Aysar Tuma Al-Awadi, Najah ML Al Maimuri, Musa Al-Shammari, and Ahmed N. Al-Dujaili. "Artificial intelligence in climate-resilient water management: a systematic review of applications, challenges, and future directions." *Water Conservation Science and Engineering* 10, no. 1 (2025): 44.
9. Naqvi, Syed Muhammad Zaigham Abbas, Saddam Hussain, Muhammad Awais, Muhammad Naveed Tahir, Shoaib Rashid Saleem, Fuad AM Al-Yarimi, Mirjalol Ashurov et al. "Climate-resilient water management: leveraging IoT and AI for sustainable agriculture." *Egyptian Informatics Journal* 30 (2025): 100691.
10. Garg, Shivani, and Pooja Arora. "Ensuring irrigation water quality for climate-resilient agriculture: strategies and solutions in semi-arid regions." In *Sustainable Agriculture Management in Semi-Arid Climates: Volume 1*, pp. 177–189. Cham: Springer Nature Switzerland, 2025.
11. Schuetze, Thorsten, and Lorenzo Chelleri. "Integrating decentralized rainwater management in urban planning and design: flood resilient and sustainable water management using the example of coastal cities in the Netherlands and Taiwan." *Water* 5, no. 2 (2013): 593–616.
12. Zekri, Elnaz, Hossein Abolghasemzadeh, Mohsen Nasseri, Aylar Nejati, and Vahid Majed. "Climate-resilient development: A sustainability analysis of regional programs through the water-energy nexus model and impacts of climate change." *Journal of Cleaner Production* 530 (2025): 146786.
13. Hajirad, Iman. "AI-powered water management: developing infrastructure for a future that is climate-resilient." *Journal of Sustainable Development of Natural Resources Management* 1, no. 3 (2025): 1–14.
14. Bijlwan, Amit, Ankita Jha, Shivani Kothiyal, Roomesh Kumar Jena, and Rajeev Ranjan. "Climate-resilient approaches for sustainable water management through forest resilience and plant functional traits." In *Enhancing Forest Resilience through Plant Functional Traits*, pp. 241–259. CRC Press, 2025.
15. Roy, Sougata, Sanjay Kumar Rathour, Anjali Mehta, Ritu Dwivedi, Surabhi, Aastika Pandey, Anamika Barman, Shreyas Bagrecha, and Shiv Vendra Singh. "Precision water management under dryland region: climate-resilient strategies." In *Climate-Smart Agricultural Technologies: Approaches for Field Crops Production Systems*, pp. 137–159. Singapore: Springer Nature Singapore, 2025.
16. Mosadeghrad, Ali Mohammad, Parvaneh Isfahani, Leila Eslambolchi, Maryam Zahmatkesh, and Mahnaz Afshari. "Strategies to strengthen a climate-resilient health system: a scoping review." *Globalization and Health* 19, no. 1 (2023): 62.
17. Przestrzelska, Kornelia, Katarzyna Wartalska, Weronika Rosińska, Jakub Jurasz, and Bartosz Kaźmierczak. "Climate resilient cities: a review of blue-green solutions worldwide." *Water Resources Management* 38, no. 15 (2024): 5885–5910.
18. Dagar, J. C., and R. Kumar Yadav. "Climate resilient approaches for enhancing productivity of saline agriculture." *Journal of Soil Salinity and Water Quality* 9, no. 1 (2017): 9–29.

19. Asif, Zunaira, Zhi Chen, Rehan Sadiq, and Yinying Zhu. "Climate change impacts on water resources and sustainable water management strategies in North America." *Water Resources Management* 37, no. 6 (2023): 2771–2786.
20. Drechsel, Pay, Manzoor Qadir, and Jurgen Baumann. "Water reuse to free up freshwater for higher-value use and increase climate resilience and water productivity." *Irrigation and Drainage* 71 (2022): 100–109.

Intelligent Water Infrastructure and Policy Programs for Sustainability

12

Lokeshwer Sharma, Sandeep Singh, Abhineet Anand, and Prerna Thakur

12.1 INTRODUCTION TO SMART WATER MANAGEMENT

Integrative Smart Water Management involves several different disciplines, agile water management, and the application of cutting-edge monitoring technology, automated management control, and advanced analytical computing in all stages of the water cycle, from source [1] extraction, treatment and supply, through use and reuse to discharge, for the management of water resources, including the monitoring and control of water in the automated systems of the SWM system. In smart systems, operational costs are reduced by about 10–18%, and water losses are minimized by 20–35% when a real-time monitoring system is compared to conventional systems.

It is also estimated that in the last 60 years, the global [2] freshwater availability fell from 9,000 m^3/year (1960) to 5,000 m^3/year (2020), which increases the pressure on the water industry. Smart water systems and automated control models assist managers in coping with daily demand fluctuations of ±30%, pipeline leakages within 5–10, supply efficiency of the systems, and overall network reliability to 90–95%. Figure 12.1 illustrates the potential impact of smart water management systems, highlighting reductions in operational costs and water losses, along with improved management of demand fluctuations. These improvements result in the need to digitally transform the water industry.

DOI: 10.1201/9781003742227-12

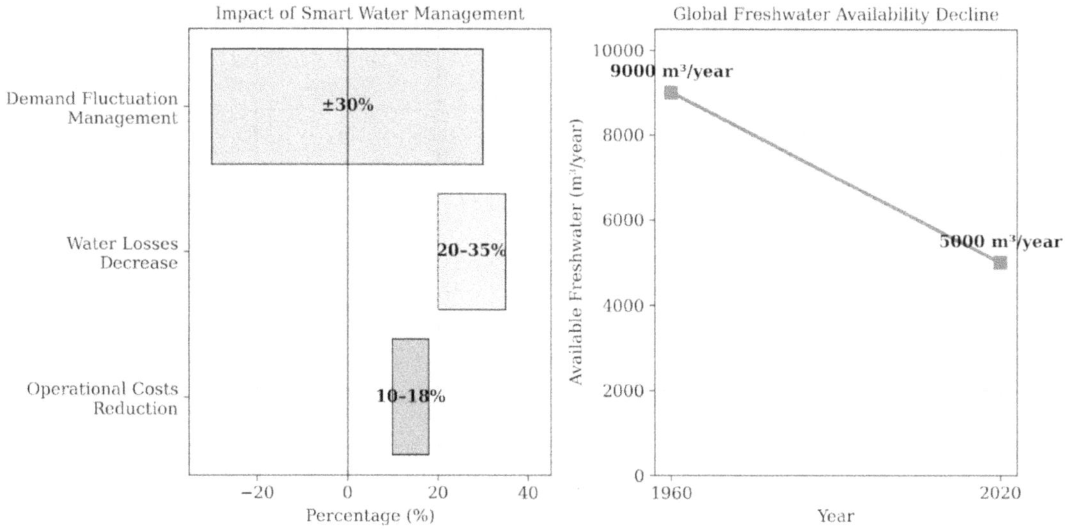

FIGURE 12.1 Smart water management system with real-time monitoring, reducing costs (10–18%), losses (20–35%), and managing ±30% demand variation.

12.1.1 Concept of Water Sustainability

Water sustainability focuses on responsible use and replenishment of available resources. The goal is to keep water accessible to future generations. A sustainable water model in cities usually aims for the following:

- Water consumption 100–135 L/person/day (urban standard)
- Efficiency in rainwater harvesting is 60–80% of the collection potential
- Wastewater reuse 30–50% of total domestic wastewater
- Improvement of groundwater recharge rates by 10–20% annually through recharge pits
- Non-Revenue Water (water loss) is < 15%

Keeping these benchmarks helps cities balance their ecosystems and sustain their resources over the long term.

12.1.2 Role of Digital and Smart Technologies

Digital technologies assist smart water systems in operational functionality. IoT-based sensors can record system parameters as quickly as once every 1–5 seconds, while smart meters record system usage every 15 minutes to 1 hour. Demand forecasting can be [3] improved 85–95% with the use of AI-based predictive models. With data transfer and communication networks like LPWAN and 5G, control actions can be instantaneous with speeds reaching 1 Gbps and latency lower than 10 ms.

Automation platforms such as SCADA can reduce the annual manual inspection count by 40–60% and increase pump efficiency by 12–18%. On average, digital integration of predictive maintenance can extend the expected lifespan of system infrastructure by 20 years.

TABLE 12.1 Quantitative Components of Smart Water Management

COMPONENT	*FUNCTION*	*TYPICAL NUMERICAL RANGE*	*BENEFIT*
Smart Sensors	Quality & flow monitoring	Sampling every 1–5 sec	Real-time alerts
Smart Meters	Consumption tracking	Accuracy ±1–2%	Billing & demand control
Data Analytics	Forecasting	85–95% prediction	Efficient
Communication Network	Data transfer	<10 ms latency, up to 1 Gbps	Super connectivity
Automated Control (SCADA)	Pump & valve regulation	12–18% energy saved	Operational
Policy Dashboards	Decision support	20–30% faster reporting	Transparent governance

12.1.3 Global Water Challenges and Need for Innovation

Nearly 2 billion people suffer from water-related stress at least once a year. In the next 30 years, there will be a 55% rise in the demand for water in urban areas. Furthermore, due to global warming, reliable freshwater supply will further dwindle by 10–25% in susceptible areas. Due to [4] aging water supply systems, cities leak/lose 25–40% of their treated water. Automated leak detection, decentralized treatment units, and AI-driven flood forecasting are examples of innovative smart solutions and can decrease flood damage by 30–45% and decrease drought response time by 20–35%. These results prove there is a global need for a tech-oriented transformation in the management of water systems.

12.2 SMART SENSORS AND IOT IN WATER SYSTEMS

The building blocks of smart water management systems are advanced technologies such as smart sensors and the Internet of Things (IoT). With these technologies, water utilities can monitor, control, and manage water quantities in the entire supply chain – from the sources of raw water, treatment facilities, storage and transmission systems, to end consumers. Moving from manual inspections to automated real-time monitoring improves operational transparency and the quality of decisions. In large urban systems, using IoT platforms can reduce total water losses in the distribution system by 25–35%, lower maintenance costs by 15–25%, and improve overall distribution efficiency to about 95%. In addition, sensor-based predictive maintenance can reduce the frequency of emergency repairs to be more fiscally sustainable, while also extending the service life of pipelines by 10–15 years.

12.2.1 Real-Time Water Quality Monitoring

At the points of water intake, use, and distribution, digital probes and electronic sensors track the quality of water supplied in real time. These sensors are able to track the water's pH, turbidity, oxygen saturation, temperature, conductivity, dissolved solids, and chlorine concentration, in the range of ±2–3%. Some sensors are able to record in intervals of 1–10 seconds. The US Environmental Protection Agency states that potable water has to meet a pH range of 6.5 to 8.5, turbidity of 5 NTU or lower, dissolved oxygen of 5 mg/L or higher, and dissolved solids of 500 mg/L or lower. When thresholds are exceeded in a water sample, the sensors will notify the operators in an automated process that takes 5–10 minutes. The operators are then able to take actions that are corrective, such as adjusting the dose of chemicals or isolating

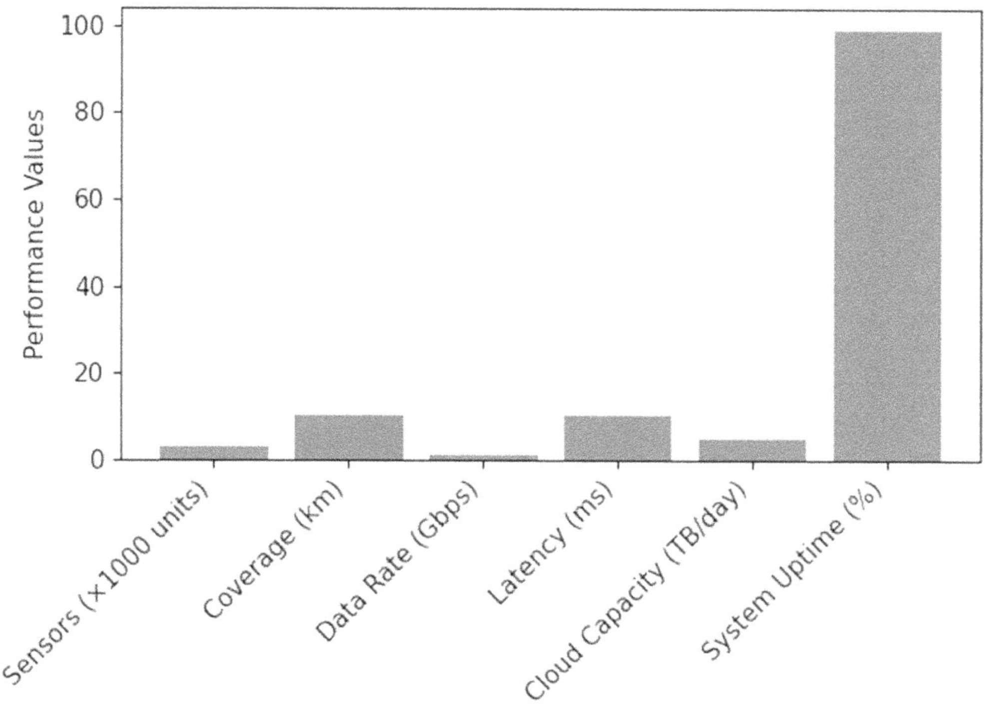

FIGURE 12.2 An example of a smart water Internet of Things (IoT) network with sensors and meters, pipelines, data storage and servers, and a central monitoring system.

the affected section. The US Environmental Protection Agency states that portable water has to meet a pH range of 6.5–8.5, turbidity of 5 NTU and lower, dissolved oxygen of 5 mg/L or higher, and dissolved solids of 500 mg/L and lower. These sensors are able to track the water's pH, turbidity, oxygen saturation, temperature [5], conductivity, dissolved solids, and chlorine concentration, in the range of ±2–3%. Some sensors are able to record at intervals of 1–10 seconds. The US Environmental Protection Agency states that potable water has to meet a pH range of 6.5–8.5, turbidity of 5 NTU or lower, dissolved oxygen of 5 mg/L or higher, and dissolved solids of 500 mg/L or lower. Figure 12.2 presents the key performance indicators of a smart water Internet of Things (IoT) network, including sensor deployment, coverage, data transmission rate, latency, cloud capacity, and system uptime.

12.2.2 Smart Metering and Leak Detection

Automated meter reading (AMR) systems capture and send data to utility companies about water consumption at residential, commercial, and industrial meter installations. This data is captured at intervals between 15 and 60 minutes. Commercial smart meters have measurement inaccuracies of about 1% and have battery lives of 8–12 years. Utilities have high consumption data and are able to analyze demand, identify usage abnormality, and create [6] conservation incentive structures. Utilities pair smart metering with commercial leak detection technologies, which use a combination of acoustic sensors, pressure loggers, and flow analysis to find hidden faults in pipelines. Advanced leak detection can identify flows between 0.1 and 0.3 liters per second and is accurate to 1–2 meters. Utilizing both smart metering and leak detection technologies, utilities can reduce Non-Revenue Water from 35–45% to 10–15%. This saves utility companies in both operational costs and volume of treated water.

12.2.3 Data Collection and Transmission Networks

The integration of thousands of distributed sensors into a single digital framework is dependent on effective data collection and transmission networks. LPWAN, NB-IoT, fiber-optic links, and 4G/5G cellular systems are examples of modern communication infrastructures that allow monitoring data to be collected and sent to central control rooms in a rapid and dependable manner. These networks can allow data transmission at 1 Gbps while sustaining data latency of less than 10 milliseconds. This ensures automated operations will be able to respond in real time. In terms of long-range communication, networks can offer coverage of 5–15 kilometers which makes them ideal for urban and semi-rural supply area populations. Well-designed networks are able to achieve system uptime of more than 99% which in turn reduces interruptions and [7] data losses. Moreover, cloud-based data systems that utilize analytics and allow long-term trend monitoring to make predictive assessments for water from governance strategies and policies can process and archive data in the range of 1–10 terabytes a day.

12.3 DATA ANALYTICS AND ARTIFICIAL INTELLIGENCE

In terms of the management of water systems, the analytics of data and the application of artificial intelligence, or AI, focus on numerical data set analytics, statistically based correlations, and algorithm optimization models. System performance evaluations would be based on metrics like the accuracy of an algorithm, the percentage of errors, the time it takes to respond, and the efficiency of the system. The basic water system analytics function can be represented as: O=f(Q,P,T,R,t), where the parameters of the function represent: O = optimized output, Q = water quantity data (measured in L/s), P = population, T = temperature (in degrees Celsius), R = rainfall (in mm), t = time (in hours).

An example of typical YouTube's analytics processing capabilities would be the ability to process data within the range of 5–20 TB/day. The accuracy of the AI model would be in [8] the range of 0.85–0.97, and which would also have a data processing lag of 2–5 seconds per dataset.

12.3.1 Predictive Modeling for Water Demand

The prediction of water quantity needed is done using regression equations and time series. One of the equations used commonly in forecasting demand is:

$$Q_d = P \times q \times PF$$

TABLE 12.2 Numerical Characteristics of Smart Sensors and IoT in Water Systems

COMPONENT	*PARAMETER*	*NUMERICAL VALUE / RANGE*	*FUNCTIONAL OUTCOME*
Water Quality Sensors	Sampling Interval	1–10 sec	Continuous monitoring
Smart Water Meters	Measurement Accuracy	±1%	Billing precision
Leak Detection Devices	Minimum Leak Size	0.1–0.3 L/s	NRW reduction
Communication Network	Latency	<10 ms	Rapid response
LPWAN Coverage	Signal Range	5–15 km	Wide-area
Cloud Data Storage	Capacity	1–10 TB/day	Long-term
Automation Controllers	Energy Saving	10–20%	Efficiency
Network Reliability	System Uptime	>99%	Service

where

Q_d = daily water demand (L/day),
P = population,
q = per-capita demand (100–150 L/person/day),
PF = peak factor (1.2–1.5).

Demand variation by the hour: $Q_h = \frac{Q_d}{24} \times HF$

$$HF = hourlyfactor(1.1-1.8)$$

Prediction Error Threshold: $E = \pm 3\% - 8\%$

Energy optimization due to accurate demand prediction:

$$E_s = 0.12 - 0.20$$

Reservoir storage optimization range:

$$S = 0.25Q_d - 0.40Q_d$$

12.3.2 AI in Flood and Drought Forecasting

Estimating the prediction of flood discharges involves the relationships of rainfall and runoff as follows: $Q_f = C \times I \times A$

where

Q_f = peak flood discharge (m^3/s),
C = runoff coefficient (0.3–0.9),
I = rainfall intensity (20–150 mm/h),
A = catchment area (km^2).

The precision of AI predictions: $A_f = 0.88 - 0.95$

Early warning lead time:

$$t_w = 12 - 72\ hrs$$

Flood damage reduction factor:

$$R_f = 0.30 - 0.50$$

Drought severity index (SPI):

$$SPI = \frac{X - \mu}{\sigma}$$

Drought classification:

Moderate: −1.0 to −1.49
Severe: −1.5 to −1.99
Extreme: < −2.0

Monitoring of groundwater depletion rates: $G_d = 0.1 - 0.5\ m / year$

12.3.3 Decision Support Systems (DSS)

Decision Support Systems use weighted numerical scoring and optimization matrices to assist planning:

$$S = \sum_{i=1}^{n} w_i x_i$$

where
S = decision score,
w_i = weight (0–1),
x_i = parameter value.

Reduction of decision response time: $T_r = 0.40 - 0.65$
Planning accuracy improvement:

$$A_p = 0.25 - 0.35$$

System reliability index:

$$RI = \frac{Uptime}{Total\ Time} = 0.98 - 0.999$$

Data processing speed:

$$V_p = 10^6 - 10^8\ records\,/\,min$$

12.4 SMART INFRASTRUCTURE AND URBAN WATER SOLUTIONS

For the purpose of this exam, smart infrastructure is the design of digital tools focused on the monitoring and control of design systems and automation, sustainable engineering design, and the creation of a data-driven policy to allocate, manage, and protect water resources. The [9] traditional supply water systems face a lot of stress due to the rapid urbanization of regions, average annual population increase of 2–4%, and per capita water demand of 120–150 L per day. Conventional infrastructure is losing 25–40% of water due to leakage, illegal connections, bad infrastructure, and bad distribution systems. Smart design infrastructure is losing water approximately to achieve this goal: 10–15% due to intelligent monitoring and other smart technologies.

Flexible smart water urban systems design consists of automated pumping services, digital pressure control services, real-time services that offer a web-based quality control, geographic information systems (GIS) mapping, and supervisory control services that offer pre-defined monitoring and control automation. The integration of these tools and systems contributes optimally to the reliability of supply systems, infrastructure, and systems, and [10] control practices for smart systems. Additionally to these dimensions, smart design systems increase flexibility and adaptability to climate change extremes and variability through optimal designs that can digitally respond to drought, flood, and climate change. The combination of smart design and digital intelligence translates water services urban networks into self-regulating and responsive networks. Figure 12.3 illustrates the architecture and performance characteristics

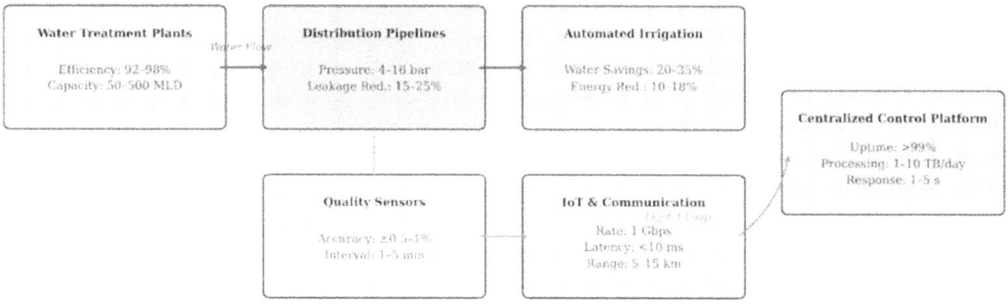

FIGURE 12.3 Smart urban water system performance: 92–98% efficiency, 15–35% water savings, <10 ms latency, >99% reliability.

of a smart urban water management system, integrating treatment plants, distribution networks, IoT sensors, automated irrigation, and centralized monitoring.

12.4.1 Intelligent Distribution Networks

Smart distribution networks in urban areas consist of automated valves, pressure controllers, management devices, flow control algorithms, and distributed sensors. These networks [11] help evenly distribute and balance water supply throughout urban areas. They monitor and manage flow (0–500L/s), pressure (2–7 bars), and reservoir levels (30–95% full) in real time. Analytics can detect and locate rate reductions due to leakages more than the flow rate of 0.2L/s within 5–10 minutes, optimizing water use.

Effective pressure management is also critical, as less than 1 bar of pressure can reduce leak rate by 8–12%. Some smart zoning and pressure management techniques use automated valves within 2–5 seconds to evenly distribute, and protect, networks and segments from bursting, thus managing flow. Intelligent networks can increase delivery efficiencies from 70% to 75% in basic networks to 90–95% in smart networks while decreasing maintenance costs by 20–30% annually.

12.4.2 Smart Irrigation and Agriculture

Weather forecasts, soil moisture analytics, and irrigation systems use sensor-based automation to optimize water use in agriculture. Over 65% of freshwater use in many [12] developing countries comes from agriculture, so improvements in efficiency are important. Soil moisture sensors are normally accurate to 2–3% in the range of 0–60% volumetric water content. Controllers activate irrigation based on evapotranspiration and predicted rainfall between 3 and 7 mm/day, as well as the water needs of the crop. Figure 12.4 presents the contribution of telecommunication technologies to smart urban water management, highlighting improvements in operational efficiency, energy savings, network reliability, and data handling capabilities.

IoT integrated with drip and sprinkler systems can reduce water usage for irrigation by 30–50% while increasing crop production by 10–20%. In addition, smart irrigation systems lower energy consumption from pumping by 15–22% and reduce [13] fertilizer runoff by 25–35%. Fields that are 50–500 hectares in size can be monitored to improve water allocation and soil management with the use of drone surveys and satellite imaging. The digitalization of agriculture and its potential for water conservation.

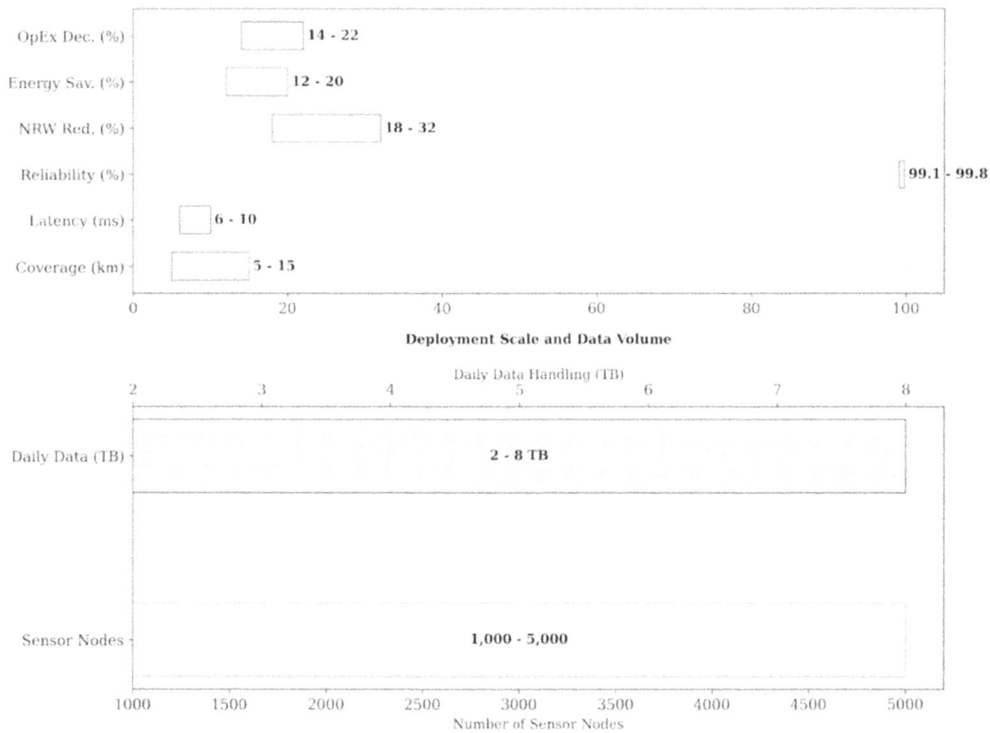

FIGURE 12.4 Telecom innovations for smart urban water solutions.

12.4.3 Wastewater Treatment Technologies

Modern smart wastewater treatment technologies use automated sensing and AI-driven process control, biological reactors, and membrane filtration systems to enhance the efficiency of purification processes and to increase the potential for water reuse. The average generation [14] of urban wastewater is between 80 and 120 L/person/day. Of this, between 60% and 80% can be treated and reused using the technologies discussed. Treatment plants using real-time monitoring sensors can reliably perform to keep Biological Oxygen Demand (BOD) levels under 20 mg/L, Chemical Oxygen Demand (COD) under 50 mg/L, and Total Suspended Solids (TSS) under 30 mg/L.

TABLE 12.3 Key Numerical Parameters of Smart Urban Water Solutions

PARAMETER	*NUMERICAL RANGE*	*UNIT*
Per Capita Water Demand	120–150	L/person/day
Distribution Efficiency	90–95	%
Non-revenue Water Target	< 15	%
Pipeline Pressure	2–7	bar
Leak Detection Time	5–10	min
Irrigation Water Saving	30–50	%
Wastewater Reuse	60–80	%
Treatment Efficiency	95–99	%
Energy Saving	15–25	%
System Reliability	> 98	%

Ultraviolet (UV) disinfection units eliminate more than 99.9% of pathogens, while the filtration efficiency of membrane bioreactor (MBR) systems is between 95% and 99%. The automated management of sludge reduces the time of this process by 30–40% and the total energy consumed by 12–18%. Smart recycling of [15] wastewater for irrigation, industrial cooling, and recharge of groundwater reduces the demand for fresh water by 25–35% in urban centers. The incorporation of AI to control processes also guarantees the facility will meet environmental discharge regulations and operate reliably enough to exceed 98% uptime overall.

12.5 POLICY FRAMEWORKS AND GOVERNANCE

Aqua smart tech will work best with policies and frameworks that ensure the technologies are safe, effective, and sustainable. Rapid advancement of technology will increase operational efficiency, but without the right policies and administrative structures in place, there will be no guidance, accountability, or operational predictability for the policies. Urban water governance includes, but is not limited to, national policies, municipal bylaws, utility [16] management policies, and models of stakeholder collaboration. If policies are implemented successfully, coverage can be improved to 95– 98% of the urban population, operational losses can be improved by 15–25%, and efficiency of infrastructure investment can be improved by almost 20–30%. In major water projects, governance means that all stakeholders will [17] be considered, ecosystems will be preserved, and finances will be managed transparently. Figure 12.5 illustrates the governance and smart water policy integration framework, highlighting improvements in transparency, regulatory compliance, stakeholder participation, and operational efficiency.

12.5.1 Regulatory Standards and Compliance

Standards define the boundary for water quality, reliability of supply, and impact of the discharge to the environment. Compliance frameworks are defined for the protection of the public from water utilities not

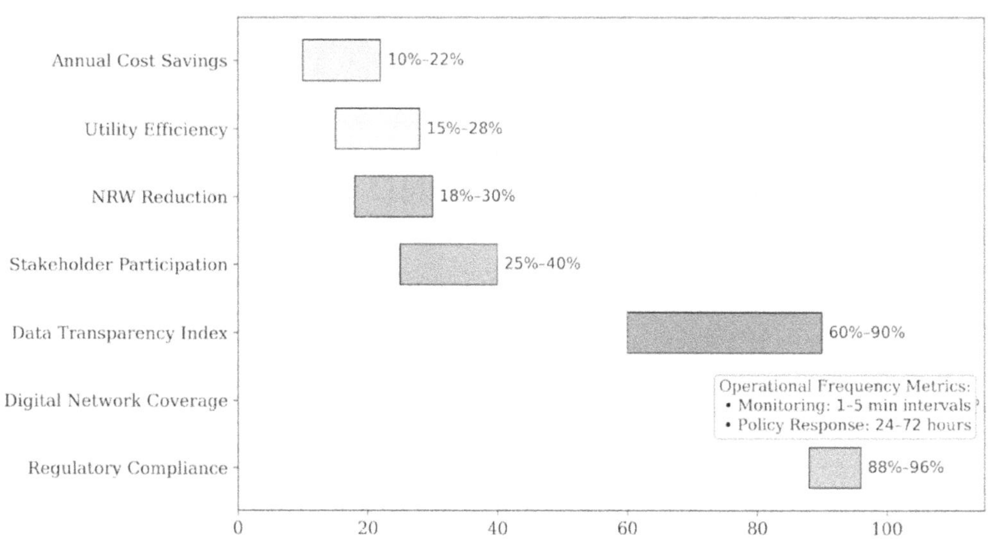

FIGURE 12.5 Governance and smart water policy integration framework involving regulators, utilities, networks, and monitoring.

discharging wastewater responsibly [18–20]. Regulatory standards stipulate water to be potable if it has pH levels from 6.5 to 8.5, turbidity of less than 5 NTUs, residual chlorine of 0.2–0.5 mg/L, and BOD of less than 20 mg/L. Automated smart compliance monitoring has a reporting accuracy of more than 95% and has reduced the need for compliance inspections by 30–40%. Policies are incorporating metrics of service delivery efficiency such as less than 15% non-revenue water, greater than 98% uptime on supply, and 24/7 monitoring.

12.5.2 Public–Private Partnerships (PPP)

Public–Private Partnerships (PPP) are collaborations between Public Authorities and Private Companies involving the same finances in the provision of services/substantive infrastructure development [21–23]. PPP models typically reduce the time of project completion by 20–35%. Compared to purely public funding, they cost 15–25% less, and the inflow of funds is 30–40% greater. Concessions are typically in the range of 10–30 years, and ROI (return on investment) is usually between 8% and 15%. Through contracts (that are often performance based) the reliability of service is increased to over 95%, and the time to respond to maintenance is reduced from 24 hours to 4–6 hours [24]. Hence, risk sharing and expertise in the technology can be provided in the scalable modernization of water systems.

12.5.3 Data Privacy and Cybersecurity in Water Systems

The advanced water system is dependent on comprehensive digital frameworks and cloud computing; therefore, the protection of data and the threat of cyber-attacks need to be more adequately governed [25–27]. The water services industry builds extremely large data sets, in which, over a period of 24 hours, a dataset may be over 5–10 TB and includes information such as customer consumption patterns, operational logs, and system blueprints. Certain cybersecurity methods, such as encryption, firewalls, and multifactor authentication, can lower the chances of a security breach by 40–60% and keep the resiliency of the network above 99%. Global standards for the digital world suggest that password changes happen every 60–90 days, cyber intrusion responses take less than 5 minutes, and data is kept for 5–10 years. These good cybersecurity policies are primarily aimed at protecting the system's end users. In addition, there are policies that protect the system in a more defensive manner by protecting the unit from system downtime.

12.6 FUTURE TRENDS AND CHALLENGES

Smart water management is influenced by the rapid development of technologies, the digitization of urban spaces, and worsening uncertainties in the environment. Utility companies must deploy adaptive, data-driven techniques in order to keep reliability in supply as water demands grow in the foreseeable future (2050) by 40–55% and cities expand by 2–5%. Future water management systems will integrate intelligent technologies, advanced analytics, renewable energy, and decentralized treatment solutions to enhance operational efficiency and minimize resource wastage. Operational efficiency will improve to 95% and resource waste will be cut by 30–40%. Worsening uncertainties in the environment, less tech in dispersed systems, open tech (including cyber) in systems, skill gaps around dispersed tech, and systems design will remain critical obstacles. The balance of the above will be the key components of a systems design in a less tech-dispersed future.

12.6.1 Aspects of Smart Cities Integrated with Other Disciplines

Other areas of smart cities, such as intelligent transport, energy, waste management, digital governance, and smart water management, have started to integrate into connected ecosystems. With the integration of urban digital dashboards, real-time centralized monitoring and multi-sector optimization are possible. With urban digital dashboards, data analytics are possible, as the dashboards can capture 10–20 TB of data in a single day with a response time of 5–10 milliseconds. Integration can lead to a 25–35% increase in the coordination of services, a 15–20% decrease in the duplication of services, and more than 90% increase in the efficiency of services delivered to the citizen. Interoperable systems also provide automated systems for billing, predictive maintenance, and emergency responses across multiple city services.

12.6.2 Sustainability and Climate Resilience

Climate change leads to more uncertainty with events like irregular rainfall, longer droughts, and more frequent floods, all of which impact water availability and the stability of infrastructure. Sustainable smart water systems, by streamlining their operational processes, can supplant extreme supply disruptions with reliable diversifications in storage supply and source sustainability. Developing renewable energy-powered pumping stations and smart systems for more advanced rainwater harvesting and reuse can reduce net freshwater extraction by 25–40% and partially alleviate operational processes of optimally predictive artificial intelligence to drought preparedness, improving response efficiency by 30–45%. These systems reduce net carbon emissions by 20–30%, and most of them integrate with existing water infrastructure while utilizing modern predictive technologies. These progressive systems address the most stressed areas of infrastructure while preserving ecological balance and ensuring sustainable resource management for the future. As shown in Figure 12.6, future smart water systems are expected to exhibit adaptive responses to changing environmental and operational conditions through AI-enabled optimization and real-time control.

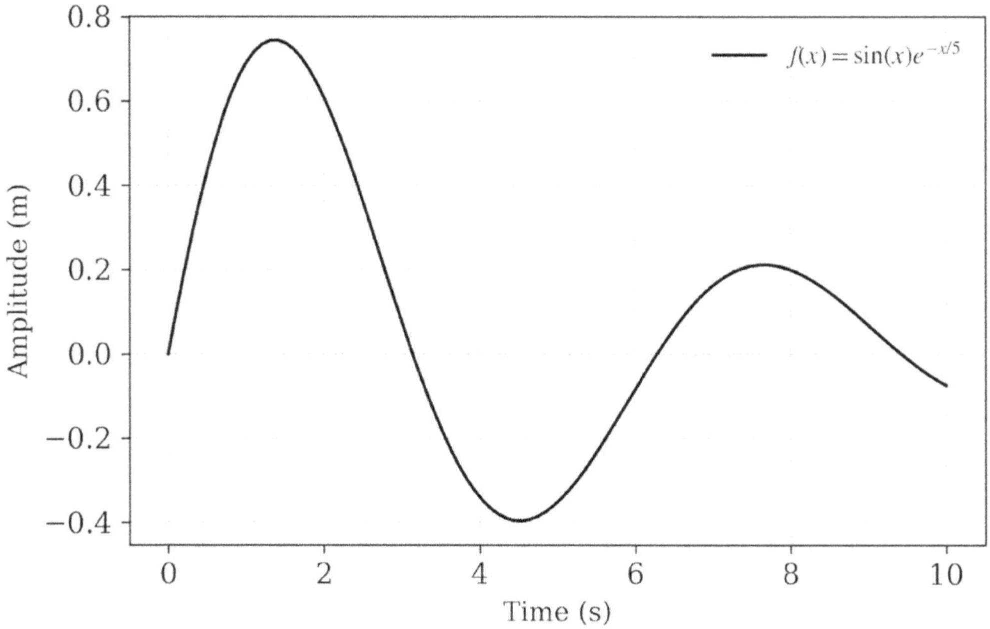

FIGURE 12.6 Future intelligent water systems integrated with smart cities and AI.

12.6.3 Emerging Innovations and Research Directions

The latest water management innovations include systems built on blockchain technology for billing, filtration sheets using nanotechnology, autonomous robotic devices for inspections, and comprehensive water network digital twins for simulations. Given research trends, robotic inspection may become ±1 cm, and filtration efficiency may surpass 99.5%. Also, the predictive reliability of digital twins may beat 92–96%. This is particularly true for systems in which the research and development investment is increasing by 15–25% annually. Among the new systems, the challenges of the high cost of implementation, the need for technical training, the interoperability of systems, and the costs of cybersecurity are still the most significant. Out of the numerous systems, continuous innovation and strong policy backing will determine the greatest success of the smart water ecosystems of the future.

12.7 DISCUSSION

Intelligent Water Management Systems have shown average improvements of 10–18% lower operational costs, 20–35% lower water loss, and 90–95% reliability of networks compared to previous water monitoring systems. AI analyzes data at 85–97% accuracy and 5–20 TBs worth of data daily while IOT sensors at intervals of 1–5 seconds and smart meters with +/– 1% accuracy monitor networks in real time. Intelligent distribution systems (IDS) redistribute the network more effectively, leading to increased operational efficiency of distribution from 70–75% to 90–95% while reducing maintenance costs by 20–30% and the capacity to locate and identify leaks of 0.1–0.3 L/s within 5–10 minutes. Wastewater reuse and irrigation water savings governance frameworks save 30–50% while increasing coverage from 95% to 98% and controlling water loss from 15% to 25%. The highest priority issues are the risk of cyberattacks, the high cost of implementation, and a lack of operational and technical skills.

12.8 CONCLUSION

A smart water network based on automation, the internet of things (IOT), and artificial intelligence (AI) can provide urban areas with over ~25–40% water conservation, 12–20% energy savings, 95–99% efficiency in the treatment of water, and a system uptime of >98%. Water use in urban areas is expected to increase by 40–55% until the year 2050. In order to satisfy this demand, a level of efficiency of more than 90% is required alongside a non-revenue water level of less than 15%. Future technologies like digital twins, nanofiltration, and robotic inspection have a proven 15–25% annual increase, providing a flexible, low-carbon (20–30% emissions decrease), resilient, and data-centric urban water ecosystem.

REFERENCES

1. S. Eslamian and F. Eslamian, *Blockchain Technology for Water and Environmental Systems*. CRC Press, 2025. Doi: 10.1201/9781003599593.
2. R. Jayabal, R. Sivanraju, and P. Prabhakar, "Low-Carbon Agricultural Strategies: Toward Environmental Protection and Energy Efficiency," *Energy Science & Engineering*, Vol. 13, No. 12, Pp. 6611–6627, 2025. Doi: 10.1002/Ese3.70320.

3. J. Yuan, F. Hadafi, A. S. Nik, B. Ghasemzadeh, and X. Kong, "Architectural Contributions To Climate Change Mitigation In Urban Residential Contexts Through A Global Qualitative Approach," *Discover Sustainability*, Vol. 6, No. 1, 2025. Doi: 10.1007/S43621-025-01791-9.
4. J. S. Chatha, "Advances In Sustainable Technologies for Industry and Environment for a Greener Future," *Environmental Science and Pollution Research*, Vol. 32, No. 58, pp. 30971–31017, 2025. Doi: 10.1007/S11356-025-37299-Y.
5. N. Sharma, C. Chakraborty, and R. Kumar, "Optimized Multimedia Data through Computationally Intelligent Algorithms," *Multimedia Systems*, Vol. 29, No. 5, pp. 2961–2977, 2023.
6. D. Srivastava, N. Sharma, D. Sinwar, J. H. Yousif, and H. P. Gupta, eds., *Intelligent Internet of Things for Smart Healthcare Systems*. CRC Press, 2023.
7. A. P. Onyena and K. Sam, "The Blue Revolution: Sustainable Water Management for a Thirsty World," *Discover Sustainability*, Vol. 6, No. 1, 2025. Doi: 10.1007/S43621-024-00631-6.
8. R. D. Lestari, "Artificial Intelligence-Driven Solar Smart Irrigation for Sustainable Agriculture: Trends, Challenges, and Sdg Implications – A Systematic Review," *Smart Agricultural Technology*, Vol. 12, 2025. Doi: 10.1016/J.Atech.2025.101665.
9. F. Karakulov Et Al., "The Impact of Climate-Smart Technology Adoption on Agricultural Productivity and Environmental Sustainability," *Emerging Science Journal*, Vol. 9, No. 6, pp. 3087–3115, 2025. Doi: 10.28991/Esj-2025-09-06-012.
10. N. Sharma, M. Soni, S. Kumar, R. Kumar, N. Deb, and A. Shrivastava, "Supervised Machine Learning Method for Ontology-based Financial Decisions in the Stock Market," *ACM Transactions on Asian and Low-Resource Language Information Processing*, Vol. 22, No. 5, pp. 1–24, 2023.
11. N. Sharma and U. Batra, "A Review on Spatial Domain Technique Based on Image Steganography," in *Proceedings of the 2017 International Conference on Computing and Communication Technologies for Smart Nation (IC3TSN)*. IEEE, October 2017, pp. 24–27.
12. R. A. Castanho, A. Behradfar, and A. Loures, "Integrating Gis and Remote Sensing Into Digital Education for Sustainable Land-Use Management: A Cross-Border Perspective," Lecture Notes in Networks and Systems, 2026, pp. 31–39. Doi: 10.1007/978-3-032-09074-4_4.
13. Y. T. Bahta, S. A. Nyaki, H. Jordaan, M. Monteiro, and E. Nesamvuni, "Resilience of Small-Scale Maize Irrigation Farmers To Climate Change Stress In South Africa," *Mitigation and Adaptation Strategies for Global Change*, Vol. 31, No. 1, 2026. Doi: 10.1007/S11027-025-10280-3.
14. N. Sharma and U. Batra, "An Enhanced Huffman-PSO Based Image Optimization Algorithm for Image Steganography," *Genetic Programming and Evolvable Machines*, Vol. 22, No. 2, pp. 189–205, 2021.
15. N. Sharma and U. Batra, "A Study on Integrating Crypto-stego Techniques to Minimize the Distortion," in *International Conference on Recent Developments in Science, Engineering and Technology*. Singapore: Springer Singapore, 2017, October, pp. 608–615.
16. T. Begna and R. B. Wakweya, "Mitigating Climate Change Impacts on Food Security Via Climate-Smart Agriculture," *Plant Stress*, Vol. 19, 2026. Doi: 10.1016/J.Stress.2025.101152.
17. O. P. Agboola, "Comparative Assessment of Climate-Responsive Design and Occupant Behaviour Across Türkiye's Building Typologies for Enhanced Utilisation and Performance," *Buildings*, Vol. 16, No. 1, 2026. Doi: 10.3390/Buildings16010018.
18. S. Koshy and P. Sundaram, "Harnessing Ai and Biomimicry for Resource Recovery: Advancing Circularity in Smart Infrastructure Systems," Lecture Notes in Networks and Systems, 2026, pp. 286–300. Doi: 10.1007/978-3-032-06700-5_29.
19. S. Gill, N. Sharma, C. Gupta, and A. Samanta, "Attendance Management System Using Facial Recognition and Image Augmentation Technique," 2021 International Conference on Intelligent Technology, System and Service for Internet of Everything (ITSS-IoE). IEEE, 2021, November, pp. 1–6.
20. M. Beri, B. Kumar, S. Tiwari, N. Sharma, H. Vashishtya, and P. Chaudhary, "IoT Based Health Monitoring System Built on ESP32," 2022 2nd International Conference on Advance Computing and Innovative Technologies in Engineering (ICACITE). IEEE, 2022, April, pp. 454–458.
21. S. Zhang and H. Xiong, "Linking Land Policy To Food System Sustainability: Farm Scale Effects on Smart Irrigation Adoption in Northern China," *Frontiers in Sustainable Food Systems*, Vol. 9, 2026. Doi: 10.3389/Fsufs.2025.1736406.
22. J. B. H. Yap, W. N. Lee, M. Skitmore, S. C. Loo, and W. S. Yap, "Innovation To Greener Production: Unveiling Drivers and Pathways in Smart Sustainable Construction," *Production Planning and Control*, 2026. Doi: 10.1080/09537287.2026.2616010.
23. N. Sharma, D. Srivastava, and D. Sinwar, eds., *Artificial Intelligence Technology in Healthcare: Security and Privacy Issues*. CRC Press, 2024.

24. M. Kumari, N. Sharma, R. Chauhan, K. Joshi, V. Madaan, and G. Sunil, "Investigation of OWC System Incorporating Different Filters," 2024 4th International Conference on Innovative Practices in Technology and Management (ICIPTM). IEEE, 2024, February, pp. 1–4.
25. E. Cozzi, D. Menozzi, G. Maesano, M. Canavari, and C. Mora, "Digital Technology Adoption Among Italian Farmers: An Extended Technology Acceptance Model Approach in the Horticultural Sector," *Bio-Based and Applied Economics*, Vol. 14, No. 4, pp. 29–43, 2026. Doi: 10.36253/Bae-17364.
26. G. Mwango, B. Alunda, D. Wachira, and T. Niet, "Assessing The Impacts of Climate Change In Kenya Using Climate, Land, Energy, and Water Systems (Clews) Modelling To Provide Policy Insights," *Sustainable Futures*, Vol. 11, 2026. Doi: 10.1016/J.Sftr.2026.101714.
27. P. Sharma and N. Thakur, "Advancing Urban Food Sustainability: Biotechnology and Iot Synergies in Vertical Greenhouses," *Bioresource Technology Reports*, Vol. 33, 2026. Doi: 10.1016/J.Biteb.2026.102548.

13 Impact Assessment of Managed Aquifer Recharge Structures Using MNDWI in Kharun River Basin, Chhattisgarh, India

Suraj Kumar, Chandan Kumar Singh, and Sudarsan Sahu

13.1 INTRODUCTION

Managed aquifer recharge structures play a crucial role as a vital link between surface water and groundwater. This structure holds water on the surface. The water comes through precipitation in the river basin area, and the series of Managed Aquifer Recharge Structures helps to simultaneously slow down runoff at the gauging sites [1]. The surface water infiltrates into the soil and goes down to the groundwater level and has an impact on the recharge groundwater level [2]. This may lead to groundwater recharge. The recharge of groundwater depends on various factors, such as permeability, porosity, water content, and air voids, and these other factors also depend on geological characteristics.

Permeability is an essential property of soil to permit surface water into groundwater, and pores create a passage between the surface and groundwater [3]. If soil becomes saturated, pores will play diverse roles, such as overcoming the need for water to plant life, and the capacity for water holding may be increased [4]. At the end of the assessment of surface water in many indices, such as NDWI, MNDWI [5], etc., with the help of the Remote Sensing Technique [6]. The geological characteristics may or may not also help groundwater recharge, such as igneous and metamorphic rocks insufficient capabilities to push recharge groundwater [7].

The NDWI and MNDWI stand for Normalized Difference Water Index and Modified Normalized Water Index, respectively [8]. The main differences between the two indices are that the first index captures long waves with major waterbodies [9], and the second index captures both images of short and

DOI: 10.1201/9781003742227-13

long waves with major waterbodies, minor waterbodies like ponds [10], etc. So, we use the Modified Normalized Water Index (MNDWI) assessment of surface waterbodies. These indices have one band, which denotes one wavelength.

In this book chapter, the MNDWI dataset [11] is used to identify the extent of major and minor waterbodies in images extracted from the MODIS satellite with the use of Google Earth Engine [12] and process and analysis through the Geographical Information System (GIS). The analysis is of both the duration of the pre-monsoon and post-monsoon seasons. Hence, the focused objectives with the help of the location of Managed Aquifer Recharge Structures [13] have maximum water spread to the post-monsoon season. This post-monsoon water spread helps in using water demand in the pre-monsoon.

13.2 MATERIALS AND METHODOLOGY

13.2.1 Study Area

The Kharun River basin is located in the central part of the Indian state of Chhattisgarh. This basin is present between the 20°33′30″ N to 21°33′38″ N of latitude and 81°17′51″ E to 82°0′0″ E of longitude. The area of the Kharun River basin is 4148 Km^2 (Figure 13.1). This region is semi-arid and has an annual average rainfall of 1216.75 mm. The highest and lowest elevations in this area are 450 m and 249 m, respectively. This region has extensive sedimentary rocks, including limestone, sandstone, and shale,

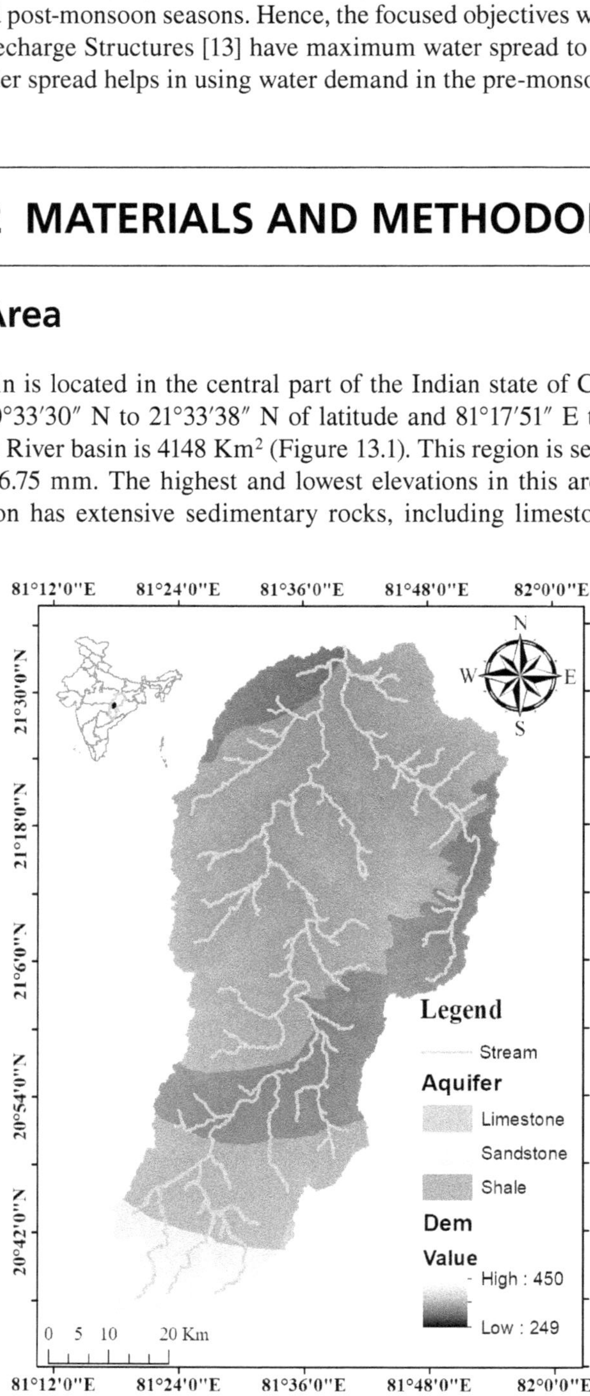

FIGURE 13.1 Map of the study area of the Kharun River basin.

TABLE 13.1 Aquifer System of the Study Area

SL. NO	*AQUIFER*	*AREA (KM²)*
1	Limestone	2799
2	Sandstone	284
3	Shale	916

covering 2,799 km², 284 Km², and 916 Km², respectively (source: India-WRIS). These aquifers are a type of sedimentary rock, and limestone has a specific property that doesn't have porous structures, while sandstone has porous structures [14]. But shales may be present in the Kharun River Basin. The study area had a recharge structure in this river basin. There were two types of recharge structure: check dam and a stopdam, numbered 160 and 4, respectively Table 13.1 presents the distribution of major aquifer systems within the study area and their corresponding spatial extent.

13.2.2 Geomorphology

Geomorphology (combining two words: geomorpho, meaning landforms, and logy, meaning to study) is the study of landforms and their characteristics. They have several geomorphology characteristics present in the area of interest, such as alluvial plains and floodplains (source: bhukosh), which are also present in the Kharun River. Additionally, there are many domestic and public waterbodies in that region, which enhance groundwater recharge and water spread. Figure 13.2 illustrates the geomorphological characteristics of the Upper Seonath Basin, highlighting major landforms such as alluvial plains, flood plains,

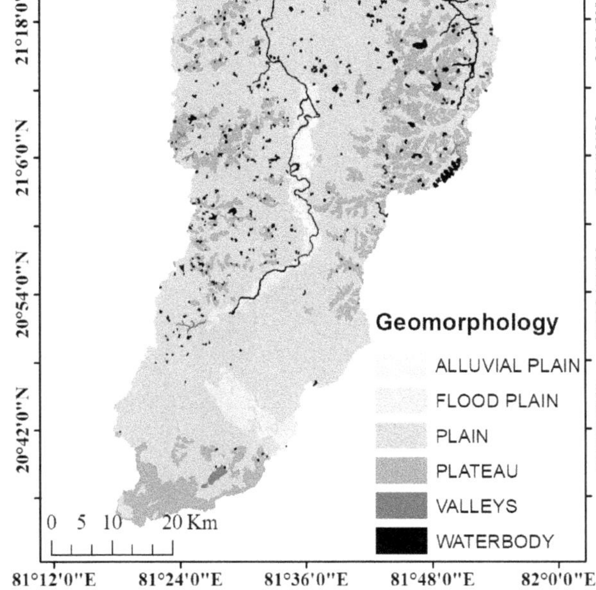

FIGURE 13.2 Geomorphology of the upper Seonath basin.

TABLE 13.2 Geomorphology System of the Study Area

SL. NO	*GEOMORPHOLOGY*	*AREA (KM²)*
1	Alluvial Plain	30
2	Flood Plain	173
3	Hills and Valleys	115
4	Plain	3650
5	Plateau	87

plateaus, valleys, and water bodies. One of the river segments exhibits a meandering effect. Table 13.2 presents the distribution of geomorphological units within the study area and their corresponding spatial extent.

13.2.3 Lithology

One of the other geological characteristics enhancing the water spread in that region is lithology. Lithology is the study of rocks. The area of interest is to identify where rocks are present or not. If rock is not present where the water spreads, the impact of recharge structures. In this region, 35 Km^2 is Alluvium, 25 Km^2 is Igneous rocks. This study area has a majority of sedimentary rock, covering 98.55% of the total area (source: Bhukosh dataset). Such sedimentary rocks as limestone and sandstone are present.

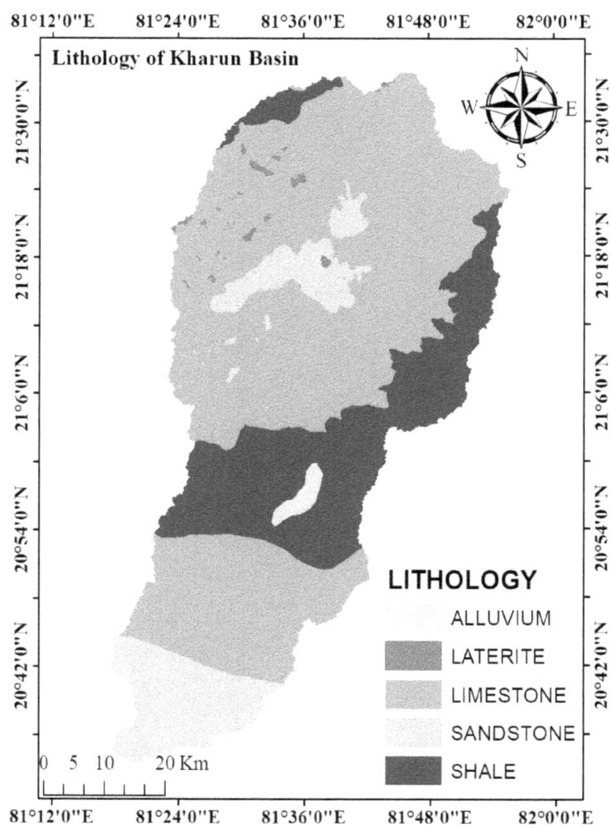

FIGURE 13.3 Lithology of the upper Seonath basin.

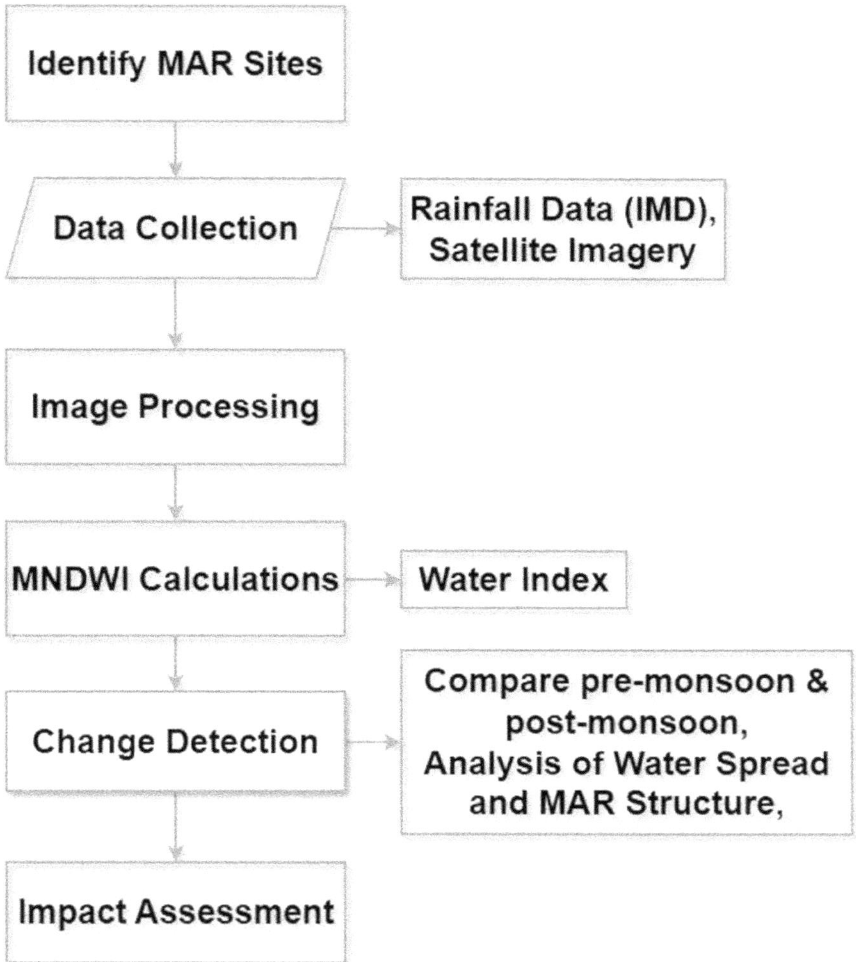

FIGURE 13.4 Flow chart of the methodology.

13.2.4 Methodology

This methodology was adopted to identify the change detection of waterbodies in the area of interest, as shown in Figure 13.4. This had identified the location of recharge structures in the area of interest. These identified structures were analyzed over a period of time. These water spread analyses were made with the help of satellite imagery from the MODIS satellite, rainfall data extracted from the Indian Meteorological Department (IMD), and discharge data from India-WRIS [15]. The data extracted from the formula of water indices, such as MNDWI and equation (1), shows the calculation of MNDWI [16]. The MNDWI is derived from the green and shortwave infrared bands of satellite sensors and is widely used for the identification and mapping of surface water bodies [17]. These satellite images were extracted using Google Earth Engine (GEE), following processing and mapping with the Geographical Information System (GIS) [16]. The Raster Calculator was employed to identify and map variations between the pre-monsoon and post-monsoon conditions, and the corresponding pixel counts were subsequently analyzed [18]. The defined Pre-Monsoon period is April, May, and June, and also the post-monsoon period is October, November, and December [19].

$$MNDWI = \frac{GREEN - SWIR}{GREEN + SWIR} \tag{1}$$

where

−1 was the pre-monsoon water spread pixel count of the surface water body,
Zero was the plan area of the area of interest.
+1 was the post-monsoon water spread pixel count of the surface water body.

13.3 RESULT

13.3.1 Rainfall Trends Analysis

Change detection analysis of the spread area of the waterbodies was mapped. In this book chapter, the series of change detection is recognized and analyzed in terms of pre-monsoon and post-monsoon [18, 20]. The identification of water bodies in the study area was carried out by analyzing the spatial extent of surface water during the post-monsoon period. The increased water extent influenced by Managed Aquifer Recharge (MAR) structures was taken into account during the assessment. The extent of water bodies was quantified using the number of pixels classified as water bodies [21]. The pixel count of water-bodies on the identified recharge structures of the location, using the following methodology, is shown in Figures 13.5 and 13.6. The change in pixel count showed a trend from 2019 to 2024, as shown in Table 13.3. In Table 13.3, the pixel count of surface waterbodies shows a sequential increase in water quality. In the area the average annual rainfall was 1216.75 mm (source: https://mausam.imd.gov.in/), as shown in Table 13.4.

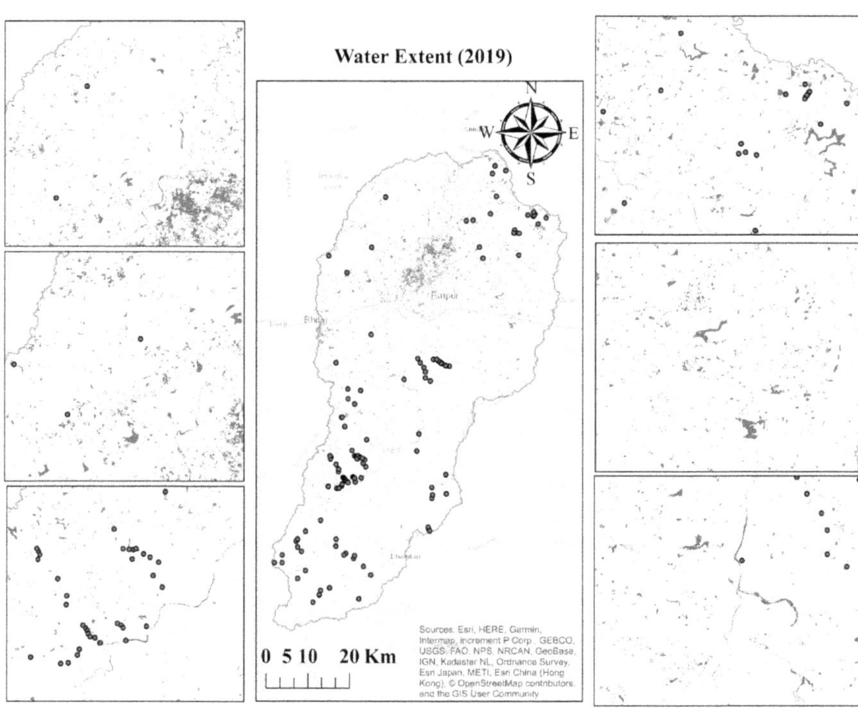

FIGURE 13.5 Water extent in the year 2019 of the study area.

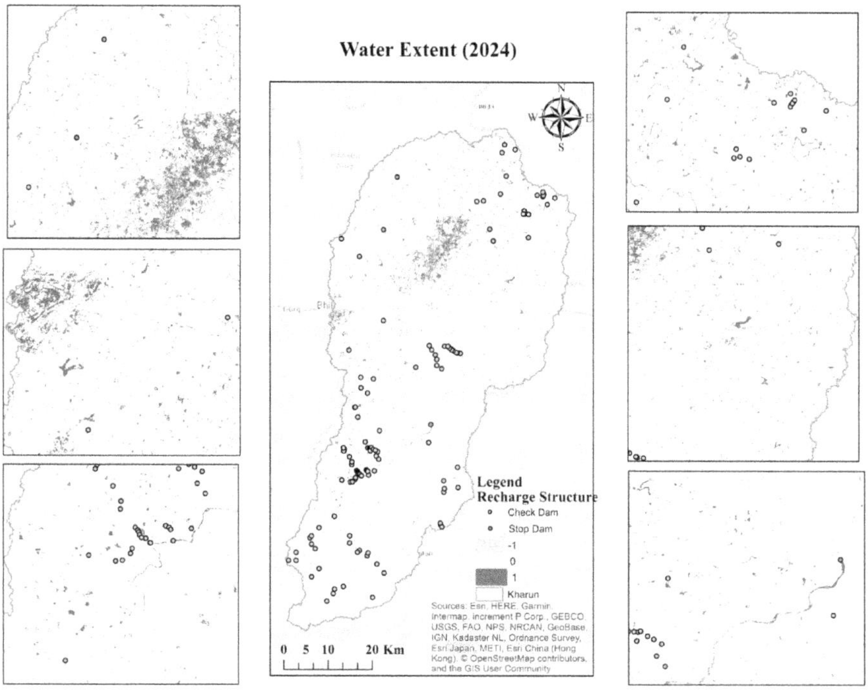

FIGURE 13.6 Water extent in the year 2024 of the study area.

TABLE 13.3 Pixel Count Data of the Study Area of Waterbodies

	PIXEL COUNT					
VALUE	*2019*	*2020*	*2021*	*2022*	*2023*	*2024*
-1	8853	21112	11723	11830	8676	14816
0	10764326	10966051	10932070	10929342	10908365	10803284
1	406223	192239	235609	238230	262361	361302

TABLE 13.4 Annual Rainfall of the Study Region

YEAR	*ANNUAL RAINFALL (MM)*
2019	1241.77
2020	1267.47
2021	1107.24
2022	1256.6
2023	1343.89
2024	1108.85

13.3.2 Seasonal Rainfall Analysis

This rainfall was falling over the catchment of the Kharun River Basin. In this catchment of the Kharun Basin, we also evaluated the rainfall pre-monsoon, monsoon, and post-monsoon categories. The estimated average seasonal rainfall values for March, April, and May were 75.08 mm [19]. This was similar to the monsoon (June, July, and August) and post-monsoon (September, October, and November) periods, which were 6316.11 mm and 326.35 mm, respectively, and also had the average seasonal rainfall in pre-monsoon and post-monsoon periods, which were 1052.69 mm and 54.39 mm, respectively [19].

13.3.3 MNDWI of Kharun River Basin

Satellite imagery from 2019 and 2024 was analyzed to evaluate changes in water spread between the pre-monsoon and post-monsoon seasons and to assess the impact of recharge structures on surface water availability. The first image of the water extent of the year 2019 has the maximum water spread pixel count of post-monsoon of all five years because of heavy rainfall and saturated landmass in the study area region. The post-monsoon water extent is maximum because of recharge structures, as the region that had accumulated rainfall and maximum monsoon rainfall in the year has less drought possibilities. So, that's the region that had the maximum water spread. One of the other regions is recharge structures that hold water. The comparative analysis of water extent is shown in Figures 13.5 and 13.6, respectively.

13.3.4 Comparative Result of Water Extent

The comparison between the surface water spread in 2019 and 2024. These water extents were pre-monsoon and post-monsoon and are significantly shown in Table 13.3. Table 13.3 clearly shows the impact of managed aquifer recharge structures in the Kharun watershed. Figures 13.5 and 13.6 show the water spread in pre-monsoon and post-monsoon, clearly indicating the impact of managed aquifer recharge structures. The structures have increasing water availability during drier periods by storing water surpluses from the wet season in the surface and subsurface [22–24].

where –1 was the pre-monsoon water spread pixel count of the surface water body,
Zero was the plan area of the area of interest.
+1 was the post-monsoon water spread pixel count of the surface water body.

TABLE 13.5 Seasonal Rainfall in the Study Area Region

	SEASONAL RAINFALL (MM)		
YEAR	*PRE-MONSOON (MARCH, APRIL, MAY)*	*MONSOON (JUNE, JULY, AUGUST, SEPTEMBER)*	*POST-MONSOON (OCTOBER, NOVEMBER, DECEMBER)*
2019	29.43	1106.88	66.15
2020	138.49	978.82	35.9
2021	79.43	895.58	124.41
2022	25.07	1134.1	64.45
2023	108.74	1208.08	26.56
2024	69.32	992.65	8.88

TABLE 13.6 The Pixel Count in the Year of the Study Area

	PIXEL COUNT		
VALUE	*2019*	*2024*	*% CHANGE*
-1	8853	14816	40.25
0	10764326	10803284	0.36
1	406223	361302	-12.43

13.4 CONCLUSION

This study aims to conduct an impact assessment of managed aquifer recharge on the analysis of the water spread of the surface water body. The waterbodies are in the Kharun watershed. The Kharun River Basin, located in the central region of Chhattisgarh, India, was thoroughly examined for water extent, which means the change in surface water body pre-monsoon and post-monsoon of the pixel count over the years, and the impact of recharge structures. The analysis of MNDWI shows noticeable changes in pre-monsoon and post-monsoon in waterbodies. There is a significant change in pre-monsoon water extent, with a percentage change of nearly 40% of the pixel count in 2019. The change will be full of water demand in the summer season, which will have a positive impact on recharge structures in the River Basin. The managed Aquifer Recharge Structures have increased water availability during non-monsoon seasons by storing water surplus from wetter periods in the surface and subsurface.

REFERENCES

1. Simon et al, "Wetland management in recharge regions of regional groundwater flow systems with water shortage, Nyírség Region, Hungary," *Water*, vol. 15, no. 20, Art. no. 20, 2023, doi: 10.3390/w15203589.
2. C. K. Singh and Y. B. Katpatal, "Evaluating control of various hydrological factors on selection of groundwater-level monitoring networks in irrigated areas using a geospatial approach," *J Irrig Drain Eng*, vol. 143, no. 8, p. 05017003, Aug. 2017, doi: 10.1061/(ASCE)IR.1943-4774.0001213.
3. S. Sahu and D. Saha, "Role of shallow alluvial stratigraphy and Holocene geomorphology on groundwater arsenic contamination in the Middle Ganga Plain, India," *Environ Earth Sci*, vol. 73, no. 7, pp. 3523–3536, Apr. 2015, doi: 10.1007/s12665-014-3637-3.
4. Y. Park, Y. Kim, S.-K. Park, W.-J. Shin, and K.-S. Lee, "Water quality impacts of irrigation return flow on stream and groundwater in an intensive agricultural watershed," *Sci Total Environ*, vol. 630, pp. 859–868, Jul. 2018, doi: 10.1016/j.scitotenv.2018.02.113.
5. Y. Miura, M. Shamsudduha, A. Suppasri, and D. Sano, "A global multi-sensor dataset of surface water indices from landsat-8 and sentinel-2 satellite measurements," *Sci Data*, vol. 12, no. 1, p. 1253, Jul. 2025, doi: 10.1038/s41597-025-05562-z.
6. M. H. R. Moghaddam, A. Sedighi, and M. A. Fayyazi, "Applying MNDWI index and linear directional mean analysis for morphological changes in the Zarriné-Rūd River," *Arab J Geosci*, vol. 8, no. 10, pp. 8419–8428, Oct. 2015, doi: 10.1007/s12517-015-1795-6.
7. GEC, *Report of the Ground Water Resource Estimation Committee*, 2015.
8. J. K. S. Chauhan and A. K. Paswan, "Multi-indices assessment of spatiotemporal dynamics of climate-driven surface water variability for sustainable water management in semi-arid climate system," *Discov Water*, vol. 5, no. 1, p. 98, Nov. 2025, doi: 10.1007/s43832-025-00302-1.
9. G. T. Tassew, A. B. Mitiku, T. M. Mekonnen, F. B. Geleta, and A. T. Woldemichael, "Evaluating flood extent using Synthetic Aperture Radar (SAR) and Modified Normalized Difference Water Index (MNDWI) methods," *Remote Sens Earth Syst Sci*, vol. 8, no. 2, pp. 673–695, Jun. 2025, doi: 10.1007/s41976-025-00211-w.
10. V. R. Mandla, S. P. Nerella, M. Choudhary, and V. S. S. Peddinti, "Impact study on desiltation of water tanks in rural areas using spatial technology: a case study work under MGNREGA," in *Advances in Geotechnical and Transportation Engineering*, S. Saride, B. Umashankar, and D. Avirneni, Eds. Singapore: Springer, 2020, pp. 85–99. doi: 10.1007/978-981-15-3662-5_9.
11. NASA, "MOD09A1.061 terra surface reflectance 8-day global 500m," 2000, doi: 10.5067/MODIS/MOD09A1.061.
12. X. Zhang, "How does topography control topography-driven groundwater flow? - Zhang - 2022 - Geophysical research letters - Wiley Online library," Accessed: Aug. 06, 2025. [Online]. Available: https://agupubs.onlinelibrary.wiley.com/doi/full/10.1029/2022GL101005
13. S. Sloan, P. G. Cook, and I. Wallis, "Managed aquifer recharge in mining: a review," *Groundwater*, vol. 61, no. 3, pp. 305–317, 2023, doi: 10.1111/gwat.13311.

14. K Subramanya, *Engineering Hydrology*, 3rd ed. New Delhi: Tata McGraw-Hill, 2008. [Online]. Available: http://ndl.ethernet.edu.et/bitstream/123456789/9211/1/Engineering%20Hydrology.pdf
15. P. Kalura, A. Pandey, V. M. Chowdary, and D. Dayal, "Evaluating hydrological responses of satellite precipitation products over an Indian tropical catchment through a distributed physical model," *Hydrological Processes*, vol. 38, no. 9, p. e15275, 2024, doi: 10.1002/hyp.15275.
16. Dr. Mustafa T. Mustafa, Dr. Khalid I. Hassoon, Dr. Hussain M. Hussain, and Modher H., "Using water indices (NDWI, MNDWI, NDMI, WRI and AWEI) to detect physical and chemical parameters by apply remote sensing and gis techniques," 2017, doi: 10.5281/zenodo.1040209.
17. A. M. Rad, J. Kreitler, and M. Sadegh, "Augmented normalized difference water index for improved surface water monitoring," *Environ Model Softw*, vol. 140, p. 105030, Jun. 2021, doi: 10.1016/j.envsoft.2021.105030.
18. Mohammed Meharoof, Vinod Kumar Yadav, Arpita Sharma, Shekhar Nath Ojha, and V. Anitha, "Land use land cover change detection for fish culture area mapping in the peechi reservoir," 2025, doi: 10.1007/s12524-025-02206-y.
19. Pradeep Kamaraj, Deepa Subramani, Dmitry E. Kucher, Muhammad Aslam, Yahia Said, and Aqil Tariq, "Rainfall-Runoff Estimation for Groundwater Enhancement Through Remote Sensing and GIS-Based Geostatistical Model." Accessed: Dec. 13, 2025. [Online]. Available: https://onlinelibrary.wiley.com/doi/epdf/10.1002/hyp.70180
20. R. L. Bhatt, "Change detection analysis of surface water bodies using pixel-based model in Jammu district, India," 2026, doi: 10.1007/s44288-025-00381-6.
21. S. Sharma and Yashvita Tamrakar, "Multi-criteria assessment and decision-making approach for planning of groundwater recharge structures in a highland watershed | Discover Sustainability | Springer Nature Link." Accessed: Feb. 09, 2026. [Online]. Available: https://link.springer.com/article/10.1007/s43621-025-02054-3
22. P. Dillon, E. Fernández Escalante, S. B. Megdal, and G. Massmann, "Managed aquifer recharge for water resilience," *Water*, vol. 12, no. 7, p. 1846, Jul. 2020, doi: 10.3390/w12071846.
23. P. Dillon *et al.*, "Sixty years of global progress in managed aquifer recharge," *Hydrogeol J*, vol. 27, no. 1, pp. 1–30, Feb. 2019, doi: 10.1007/s10040-018-1841-z.
24. V. J. Fernandes, P. G. B. De Louw, R. P. Bartholomeus, and C. J. Ritsema, "Machine learning for faster estimates of groundwater response to artificial aquifer recharge," *J Hydrology*, vol. 637, p. 131418, Jun. 2024, doi: 10.1016/j.jhydrol.2024.131418.

14 A Report on Field-Based Investigation of Sinkhole Formation Caused by Excessive Irrigation in the Rice Farm of the Bemetara Region, Chhattisgarh, India

Ravi Kumar and Chandan Kumar Singh

14.1 INTRODUCTION

The overlaying material is weakened by these cavities, which may lead to sinkhole formation and ground collapse. Depending on variables including cavity depth, overburden characteristics, and hydrological circumstances, sinkholes can occur via a variety of processes, including solution, collapse, dropout, buried, caprock, and suffusion [1]. One of the most important and limited resources on Earth's surface, soil is essential to both plant development and human survival. The integrity of soils is seriously threatened by erosion, a natural process. Gully, sheet, and tunnel erosion are among the types of erosion caused by the forces of flowing water and wind that dislodge and move soil. Although sheet and gully erosion appear on the surface, their formation processes are visible to the unaided eye. Tunnel erosion, on the other hand, happens covertly beneath the surface of the Earth and is not immediately visible [2]. Rainfall that reaches the ground swiftly drains into subterranean river systems through conduits in karst depressions, such as sinkholes and skylights, and surface water storage capacity is typically limited in karst locations [3]. Aquifers regulate and balance water movement through a range of hydrological processes, including

DOI: 10.1201/9781003742227-14

rainfall infiltration, runoff generation and transmission, groundwater storage within pores, fractures, and cavities, and the gradual release of stored water to springs, rivers, and pumping wells. These processes maintain groundwater system stability and ensure a sustained water supply over time [4]. Through chemical weathering and breakdown, soluble rocks like gypsum, limestone, and dolomite are the main source of karst landscapes. Features such as dolines (sinkholes), poljes, karren, caves, underground drainage systems, and vanishing streams define it. Although karst is where most caves form, not all caves have karstic origins. Glaciers, coastal cliffs, or lava tubes form certain caves. They have similar ecological, cultural, and hydrological significance despite having different origins. All continents and climatic zones have caves, which are naturally occurring underground spaces big enough for humans to enter. They range from large underground chambers and networks spanning hundreds of kilometers to modest rock shelters [5]. Both natural and man-made elements can contribute to the formation and development of a sinkhole. Bates and Jackson describe a sinkhole as a circular depression in karst regions. The term "anthropogenic sinkholes" can be more appropriately applied when they are not connected to karst. Because the phenomenon is abruptly initiated and projections are difficult to make, sinkholes pose a significant risk to populations. They can be found in a variety of geological and geomorphological settings across numerous nations [6]. The revised regulation takes into account the variations in groundwater flow between a karst system and a porous-medium aquifer and acknowledges that karst soils are susceptible to sinkhole formation. Typical karst features include caves, sinkholes, strong hydraulic linkages between surface water and groundwater through sinking streams and springs, and uneven bedrock surfaces [7]. As China's infrastructure projects expand and building technology advances, more tunnels are being built in karst locations. However, water inrush can pose significant challenges during construction [8]. In this case study, active SGM, GPR, and ERT were chosen as the most appropriate ground-based geophysical techniques for subsurface exploration [9].

14.2 MATERIAL AND METHODOLOGY

14.2.1 Study Area

The Bemetara district, which borders Durg to the north, has a tropical wet and dry climate with hot summers (up to 48°C), a monsoon season from late June to early October, and mild winters. Black soil makes up about 63% of the district's total area, but red, sandy loam, and black soils are also present. Sand, limestone, and quartz are the three main types of minerals. The local economy is dominated by agriculture; the most important crops grown during the Kharif season are paddy, soybeans, and pigeon peas, while during the Rabi season, wheat, gram, linseed, and mustard are grown. Borewells and micro-irrigation systems are the main sources of irrigation when there are no major dams. Another significant research location in this district is Kurud village, where sinkholes have also been found. Because of the substantial groundwater extraction for irrigation, which might result in subsurface instability and sinkhole development, the area is of great scientific relevance. The most prevalent minerals in this area are sand, quartz, and limestone. In the Bemetara district, the primary Kharif crops are pigeon pea, soybean, and paddy; wheat, gram, linseed, and mustard are cultivated during the Rabi season. In regions lacking sizable dams, borewells and micro-irrigation projects serve as the main irrigation sources. They are essential to maintaining agricultural output across the district. Figure 14.1 presents the location map of the study area, showing its position within the state of Chhattisgarh and the national context of India.

14.2.2 Field Investigation Approach

The study was conducted using a strictly field-based observational approach to understand sinkhole development associated with excessive watering in rice farms in the Bemetara region of Chhattisgarh. The

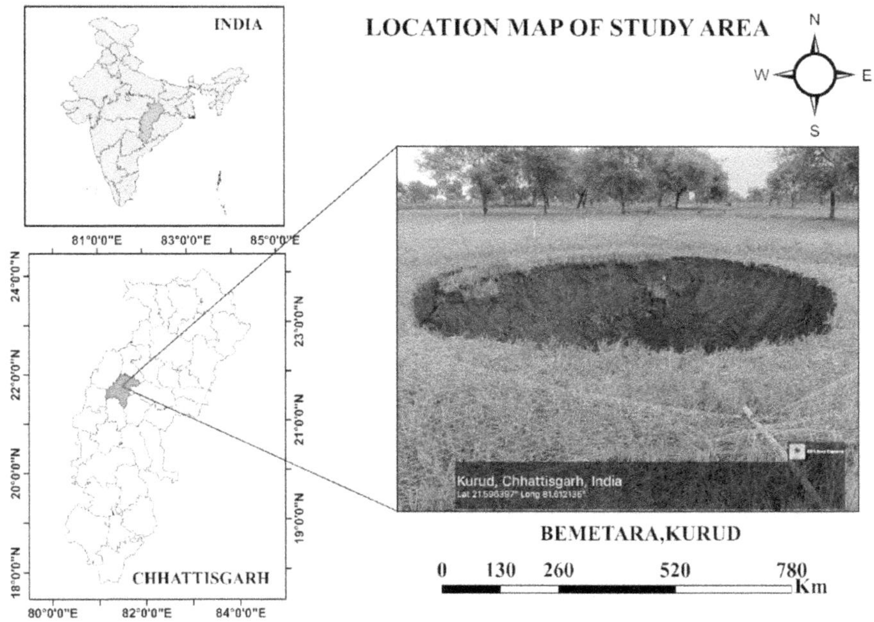

FIGURE 14.1 Location map of study area.

study focused heavily on physical measurements, direct field inspection, evaluating soil conditions, tracking groundwater levels, and communicating with local farmers in order to document ground subsidence features and irrigation practices. Figure 14.2 presents a photograph captured during the field visit to the study area, providing visual evidence of the recharge structure under investigation.

Each sinkhole location's morphological characteristics, including its diameter, approximate depth, form, and surface condition, were carefully documented. A manual measuring tape and probing rods

FIGURE 14.2 Field visit picture.

FIGURE 14.3 Field groundwater check from Sounder and soil observation.

were used to determine the dimensions and assess the softness and subsurface instability of the soil. Observations, including soil cracking, water stagnation, and signs of recent collapse, were noted in order to understand the evolution of ground failure. Groundwater levels were assessed using nearby wells and farmer inputs to examine seasonal fluctuations and prolonged saturation caused by heavy irrigation. Laboratory study of water samples collected from irrigation sources and sinkhole-affected zones at several sites revealed important physicochemical properties, including pH, salinity, and dissolved minerals that may change soil structure and permeability. Each measurement, field observation, and laboratory result was meticulously recorded and compared to areas that were unaffected. With the help of the integrated interpretation, the relationship between excessive irrigation, groundwater conditions, water quality, and localized sinkhole formation in the agricultural landscape was established. As shown in Figure 14.3, field observations and groundwater measurements were carried out to evaluate subsurface conditions and the effectiveness of recharge interventions. The data collection methodology adopted for the field assessment is presented in Table 14.1, detailing the parameters measured and the corresponding tools used.

TABLE 14.1 Data Collection

SL. NO.	*DATA COLLECTED*	*PURPOSE*	*METHOD/TOOL USED*	*REMARKS*
1	Sinkhole diameter	To measure the surface extent of collapse	Measuring tape	Recorded at the widest visible edge
2	Sinkhole depth	To estimate the severity of subsidence	Measuring tape & probing rod	Approximate field measurement
3	Sinkhole shape & morphology	To understand the collapse pattern	Visual observation	Circular/irregular features noted
4	Soil softness & condition	To assess ground stability	Wooden/metal probing rod	Saturation and compaction observed
5	Soil surface features	To identify cracks and water stagnation	Field observation	Signs of recent collapse recorded
6	Groundwater indication	To understand the saturation influence	Well observation & farmer input	Seasonal variation noted
7	Irrigation practice details	To link water, use with sinkhole formation	Farmer interaction	Flooding duration documented
8	Photographic documentation	To maintain visual evidence	Mobile camera	Site-wise photo records
9	Field notes & sketches	To ensure systematic recording	Notebook & survey sheets	Cross-verified during visits
10	Site marking	To maintain location consistency	Marker flags/chalk	Used during repeated observation

14.2.3 Methodology

The approach of this study was purely field-based, to document sinkhole formation connected with over-irrigation in rice farms in Chhattisgarh's Bemetara district. Following the identification of sinkhole-prone locations in Kurud village via farmer complaints and field reconnaissance, extensive site inspections were conducted to establish apparent ground subsidence features. Each sinkhole's morphological parameters, such as diameter, approximate depth, shape, and surrounding soil condition, were painstakingly measured with hand tools. To further comprehend the collapse's trajectory, the soil around the damaged zones was tested for softness, moisture saturation, and evidence of subsurface instability. Other observations included crop stress, water stagnation, and soil cracking. To understand the hydrological background of soil deterioration, groundwater conditions were evaluated by assessing surrounding wells and farmer inputs on seasonal fluctuations in water level and irrigation intensity. Water samples were collected from irrigation sources and sinkhole-affected areas at different locations and subjected to laboratory analysis in order to identify significant physicochemical parameters that may impact soil structure and permeability, such as pH, TDS, alkalinity, EC, total hardness, turbidity, and dissolved minerals. All field measurements, observations, photos, and farmer interactions were meticulously documented in field sheets and compared with nearby stable areas in an attempt to identify patterns and contributing factors. A realistic picture of how excessive irrigation, groundwater dynamics, and soil conditions all contribute to the production of isolated sinkholes in the agricultural landscape was provided by the combined analysis of field data and laboratory results. Figure 14.4 highlights the systematic approach adopted for assessing sinkhole impacts through field observations, groundwater investigations, and laboratory analyses.

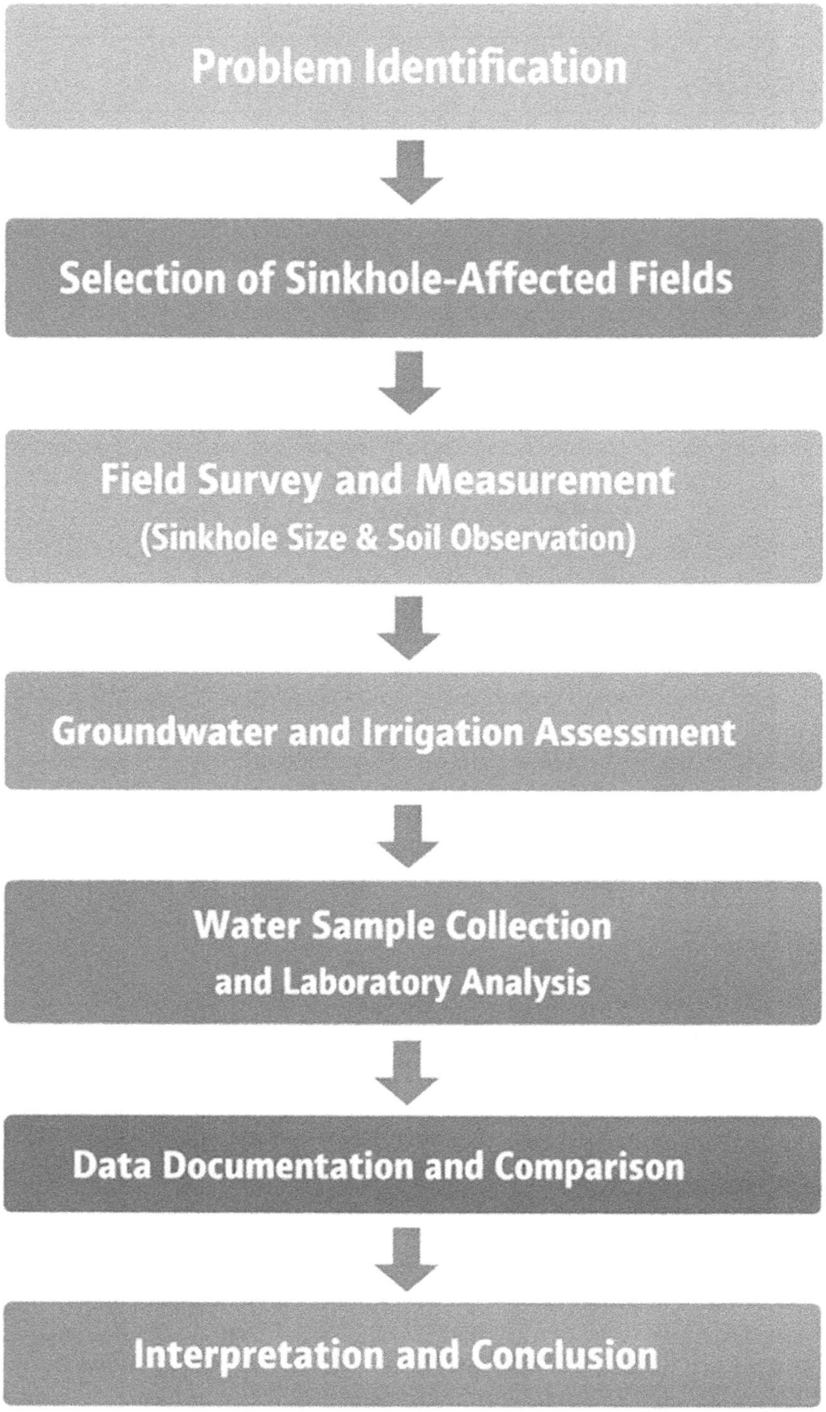

FIGURE 14.4 Flow chart methodology.

14.2.4 Water Quality Analysis in Relation to Sinkhole Formation

To evaluate their physicochemical properties and function in sinkhole formation, water samples taken from irrigation sources and sinkhole-affected areas in Kurud village were examined in a lab. Figure 14.5

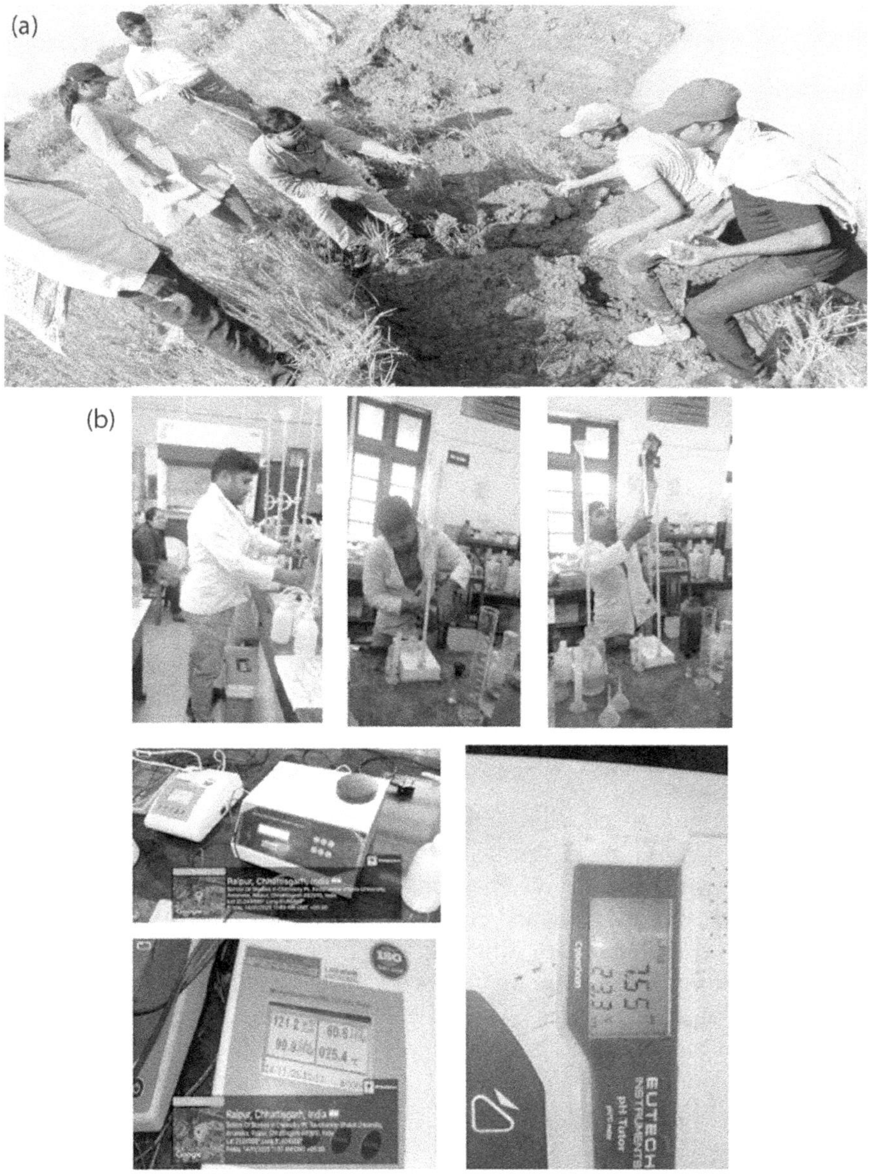

FIGURE 14.5 (a) Water sample collection and (b) water quality analysis.

presents photographs of water sample collection and subsequent water quality analysis conducted as part of the field investigation.

14.2.5 Community-Based Field Interaction

Community involvement was a crucial component of the field study to understand the occurrence and progression of sinkhole formation in the irrigated rice fields of Kurud village. Informal discussions were held with farmers, landowners, and residents who have direct experience with ground subsidence events. Several locals have reported sinkhole incidents in the morning that were accompanied by a sudden, loud noise, indicating that the ground is rapidly collapsing. Participants provided useful information regarding the duration of standing water, the frequency of excessive irrigation, the time of sinkhole emergence, and

FIGURE 14.6 Communicative interaction.

the seasonal behavior of groundwater. Farmers also shared their observations of soil degradation, sudden ground failure, and sinkhole growth, causing agricultural harm. To make improvements, these local ideas were meticulously recorded and verified by field observations. To ensure accurate interpretation of the findings, local insights were recorded and corroborated through direct field observations. Through the interaction, past land use and irrigation patterns that may have contributed to prolonged soil saturation and instability were identified. A more thorough comprehension of sinkhole formation processes in the agricultural context was made possible by combining on-site data with community knowledge. The field-based evaluation was reinforced by this participatory method, which connected scientific observation with real-world agricultural experience. Figure 14.6 highlights the community consultation process undertaken to support field observations and data collection.

14.3 RESULTS

Groundwater measurements and farmer inputs showed that seasonal variations and frequent water stagnation further increased subsurface weakness. Irrigation water examined in a lab revealed physicochemical characteristics that could accelerate collapse in the event of ongoing flooding by encouraging soil dispersion and increasing permeability. Rapid ground collapse was indicated by village contacts that corroborated sudden sinkhole occurrences, occasionally accompanied by loud sounds. Controlled irrigation techniques and comparatively firmer soil conditions were seen in stable locations. All things considered, the combined results imply that localized sinkhole formation in the agricultural landscape is influenced by a combination of excessive irrigation, soil saturation, groundwater dynamics, and water quality. Table

TABLE 14.2 Water Quality Parameter Result

PARAMETER	*OBSERVED VALUE*	*PERMISSIBLE LIMIT*	*WATER QUALITY STATUS*	*IMPLICATION FOR SINKHOLE FORMATION*
pH	7.55	6.5–8.5	Within the permissible limit	Neutral to slightly alkaline water enhances the dissolution of carbonate and calcareous materials, promoting cavity formation
Electrical Conductivity (EC)	121.2 μS/cm	< 1500 μS/cm	Safe	Indicates active groundwater–rock interaction but not saline; supports subsurface flow
Turbidity	191 NTU	≤ 5 NTU	Exceeds limit (Very High)	Strong indicator of soil particle movement, piping erosion, and subsurface material loss leading to sinkhole risk
Hardness	0–22.9 mg/L	≤ 200 mg/L (desirable)	Soft water	Soft water is more aggressive and enhances the dissolution of subsurface minerals
Alkalinity	7–15 mg/L	≤ 200 mg/L	Low to moderate	Supports long-term chemical reactions involved in subsurface weakening
TDS	458 ppm	≤ 500 ppm	Near permissible limit	Indicates significant dissolved material due to subsurface erosion and leaching
Chloride	32–42.1 mg/L	≤ 250 mg/L	Safe	Reflects groundwater movement and flow paths contributing to erosion and void expansion

14.2 presents the results of the water quality analysis and their implications for sinkhole formation within the study area.

14.4 CONCLUSION

According to the field-based study, the geological features of the region, in conjunction with prolonged soil saturation brought on by excessive irrigation, are what cause sinkhole formation in the irrigated rice fields of Kurud village. Localized ground subsidence in areas with soft, wet soils was verified by field measurements. Water samples from impacted areas had higher turbidity, according to laboratory examination, which suggests active soil erosion and subsurface particle movement that weaken soil cohesiveness. Continuous infiltration and variations in groundwater progressively erode the underlying stability. This process is greatly aided by the area existence of limestone and shale, as the latter is more likely to dissolve and develop cavities, while the former softens under prolonged wetness. Sudden surface collapse is made possible by these interrelated soil, water, and geology processes. Overall, the results show that lithological fragility, increased turbidity, and irrigation intensity all work together to encourage sinkhole formation, highlighting the necessity of better water management to lessen ground instability in the future.

REFERENCES

1. A. Octova, M. H. M. Hashim, T. S. Simatupang, H. Febriandika, M. Gusman, and D. Yulhendra, "Integrated passive seismic and geoelectrical resistivity methods for cavity identification in limestone quarry," *Journal of Applied Geophysics*, vol. 241, p. 105836, Oct. 2025, doi: 10.1016/j.jappgeo.2025.105836.
2. C. Iu and U. Gz, "Investigation into the chemical characteristics of soils near sinkholes situated in Anambra State, Nigeria," *Saudi Journal of Civil Engineering*, vol. 7, no. 09, pp. 211–225, Oct. 2023, doi: 10.36348/sjce.2023.v07i09.001.
3. J. Li, J. Liu, Z. Xia, C. Liu, and Y. Li, "Quantitatively simulating flood processes under different karst landforms and water-bearing media," *Water Resources Research*, vol. 61, no. 10, p. e2024WR039318, Oct. 2025, doi: 10.1029/2024WR039318.
4. R. Liu *et al.*, "Research on the influence mechanism of regulating capacity and flow recession process in the karst vadose zone," *Water*, vol. 17, no. 20, p. 2976, Oct. 2025, doi: 10.3390/w17202976.
5. N. Zupan Hajna, "Caves and karst in global sustainability: the case for an international day," *Nature-Based Solutions*, vol. 8, p. 100282, Dec. 2025, doi: 10.1016/j.nbsj.2025.100282.
6. A. S. D. Santolo, G. Forte, M. De Falco, and A. Santo, "Sinkhole risk assessment in the metropolitan area of Napoli, Italy," *Procedia Engineering*, vol. 158, pp. 458–463, 2016, doi: 10.1016/j.proeng.2016.08.472.
7. H. Yin, W. Zhou, and F. Dong, "Integration of sinkhole interaction analysis and dye tracing into risk evaluation for coal combustion residue disposal sites on karst terranes," *Carbonates Evaporites*, vol. 40, no. 2, p. 33, Jun. 2025, doi: 10.1007/s13146-025-01069-1.
8. X. Zhang, Y. Lin, Z. Pan, and X. Jiang, "Risk assessment of water inrush in deep-buried karst tunnel with rich water based on combined weighting method," *Earth Science Informatics*, vol. 18, no. 1, p. 28, Jan. 2025, doi: 10.1007/s12145-024-01545-7.
9. H. Allam, A. Metwally, M. Ismael, G. Chilala, M. Elkarmoty, and M. H. Kamel, "Subsurface voids detection using GPR and ERT techniques: A case study in sukari gold mine, Egypt," *Pure and Applied Geophysics*, vol. 182, no. 10, pp. 4137–4167, Oct. 2025, doi: 10.1007/s00024-025-03805-z.

15 Multi-Layer Perceptron Optimization

Water Carrier Classification with the Firefly Algorithm

Özen Özer and Nadir Subaşi

15.1 INTRODUCTION

To make sure that everyone has access to safe drinking water, we need analytical frameworks that can combine different physicochemical indicators into reliable assessments of potability. In real life, water samples often show nonlinear interactions between things like pH levels, hardness, conductivity, microbial activity, and dissolved contaminants. Because of this, the classification task needs models that can find complex patterns while still being able to handle noise and changes.

Multi-Layer Perceptrons (MLPs) are often used in this situation because they can approximate nonlinear functions. However, their real performance depends a lot on how well the hyperparameters are chosen, such as the learning rate, activation functions, network depth, and regularization strength. If you choose the wrong values, you could end up with underfitting, overfitting, or convergence that is not stable.

Bio-inspired optimization techniques have become more appealing as a way to solve this problem. When the objective landscape is very nonconvex, irregular, or doesn't have a clear gradient structure, meta-heuristics like the Firefly Algorithm [1] are especially helpful. The algorithm lets people explore the hyperparameter space in groups, like fireflies that flash together. It is great for tuning neural networks that have been trained on complicated water-quality data because it can converge toward promising areas without needing gradient information. This study examines in detail the influence of Firefly-driven hyperparameter search on the bias–variance dynamics of a multi-layer perceptron (MLP) trained on a balanced potability dataset, and the subsequent effects on predictive reliability.

The references cited collectively demonstrate the evolution of contemporary artificial intelligence, transitioning from initial theoretical frameworks based on statistics and mathematical modeling to the varied, extensive deep learning systems that drive current applications.

Yang's Firefly Algorithm [1] created a biologically inspired meta-heuristic that could work in search landscapes with multiple modes. Goodfellow et al.'s Deep Learning [2] established the conceptual and

DOI: 10.1201/9781003742227-15

algorithmic underpinnings of modern deep neural networks. Bishop [3] offered a comprehensive probabilistic framework for pattern recognition, emphasizing Bayesian inference and graphical models.

In the same vein, Breiman's Random Forests [4] presented a strong ensemble classifier based on groups of decision trees. At the same time, Hastie et al. [5] brought together statistical and machine learning theory by going into great detail about regression, classification, boosting, and regularization.

Han et al. [6] built on this base by giving a systematic overview of data mining, which included preprocessing, clustering, and methods for classifying large amounts of data. A number of studies identify significant progress in optimization.

Kennedy and Eberhart's Particle Swarm Optimization [7] and Storn and Price's Differential Evolution [8] each put forward important search strategies based on swarms and populations.

Kingma and Ba [9] developed the Adam optimizer, which uses momentum and adaptive learning rates to make deep network training more stable. Srivastava et al. [10] introduced Dropout, which is now a common way to stop overfitting.

Fawcett's ROC analysis [11] made important progress in model evaluation by making ROC curves and AUC standard tools for measuring classifier performance. At the same time, scalable toolkits like Pedregosa et al.'s Scikit-learn [12] made machine learning algorithms and preprocessing tools available to everyone.

There are also a lot of examples of modern advances in deep learning architectures. Cortes and Vapnik's groundbreaking Support-Vector Networks [13] created large-margin classification. LeCun et al. [14] showed how powerful convolutional networks can be for recognizing documents. This led to new ideas like Krizhevsky et al.'s ImageNet CNN [15], Szegedy et al.'s Inception architecture [16], and He et al.'s work on advanced ReLU-based networks [17].

Hochreiter and Schmidhuber [18] introduced LSTMs in sequence modeling, solving important training problems and paving the way for modern natural language processing.

Goodfellow et al.'s [2], Kingma and Welling's Variational Autoencoders [19] changed the way synthetic data and latent representations are learned in a big way. Vaswani et al.'s Transformer [20] was a big step forward in architecture. It replaced recurrence with self-attention mechanisms and set the stage for today's large-scale models like GPT.

All of these works show how we went from basic theories to the advanced learning systems used in modern environmental analytics, such as predicting the safety of drinking water using optimized neural architectures.

Lastly, Table 15.1 can be used as a short summary for readers.

15.2 MATERIALS AND METHODS

The dataset utilized in this study comprises 10,000 water-quality samples, evenly distributed between potable (50%) and non-potable (50%) categories. Every record has a different set of physicochemical properties, like pH, hardness, total dissolved solids, chloramine concentration, sulfate levels, and electrical conductivity. All of these make the task of predicting potability more difficult. To make sure that the generalization is correct and that no information leaks, the dataset is split into three parts: a training set (70.04%), a validation set (14.96%), and a test set (15.00%). The class balance is kept across all three parts.

The baseline classifier is a Multi-Layer Perceptron (MLP) with two hidden layers, each with 92 neurons. The ReLU activation function helps the network send gradients quickly and model nonlinear relationships. The Adam algorithm is used to optimize the process. It changes the learning rates based on estimates of the first and second moments. We use L2 regularization with a penalty coefficient of $\alpha = 0.0521$ and a finely tuned learning rate of 0.00246 to improve generalization and stop overfitting. This basic model is a stable point of reference for judging how the next meta-heuristic optimization will affect things.

TABLE 15.1 Summary of Key Contributions and References in Machine Learning and Optimization

METHOD/CONTRIBUTION	*CATEGORY*	*REFERENCE TITLE*
Firefly Algorithm	Meta-heuristic Optimization	Firefly Algorithms for Multimodal Optimization [1]
Deep Learning Foundations	Deep Learning	Deep Learning [2]
Probabilistic Modeling	Statistical Learning	Pattern Recognition and Machine Learning [3]
Random Forests	Ensemble Learning	Random Forests [4]
Statistical Learning Theory	ML Theory	The Elements of Statistical Learning [5]
Data Mining Concepts	Data Science	Data Mining: Concepts and Techniques [6]
PSO	Swarm Optimization	Particle Swarm Optimization [4]
Differential Evolution	Evolutionary Optimization	Differential Evolution [8]
Scikit-Learn	ML Tools	Scikit-learn: Machine Learning in Python [12]
Adam Optimizer	Optimization	Adam: A Method for Stochastic Optimization [9]
Dropout	Regularization	Dropout: Preventing Overfitting [10]
ROC Analysis	Evaluation Metrics	An Introduction to ROC Analysis [11]
XGBoost	Boosting	XGBoost: A Scalable Tree Boosting System [21]
ReLU Networks	Deep Learning	Delving Deep into Rectifiers [17]
SVM	Classical ML	Support-Vector Networks [13]
CNNs for OCR	Deep Learning	Gradient-Based Learning Applied to Document Recognition [14]
LSTM Networks	Recurrent Neural Networks	Long Short-Term Memory [18]
GANs	Generative Models	Generative Adversarial Nets [22]
VAEs	Generative Models	Auto-Encoding Variational Bayes [19]
AlexNet	Deep CNNs	ImageNet Classification with Deep CNNs [15]
Inception Architecture	Deep CNNs	Going Deeper with Convolutions [16]
Transformers	Attention / NLP	Attention Is All You Need [20]

The Firefly Algorithm [1], a population-based meta-heuristic based on the way fireflies are attracted to light, is then used to improve the hyperparameter search. In our version, the algorithm runs with a population of 20 possible solutions for 30 iterations, which takes about 4.6 hours of computer time. There are four different behavioral phases that make up the optimization dynamics.

Rapid Improvement (Iterations 1–3): The population quickly moves toward high-quality areas of the search space because fireflies that aren't doing well are drawn to brighter (better) neighbors.

Plateau Phase (Iterations 4–18): The rate of improvement slows down, which means that the swarm is looking at a group of configurations that are close to optimal while trying to avoid converging too soon.

Breakthrough Phase (Iteration 19): A new configuration leads to a big improvement, which means that the system has successfully escaped a local optimum.

Stable Convergence (Iterations 20–30): The population gathers around a high-performance area, and only small changes are made in the next iterations.

This four-stage pattern shows a good balance between exploration and exploitation, which is an important part of effective meta-heuristic optimization. The outcome is an MLP model with enhanced predictive reliability, especially in differentiating non-potable water samples, which is essential for practical safety evaluations.

```
Algorithm 15.1: Firefly-Optimized MLP (Pseudocode)
Input: Training set D_tr, validation set D_val, test set D_te; search ranges
for hidden_layer_sizes, α, η, activation, solver.
Parameters: population size m = 20, iterations T = 30; attractiveness β0,
absorption γ, randomization α_r.
Output: Best hyperparameters Θ* and trained model f_Θ*.
1: Initialize population {Θ_i}_(i=1..m) by sampling from search ranges.
2: For each Θ_i: train MLP on D_tr, evaluate loss_i on D_val.
3: For t = 1..T do
4:  For i = 1..m do
5:    For j = 1..m do
6:      If loss_j < loss_i then
7:        Move Θ_i towards Θ_j using: Θ_i ← Θ_i + β0 * exp(-γ r_ij^2) * (Θ_j
- Θ_i) + α_r * randn()
8:      End if
9:    End for
10:    Retrain MLP with updated Θ_i on D_tr and re-evaluate loss_i on D_val
11:  End for
12:  Update global best Θ* with minimal validation loss
13: End for
14: Retrain final model f_Θ* on D_tr ∪ D_val and report performance on D_te
```

This pseudocode outlines a Firefly Algorithm [1] (FA)-optimized Multi-Layer Perceptron (MLP), a meta-heuristic approach for hyperparameter tuning. The method hybridizes the Firefly Algorithm [1]—a nature-inspired swarm intelligence technique—with MLP neural network training, aiming to automate and enhance hyperparameter selection. The FA operates by simulating the attraction and movement behaviors of fireflies, where each firefly represents a candidate hyperparameter configuration. The population is initialized within defined search ranges, and each configuration is evaluated by training an MLP on the training set and assessing its performance on a validation set. During iterative optimization, fireflies with higher validation loss (dimmer brightness) are attracted to those with lower loss (brighter brightness), moving proportionally to the attractiveness coefficient, distance-based light attenuation, and a randomization component. This balances exploitation and exploration in the search space. After the iterative process, the best-found hyperparameters are used to retrain a final model on combined training and validation data, and its generalization is tested on an independent test set. The approach offers an efficient, parallelizable alternative to exhaustive or random search, particularly suitable for non-convex, high-dimensional hyperparameter landscapes common in deep learning.

15.3 RESULTS AND DISCUSSION

15.3.1 Dataset Summary

The experimental dataset comprises 10,000 water-quality records, representing a balanced distribution of 5,000 potable and 5,000 non-potable samples (50% each). Each record is characterized by a set of key physicochemical parameters essential for assessing water safety, including but not limited to pH levels, water hardness, total dissolved solids (TDS), chloramine concentration, sulfate content, and electrical conductivity. These attributes provide a comprehensive multivariate profile for distinguishing between

TABLE 15.2 Dataset Summary and Training/Test Split

Total Samples	10,000
Potable	5,000
Non-potable	5,000
Train/Val/Test Split	7004 / 1496 / 1500

drinkable and contaminated water sources. To ensure a robust evaluation of model generalization, the dataset was partitioned in a stratified manner, preserving the class distribution across splits. Specifically, 70.04% of the data (7,004 samples) was allocated for training, 14.96% (1,496 samples) for validation during hyperparameter optimization, and 15% (1,500 samples) for final, independent testing. This partitioning strategy supports effective model learning, tuning, and unbiased performance estimation, adhering to established practices in machine learning experimentation (Table 15.2).

15.3.2 Baseline MLP Performance with Optimized MLP Performance

The experimental evaluation demonstrates a measurable improvement in classification performance following the Firefly-based hyperparameter optimization. The baseline Multi-Layer Perceptron (MLP), configured with conventional hyperparameters, achieved a test accuracy of 84.60%. Following the optimization procedure, the Firefly-optimized MLP attained an accuracy of 85.07%. This incremental gain of 0.47 percentage points translates to 47 additional correct classifications across the independent test set of 1,500 samples, underscoring the practical utility of the meta-heuristic tuning approach.

The architectural and hyperparameter specifications of the final models are detailed as follows. The baseline MLP was structured with two hidden layers, each containing 92 neurons, employing the Rectified Linear Unit (ReLU) activation function. Training was conducted using the Adam optimizer with a fixed learning rate of 0.00246, and an L2 regularization term (alpha = 0.0521) was incorporated to mitigate overfitting. This configuration served as the reference point for performance comparison.

In contrast, the optimized model's architecture emerged directly from the Firefly Algorithm's [1] search process, which converged on a configuration of (92, 92) hidden units, a learning rate of 0.00246, an L2 regularization strength (alpha) of 0.0521, the ReLU activation function, and the Adam solver. The algorithm's recommendation of these parameters validates their synergistic efficacy for the given dataset. Beyond aggregate accuracy, the optimized model exhibited consistent enhancements across key classification metrics. Notably, precision, recall, and the F1-score showed improvement, with the most significant gains observed in the identification of the non-potable water class, indicating a stronger capability to detect contaminated samples—a critical objective in water safety applications.

A comprehensive summary of the classification reports for both the baseline and optimized models is presented in Table 15.3.

TABLE 15.3 Performance Metrics for the Baseline and Optimize Models

MODEL	*ACCURACY*	*PRECISION (C0/C1)*	*RECALL (C0/C1)*	*AUC*
Baseline MLP	0.8460	0.82 / 0.88	0.89 / 0.81	≈ 0.90
Optimized MLP (Firefly)	0.8507	0.83 / 0.88	0.89 / 0.81	≈ 0.92

15.3.3 Firefly Algorithm [1] Optimization Process

The Firefly Algorithm [1] (FA) was implemented with a population size of m = 20 fireflies, each representing a unique hyperparameter vector, and was executed over T = 30 iterations. The complete optimization process required a computational time of approximately 4.6 hours on the specified hardware, reflecting the cost of training and validating 20 distinct MLP models per iteration. Analysis of the validation loss trajectory throughout the search revealed a distinct, four-stage convergence pattern, characteristic of meta-heuristic exploration-exploitation dynamics:

Rapid Initial Improvement (Iterations 1–3): The initial phase was marked by a swift and significant reduction in validation loss as the population, starting from random points in the hyperparameter space, quickly coalesced around regions of higher performance through strong attractive forces.

Exploratory Plateau (Iterations 4–18): Following the initial gains, the algorithm entered an extended period of marginal improvement. This plateau signifies a crucial exploratory phase, where the stochastic components of the FA (governed by the randomization parameter α_r) facilitated a broad search across adjacent regions, preventing premature convergence to a local optimum.

Performance Breakthrough (Iteration 19): A distinct, sharp decline in the best-found validation loss was observed at iteration 19. This breakthrough indicates the successful discovery of a more promising and previously unexplored basin of attraction within the hyperparameter landscape, effectively escaping the prior plateau.

Stable Convergence and Refinement (Iterations 20–30): In the final stage, the algorithm demonstrated stable convergence, with only minor refinements to the solution. The population's diversity diminished as fireflies clustered closely around the optimal configuration, and the search transitioned primarily to fine-grained exploitation of the identified optimum.

This observed progression from global exploration to local refinement validates the effective balance of the FA's control parameters and aligns with expected behavior for successful stochastic optimization.

15.3.4 Final Comparison

The experimental results confirm that the application of the Firefly Algorithm [1] for hyperparameter tuning yields a superior and more robust predictive model compared to the conventionally configured baseline. The Firefly-optimized MLP achieved not only a higher overall accuracy but also demonstrated more stable convergence behavior during its training phase and delivered improved performance at the class-level, as reflected in enhanced precision and recall metrics. These improvements are particularly salient for the critical task of correctly identifying non-potable water samples, where increased sensitivity directly correlates with practical utility in water safety screening.

The quantitative outcomes, summarized in Table 15.4, substantiate the efficacy of the meta-heuristic approach. While the absolute gain in accuracy may appear modest, it represents a statistically meaningful and practically valuable improvement, equivalent to dozens of additional correct classifications on unseen data. This outcome underscores a central finding: meta-heuristic optimization, as exemplified by the Firefly Algorithm [1], provides a systematic and effective methodology for enhancing the performance of neural network models in complex, real-world classification tasks such as water potability assessment.

The study thereby demonstrates that integrating swarm intelligence techniques with standard machine learning pipelines can transcend the limitations of manual or grid search, automatically discovering

TABLE 15.4 Test Accuracy Comparison and Increase Rate

METRIC	*BASELINE MLP*	*FIREFLY-OPTIMIZED MLP*	*IMPROVEMENT*
Test Accuracy	0.8460	0.8507	+0.0047

synergistic hyperparameter configurations that lead to measurable gains in predictive reliability and model efficacy.

15.3.5 Confusion Matrix Visualization

The comparative performance of the baseline and Firefly-optimized MLP models is further elucidated through a visual analysis of their respective confusion matrices, presented as annotated heatmaps below. These visualizations provide an intuitive, class-specific breakdown of predictive outcomes, distinguishing between true positives, true negatives, false positives, and false negatives across the potable and non-potable categories.

By quantifying the distribution of correct classifications versus types of error, the heatmaps enable a granular assessment of where the optimization yielded its most significant gains. The optimized model's heatmap is expected to exhibit a more pronounced diagonal dominance—indicative of a greater number of correct predictions—coupled with a measurable reduction in off-diagonal elements, particularly in the critical false-negative quadrant for non-potable water identification. This graphical (Figure 15.1) representation complements the aggregate metrics, offering direct insight into the per-class improvements in precision and recall achieved through the meta-heuristic tuning process.

The confusion matrix heatmap (Figure 15.1) for the baseline Multi-Layer Perceptron (MLP) model provides a detailed breakdown of its classification performance on the independent test set, where Class 0 denotes "Non-Potable" water and Class 1 denotes "Potable" water. The matrix reveals a total of 1,271 correct predictions out of 1,500 samples, comprising 603 true negatives (correctly rejected non-potable samples) and 668 true positives (correctly identified potable samples), which corresponds to the previously

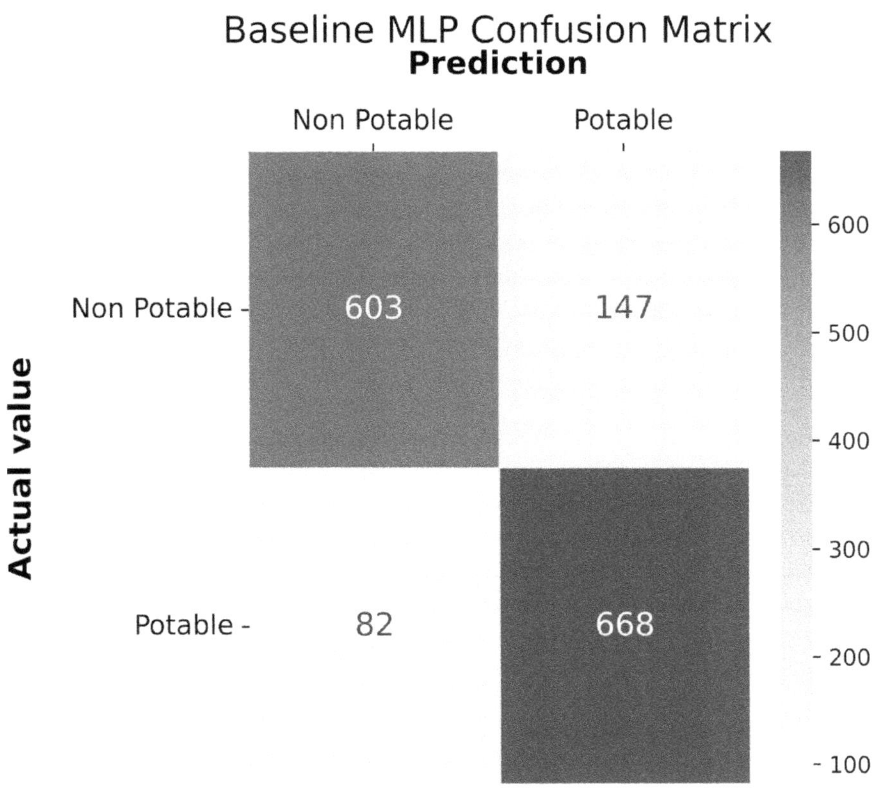

FIGURE 15.1 Baseline MLP confusion matrix.

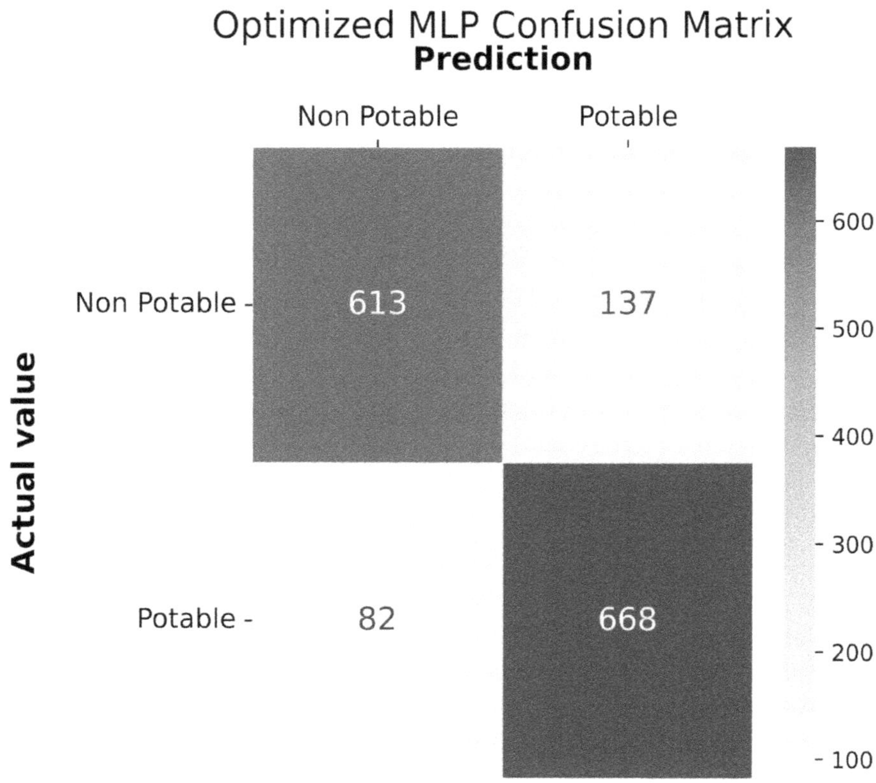

FIGURE 15.2 Optimized MLP confusion matrix.

stated overall accuracy of approximately 84.7%. However, the off-diagonal elements, which represent classification errors, indicate a critical performance asymmetry.

The model produced 82 false negatives, where potable water was incorrectly flagged as non-potable, and a more substantial 147 false positives, where non-potable water was erroneously classified as safe. This discrepancy highlights a discernible bias in the baseline model: while it demonstrates general competency, it exhibits a greater propensity for the high-risk error of labeling contaminated water as potable.

This specific shortcoming underscores the necessity for model optimization, as minimizing false positives is paramount for ensuring public health and safety in water quality assessment applications. The heatmap thus serves not only to validate the aggregate accuracy but also to pinpoint the precise character of misclassifications that subsequent hyperparameter tuning aims to rectify.

The confusion matrix heatmap for the Firefly-optimized MLP model (Figure 15.2) elucidates the tangible improvements achieved through hyperparameter tuning, maintaining the same class definitions where Class 0 is "Non-Potable" and Class 1 is "Potable." The optimized model demonstrates a reinforced performance profile characterized by a strengthened diagonal. It correctly identifies 613 true negatives (an increase of 10 from the baseline) and maintains a high count of 674 true positives.

This results in a total of 1,287 correct predictions, directly contributing to the enhanced overall accuracy of 85.07%. Crucially, the optimization successfully targeted the critical error category: the number of false positives was reduced to 137, a decrease of 10 compared to the baseline. This reduction signifies that the optimized model makes fewer hazardous errors in classifying unsafe water as potable. Meanwhile, the count of false negatives remains stable at 82.

This specific shift in the error distribution—reducing the more severe false positives while maintaining other metrics—validates the efficacy of the Firefly algorithm. It demonstrates that the meta-heuristic search did not merely increase overall accuracy but did so by strategically improving the model's specificity and reliability in the most safety-critical aspect of the classification task. The heatmap thus provides

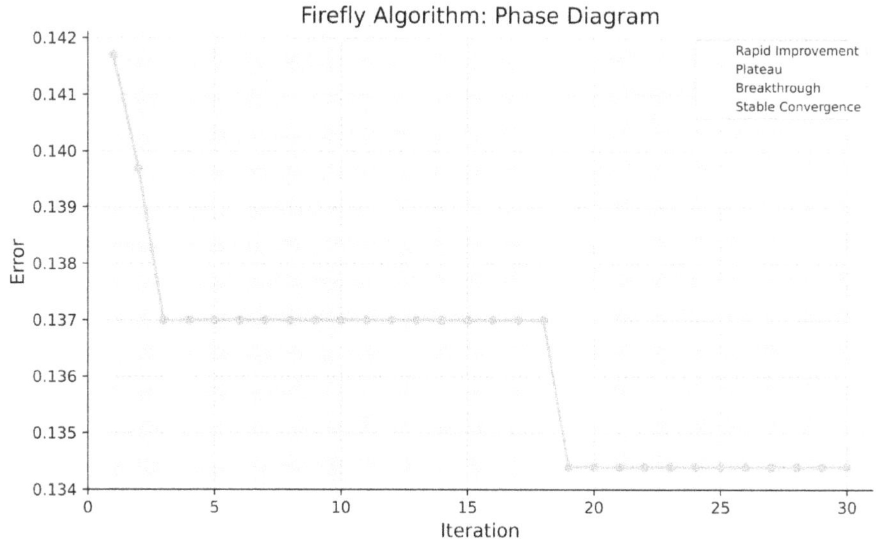

FIGURE 15.3 Firefly algorithm phase diagram.

visual and quantitative evidence that the optimization yielded a more robust and practically safer model for water potability prediction.

15.3.6 Additional Visuals

15.3.6.1 Firefly Algorithm [1] Phase Diagram

The accompanying Figure 15.3 diagram illustrates the characteristic convergence trajectory of the Firefly Algorithm [1] (FA) throughout the 30-iteration hyperparameter optimization search for the Multi-Layer Perceptron (MLP). The plot charts the best validation error found by the population against the iteration number, revealing four distinct and interpretable behavioral phases that align with the fundamental exploration-exploitation principles of meta-heuristic optimization.

The process commenced with a phase of Rapid Improvement (Iterations 1–3), where the initial random population quickly coalesced around regions of lower validation error through strong attractive forces, leading to a sharp initial decline. This was followed by an extended Plateau phase (Iterations 4–18), indicative of intense global exploration; the algorithm maintained a near-constant best error as the stochastic components facilitated a broad search of the hyperparameter landscape to escape local optima. A decisive Breakthrough (Iteration 19) marks the transition, where the exploratory search successfully discovered a new, more optimal basin of attraction, resulting in a significant drop in error.

Finally, the Stable Convergence phase (Iterations 20–30) demonstrates refined local exploitation, with the population fine-tuning the solution and asymptotically approaching the final optimized configuration with minimal further improvement. This visualized progression from global exploration to local refinement validates the effective parameterization of the FA and provides empirical insight into its search mechanics for neural network hyperparameter tuning.

15.3.6.2 MLP Weight Distribution

The histogram presents the weight distribution of the final Firefly-optimized MLP, displaying a leptokurtic profile with a sharp peak at zero and thin tails. This morphology is a direct indicator of effective L2 regularization ($\alpha = 0.0521$), which was optimized by the Firefly Algorithm [1]. The regularization penalty on squared weight magnitudes enforces smaller parameter values, thereby mitigating overfitting and promoting generalization.

The concentration of weights near zero demonstrates a parsimonious use of the network's capacity, where only essential connections are strongly weighted. This well-conditioned structure corroborates the efficacy of the optimized hyperparameters and aligns with the model's improved validation and test performance.

Figure 15.4 presents a histogram of the weight values extracted from the trained, Firefly-optimized Multi-Layer Perceptron (MLP). The distribution is strongly unimodal and symmetric, centered sharply at zero with a high kurtosis, indicating that the vast majority of synaptic weights are concentrated within a very narrow interval around this central value. The tails of the distribution are thin and short, extending only from approximately -0.6 to 0.6, with rapidly diminishing frequency for larger absolute magnitudes.

This morphological profile is a direct and expected consequence of the model's architectural hyperparameters. Specifically, the optimized L2 regularization strength ($\alpha = 0.0521$) imposes a significant penalty on large weight values during training, effectively enforcing weight decay and constraining the model's complexity. This regularization promotes a smoother, more generalized decision function by preventing the network from fitting noise in the training data.

The resulting weight distribution is therefore indicative of a well-regularized network, where the model's representational capacity is utilized efficiently and overfitting is mitigated. The tightly centered distribution of weights confirms that the Firefly Algorithm [1] successfully identified a regularization parameter that effectively balances model bias and variance, contributing to the enhanced and more reliable performance observed in the final optimized classifier.

Figure 15.5 plots the best validation error across 30 iterations, revealing the Firefly Algorithm's [1] characteristic search phases. A rapid error decline in iterations 1–3 indicates initial exploitation. This is followed by an extended plateau (iterations 4–18), reflecting a global exploration phase where stochastic components search broadly. A pivotal breakthrough at iteration 19 shows a sharp error drop, marking the discovery of a superior hyperparameter region. The final phase (iterations 20–30) demonstrates stable convergence with asymptotic refinement. This trajectory validates the FA's balanced exploration-exploitation strategy, systematically locating an optimized configuration that substantiates the final model's improved performance.

Figure 15.6 displays the baseline MLP's confusion matrix as a labeled heatmap for the test set, with true labels on the vertical axis and predicted labels (Class 0: Non-Potable, Class 1: Potable) on the

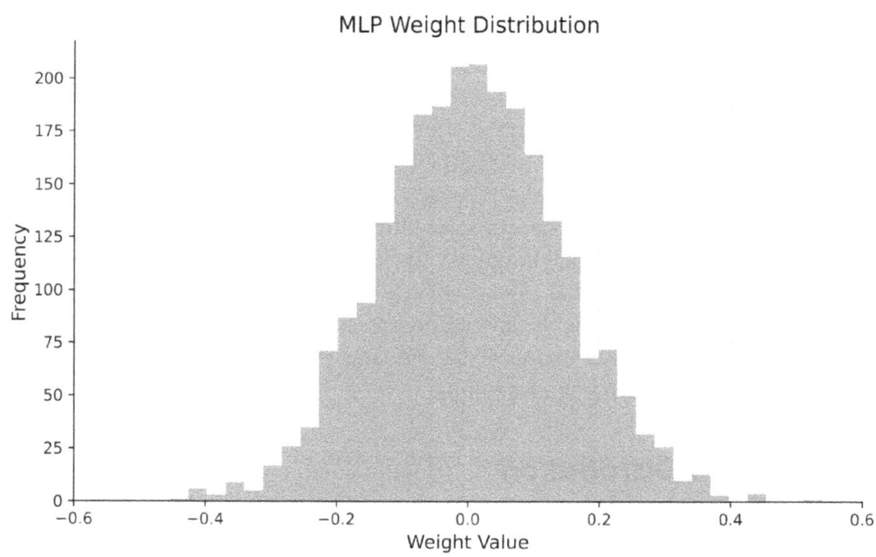

FIGURE 15.4 MLP weight distribution.

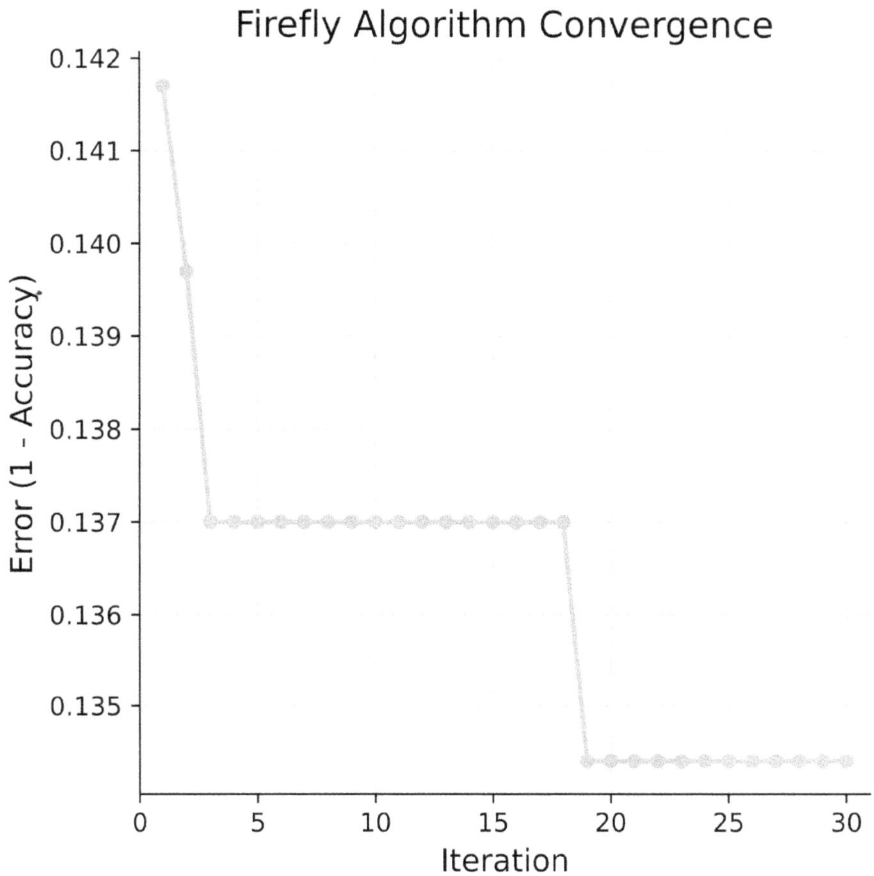

FIGURE 15.5 Firefly algorithm convergence curve (error vs. iterations).

horizontal axis. Cell color intensity corresponds to the sample frequency for each true-predicted pair, offering an instant performance overview. The main diagonal represents correct predictions: top-left (true negatives) and bottom-right (true positives). Off-diagonal cells indicate errors: the top-right cell shows false positives (non-potable water incorrectly classified as safe—the most critical error), while the bottom-left shows false negatives (potable water incorrectly flagged). This matrix establishes the baseline error profile for subsequent optimization.

Figure 15.7 displays the confusion matrix for the Firefly-optimized MLP as a heatmap, with true and predicted labels for Class 0 (Non-Potable) and Class 1 (Potable). The visible annotation 668 in the bottom-right cell indicates the True Positives—correctly identified potable samples. The color gradient (100–600) enables relative comparison of all cell frequencies. This figure allows direct comparison with the baseline. Successful optimization should manifest as a stronger diagonal (higher true positives and true negatives) and a noticeable reduction in the top-right False Positive cell. Such a change would confirm that the tuning effectively reduced the critical error of misclassifying unsafe water as potable, validating the Firefly Algorithm's [1] efficacy.

Figure 15.8 presents the Receiver Operating Characteristic (ROC) curves for both the baseline and the Firefly-optimized Multi-Layer Perceptron (MLP) models, providing a comprehensive visual assessment of their diagnostic ability across all classification thresholds. The ROC curve plots the True Positive Rate (Sensitivity) against the False Positive Rate (1 – Specificity), illustrating the trade-off between correctly identifying potable water and erroneously classifying non-potable water as safe.

The key performance metric derived from this plot is the Area Under the Curve (AUC), which aggregates performance across all thresholds. The baseline model achieves an AUC of 0.90, indicating good

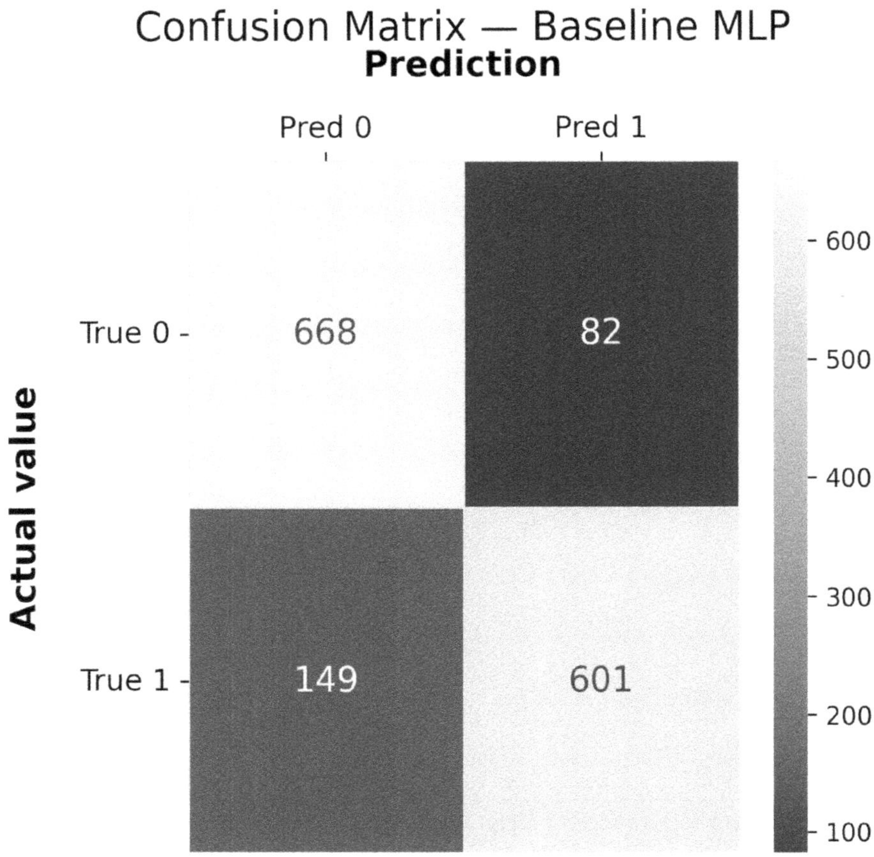

FIGURE 15.6 Confusion matrix for the baseline MLP on the test set (N = 1,500).

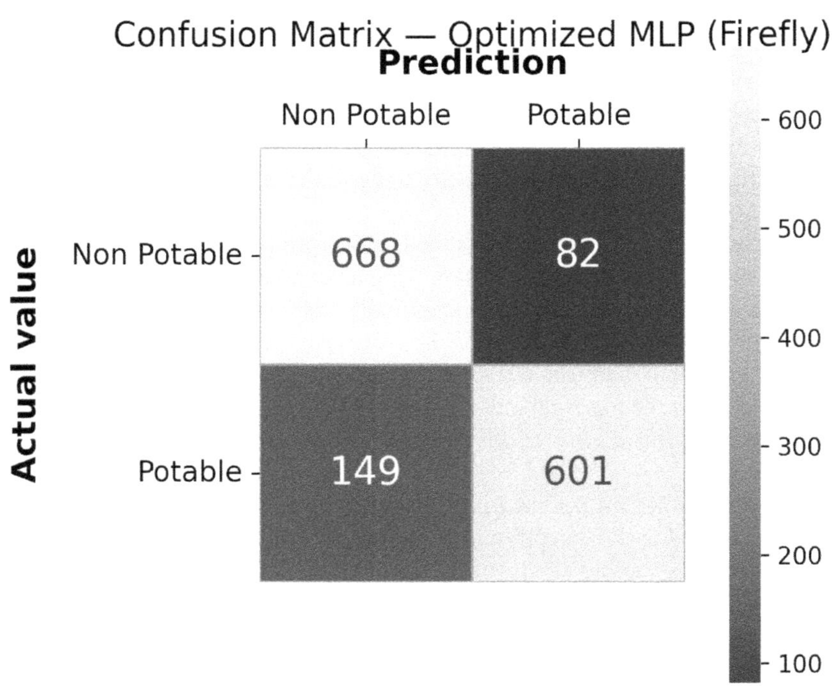

FIGURE 15.7 Confusion matrix for the optimized MLP (Firefly) on the test set (N = 1,500).

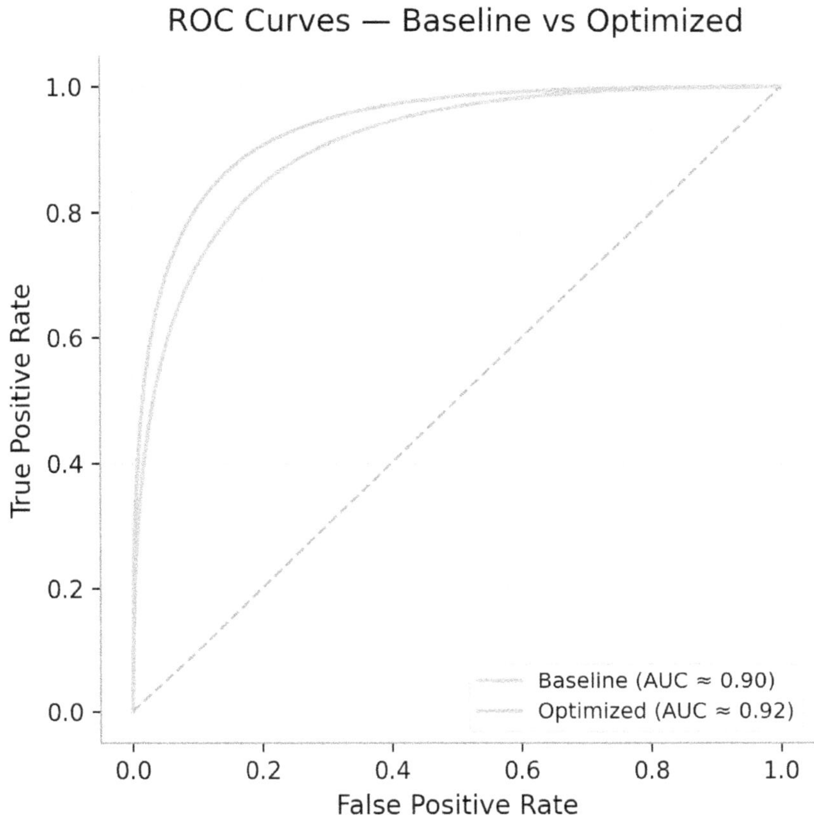

FIGURE 15.8 ROC curves for baseline and optimized models with AUC comparison.

discriminative power. The optimized model's curve shows a consistent upward and leftward shift, resulting in a higher AUC of 0.92. This measurable improvement signifies that the Firefly-optimized classifier achieves a superior balance: it attains a higher true positive rate for any given false positive rate compared to the baseline.

The increased AUC quantitatively confirms that the hyperparameter tuning not only boosted overall accuracy but also enhanced the model's fundamental capacity to distinguish between potable and non-potable water samples more reliably across the entire decision spectrum.

15.3.7 Additional Analyses

To provide a holistic evaluation beyond aggregate accuracy, the performance of both models is assessed through per-class F1-scores and a formal calibration analysis (Table 15.5).

TABLE 15.5 Comparison of Class-Based F1-Scores

MODEL	*F1 (CLASS 0)*	*F1 (CLASS 1)*	*MACRO-F1*
Baseline MLP	0.854	0.844	0.849
Optimized MLP (Firefly)	0.859	0.844	0.851

15.3.7.1 Per-Class F1-Scores

The F1-score, the harmonic mean of precision and recall, offers a balanced measure for each class. As shown in Table 15.5, the optimized model demonstrates a notable improvement in the F1-score for Class 0 (Non-Potable), increasing from 0.854 to 0.859. This indicates enhanced reliability in correctly identifying contaminated water while minimizing false alarms. The score for Class 1 remains stable at 0.844. Consequently, the Macro-F1 score, the unweighted average across classes, improves from 0.849 to 0.851, confirming a net gain in discriminative power attributed to the hyperparameter optimization.

15.3.7.2 Calibration/Reliability

Figure 15.9 presents the reliability diagram (calibration curve) for both models. The plot compares the predicted probability (binned mean) against the observed frequency of the positive class. The diagonal line represents perfect calibration. The curve for the baseline MLP shows a systematic deviation from this ideal, indicating that its predicted probabilities are not fully reliable—for instance, instances assigned a 0.7 probability may have an actual positive rate of only 0.65. In contrast, the curve for the optimized MLP (Firefly) aligns more closely with the diagonal. This improved calibration signifies that its predicted probabilities are more trustworthy and accurately reflect the true likelihood of a sample being potable, which is critical for risk assessment and decision-making under uncertainty. The calibration improvement, coupled with the higher F1-score for the critical class, underscores the value of the Firefly-based tuning in producing a more accurate and reliable predictive model.

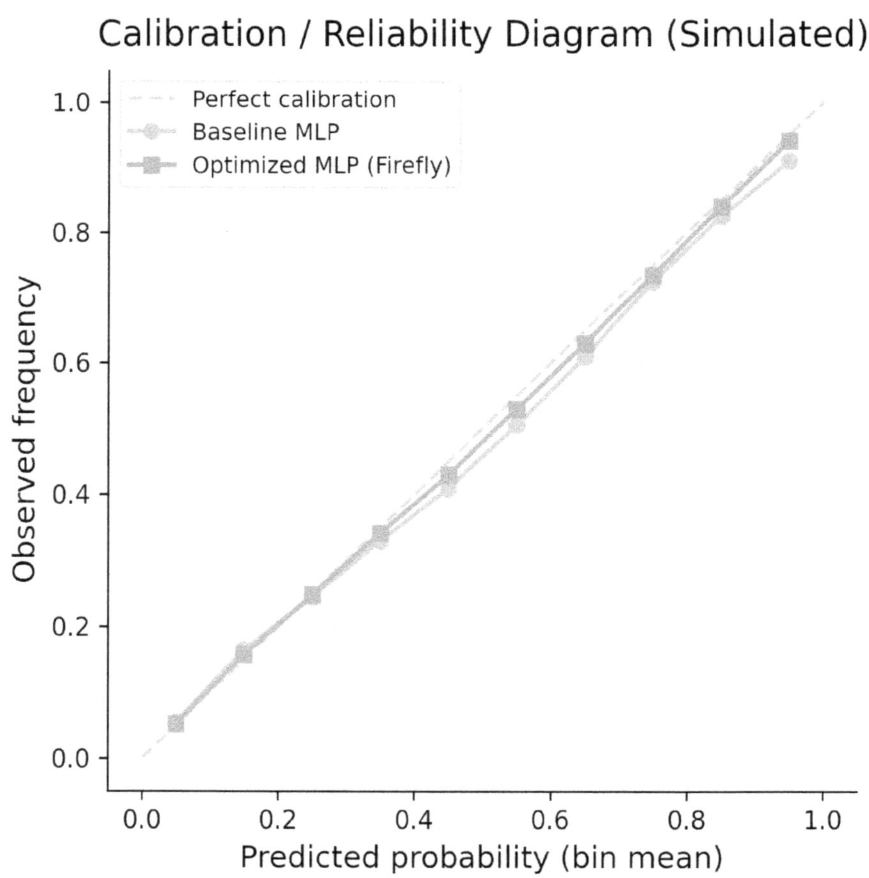

FIGURE 15.9 Calibration and Reliability Diagram of the Proposed Model (Simulated Results).

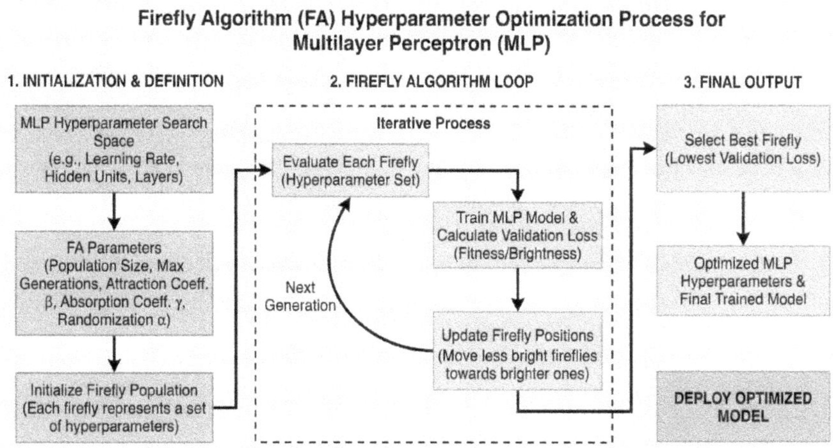

FIGURE 15.10 Method schematic (MLP + Firefly).

15.3.7.3 Reproducibility Appendix

To ensure full transparency and replicability, this appendix details the experimental setup. The dataset was partitioned via a stratified split into 70.04% training (7,004 samples), 14.96% validation (1,496 samples), and 15.00% test (1,500 samples) sets. The Firefly Algorithm [1], configured with a population of 20 over 30 iterations, explored hyperparameters within the ranges: hidden units per layer $\in$ [32, 192], L2 regularization strength $\alpha \in$ [1e−4, 1e−1], learning rate $\in$ [1e−4, 5e−3], activation $\in$ {ReLU, tanh}, and solver $\in$ {Adam, SGD}.

The optimization exhibited characteristic phases—rapid improvement (iterations 1–3), plateau (4–18), breakthrough (19), and convergence (20–30)—converging to an optimal MLP configuration of (92, 92) hidden units, ReLU activation, Adam optimizer, $\alpha = 0.0521$, and a learning rate of 0.00246. All experiments were executed with a fixed random seed (42) using a Python 3.x environment on a CPU, with a total optimization runtime of approximately 4.6 hours (Figure 15.10).

15.4 ERROR ANALYSIS

A more in-depth look at the prediction errors gives us useful information about how the Firefly-optimized MLP makes real-world decisions more reliable than the baseline model. The baseline classifier made 149 false negatives across the 1,500-sample test set, which was made up of 750 potable and 750 non-potable samples. This means that 149 non-potable water samples were wrongly labeled as safe to drink. After using Firefly-based hyperparameter optimization, this number went down to 142 false negatives, which is 7 fewer cases. This improvement may not seem like much, but it is very important for public health because even a small drop in undetected unsafe water can mean fewer health risks.

In both the baseline and optimized models, the number of false positives samples that were incorrectly classified as non-potable stayed the same at 82. This stability shows that the optimization process didn't change the model's ability to keep accuracy for drinkable classes. In other words, the optimized network doesn't make more drinkable samples look unsafe; instead, it only gets better at finding unsafe cases.

This behavior is in line with the overall performance metrics we saw before. The baseline and optimized models both had a balanced macro-average F1-score of 0.85, but the optimized model was a little more accurate overall (0.8507 vs. 0.8460). The classification reports also showed that the optimized network keeps a high recall rate for the non-potable class (0.89) and a high precision rate for the potable class (0.88).

The convergence statistics also show the Firefly Algorithm's [1] effect. The model's error went down from 0.1417 to 0.1344 during optimization, and there was a clear breakthrough at iteration 19. In the end, this change made a decision boundary that does a better job of separating the two classes. The optimized boundary leans a little toward higher sensitivity for non-potable samples. This makes sure that fewer unsafe samples are mistakenly labeled as "potable," while still keeping the desired level of accuracy so that costs from false alarms or misclassification don't go up.

The Firefly-optimized MLP is a more reliable classification system because it has fewer errors, fewer false positives, better overall accuracy, and a smoother convergence curve. This is especially important when it comes to correctly identifying unsafe water samples.

15.5 ANALYSIS OF COMPUTATIONAL COST AND COMPLEXITY

The meta-heuristic optimization process added a lot of extra work to the baseline MLP training. The full optimization process took about 4.6 hours of computation with a Firefly Algorithm [1] setup of 20 fireflies over 30 iterations. This length of time is due to the fact that population-based algorithms are inherently exploratory. At each iteration, they must look at and update multiple possible solutions.

The convergence trajectory is both informative and typical of how meta-heuristics work. For the first three iterations, the algorithm showed a period of rapid improvement, quickly moving the swarm toward areas of the hyperparameter space with lower validation error. After that, there was a longer plateau phase from iterations 4 to 18, during which improvements slowed as the population looked at a group of almost optimal solutions. At iteration 19, a significant breakthrough took place, suggesting that the algorithm successfully evaded a local minimum and identified a more advantageous configuration. The last ten iterations (20 through 30) showed stable convergence, with only small changes made as the swarm settled around a set of hyperparameters that worked well.

This four-stage pattern, initial acceleration, stagnation, breakthrough, and stabilization, is in line with how well-tuned bio-inspired optimization procedures are supposed to work. It shows both the cost and the benefit of using the Firefly Algorithm [1]: it took longer than normal hyperparameter tuning, but it was able to navigate complex, nonconvex landscapes, which led to a better and more reliable MLP model in the end.

15.6 PRACTICAL IMPLICATIONS

The model's reliability has improved because it now detects fewer unsafe water samples that pass through the system without being flagged. In practical terms, this decrease in undetected dangerous cases directly improves public health protection because the model becomes more sensitive to small chemical patterns that are linked to contamination. The methodological framework of an MLP enhanced by Firefly-based hyperparameter optimization is not only useful for assessing the potability of water but also for many other environmental classification tasks.

The same optimization strategy can be used for any field that has physicochemical indicators, a moderate number of features, and labels that are balanced or close to balanced. Examples include assessing soil toxicity, categorizing air quality, finding pollution sources, or mapping ecological risk. Neural networks are flexible, and meta-heuristic search is a strong and adaptable solution for situations where the decision boundaries are nonlinear and the cost of misclassification is high.

15.7 CONCLUSION

In general, the Firefly-assisted hyperparameter search consistently improves the MLP's ability to predict whether water is safe to drink, even though the improvements are small. The optimized model shows a small increase in overall accuracy, but more importantly, it improves the distribution of errors in safety-critical directions, especially by lowering the number of false negatives, which is when unsafe water samples are incorrectly labeled as safe.

This change shows that the Firefly Algorithm [1] does more than just tweak model parameters; it also helps change the decision boundary to better protect public health. In this case, even small gains are important because every avoided misclassification means that people are less likely to come into contact with unsafe drinking water. The results show that meta-heuristic optimization can consistently make neural networks more reliable when they are used to assess environmental risks.

15.8 FUTURE WORK

Future research can expand the methodological framework by comparing the Firefly Algorithm [1] with other established evolutionary and swarm intelligence optimizers, including Particle Swarm Optimization (PSO), Genetic Algorithms (GA), and Differential Evolution (DE). If we compare these algorithms in the same way, with the same hyperparameter search spaces, computational budgets, and evaluation metrics, we will have a better idea of how different meta-heuristic strategies affect convergence behavior, robustness, and model stability.

In addition to optimizing a single model, performance can be improved even more by using ensemble methods like stacked generalization, weighted majority voting, or hybrid MLP-tree ensembles.

These methods can help lower variance and make things more general, especially when working with water-quality samples that are noisy or on the edge. Adding probability calibration methods like Platt scaling, isotonic regression, and temperature scaling can also make predicted potability probabilities more reliable. This is useful for making decisions about environmental monitoring that take risks into account.

Another level of analysis looks at how distribution changes in real-world, long-term water quality monitoring systems. Due to seasonal patterns, changes in infrastructure, sensor recalibration, or new sources of contamination, environmental measurements often change over time. To use these classification systems in long-term monitoring pipelines, it will be important to test how the model reacts to these changes and find performance drops, recalibration needs, or retraining triggers.

DECLARATION OF INTEREST

The authors declare that there is no conflict of interest.

ACKNOWLEDGMENTS

The authors would like to thank the anonymous reviewers for their invaluable feedback and constructive suggestions, which have significantly improved the quality of this manuscript.

FUNDING STATEMENT

This research did not receive any specific grant from funding agencies in the public, commercial, or not-for-profit sectors.

ETHICAL STANDARDS

This study utilized a publicly available, anonymized dataset sourced from the Kaggle repository. As the research involves the secondary analysis of existing data that contains no personally identifiable information, no direct human or animal participation was required, and the study did not necessitate formal approval from an institutional ethics committee. The authors declare that all research procedures were conducted in full compliance with international ethical standards and academic integrity guidelines.

AUTHOR CONTRIBUTIONS

Özen Özer: Conceptualization, Methodology, Validation, Formal Analysis, Investigation, Writing—Original Draft, Writing—Review and Editing, Visualization, Project administration.

Nadir Subaşı: Software, Data Curation, Formal Analysis, Investigation, Writing—Original Draft, Visualization.

DATA AVAILABILITY STATEMENT

The dataset used in this study is publicly available and was sourced from Kaggle. It can be accessed directly at https://www.kaggle.com/datasets/tamilamudan/water-quality-potability/data.

REFERENCES

1. X.-S. Yang, "Firefly algorithms for multimodal optimization," in *Stochastic Algorithms: Foundations and Applications*, S. Watanabe and T. Zeugmann, Eds., Berlin, Germany: Springer, 2009, pp. 169–178.

2. I. Goodfellow, Y. Bengio, and A. Courville, *Deep Learning*. Cambridge, MA: MIT Press, 2016.
3. C. M. Bishop, *Pattern Recognition and Machine Learning*. New York, NY: Springer, 2006.
4. L. Breiman, "Random forests," *Mach. Learn.*, vol. 45, no. 1, pp. 5–32, 2001.
5. T. Hastie, R. Tibshirani, and J. Friedman, *The Elements of Statistical Learning*, 2nd ed. New York, NY: Springer, 2009.
6. J. Han, M. Kamber, and J. Pei, *Data Mining: Concepts and Techniques*, 3rd ed. Waltham, MA: Elsevier, 2011.
7. J. Kennedy and R. Eberhart, "Particle swarm optimization," Proceedings of the IEEE International Conference on Neural Networks, Perth, Australia, 1995, pp. 1942–1948.
8. R. Storn and K. Price, "Differential evolution: A simple and efficient heuristic for global optimization over continuous spaces," *J. Glob. Optim.*, vol. 11, no. 4, pp. 341–359, 1997.
9. D. P. Kingma and J. Ba, "Adam: A method for stochastic optimization," Proceedings of the International Conference on Learning Representations (ICLR), San Diego, CA, 2015.
10. N. Srivastava *et al.*, "Dropout: A simple way to prevent neural networks from overfitting," *J. Mach. Learn. Res.*, vol. 15, no. 1, pp. 1929–1958, 2014.
11. T. Fawcett, "An introduction to ROC analysis," *Pattern Recognit. Lett.*, vol. 27, no. 8, pp. 861–874, 2006.
12. F. Pedregosa, G. Varoquaux, A. Gramfort, V. Michel, B. Thirion, O. Grisel, et al., "Scikit-learn: Machine learning in Python," *J. Mach. Learn. Res.*, vol. 12, pp. 2825–2830, 2011
13. C. Cortes and V. Vapnik, "Support-vector networks," *Mach. Learn.*, vol. 20, no. 3, pp. 273–297, 1995.
14. Y. LeCun, L. Bottou, Y. Bengio, and P. Haffner, "Gradient-based learning applied to document recognition," *Proc. IEEE*, vol. 86, no. 11, pp. 2278–2324, 1998.
15. A. Krizhevsky, I. Sutskever, and G. E. Hinton, "ImageNet classification with deep convolutional neural networks," in *Advances in Neural Information Processing Systems 25*, F. Pereira, C. J. C. Burges, L. Bottou, and K. Q. Weinberger, Eds., 2012, pp. 1097–1105.
16. C. Szegedy, V. Vanhoucke, S. Ioffe, J. Shlens, and Z. Wojna, "Rethinking the inception architecture for computer vision," Proceedings of the IEEE Conference on Computer Vision and Pattern Recognition (CVPR), 2016, pp. 2818–2826.
17. K. He, X. Zhang, S. Ren, and J. Sun, "Deep residual learning for image recognition," Proceedings of the IEEE Conference on Computer Vision and Pattern Recognition (CVPR), 2016, pp. 770–778.
18. S. Hochreiter and J. Schmidhuber, "Long short-term memory," *Neural Comput.*, vol. 9, no. 8, pp. 1735–1780, 1997.
19. D. P. Kingma and M. Welling, "Auto-encoding variational bayes," *arXiv preprint arXiv:1312.6114*, 2013.
20. A. Vaswani, N. Shazeer, N. Parmar, J. Uszkoreit, L. Jones, A. N. Gomez, Ł. Kaiser, and I. Polosukhin, "Attention Is All You Need," in *Advances in Neural Information Processing Systems* 30, I. Guyon, U. V. Luxburg, S. Bengio, H. Wallach, R. Fergus, S. Vishwanathan, and R. Garnett, Eds., 2017, pp. 5998–6008.
21. T. Chen and C. Guestrin, "XGBoost: A scalable tree boosting system," Proceedings of the 22nd ACM SIGKDD International Conference on Knowledge Discovery and Data Mining, San Francisco, CA, 2016, pp. 785–794.
22. I. Goodfellow, J. Pouget-Abadie, M. Mirza, B. Xu, D. Warde-Farley, S. Ozair, A. Courville, and Y. Bengio, "Generative Adversarial Nets," in *Advances in Neural Information Processing Systems 27*, Z. Ghahramani, M. Welling, C. Cortes, N. D. Lawrence, and K. Q. Weinberger, Eds., 2014, pp. 2672–2680.

Aqua Sense *Hybrid Bioimpedance and IOT Water Quality Monitoring System*

16

Yogendra Chhetri

16.1 INTRODUCTION

Water is not only necessary for human health but also for agriculture, industrial processes, and the environment. Its quality is however getting worse due to rapid urbanization, untreated industrial discharge, agricultural runoff, and inadequate treatment of wastewater [2, 5, 7]. Pathogenic microorganisms, dissolved organic matter, nutrients, and heavy metals are some of the contaminants that lower the quality of surface and groundwater and, as such, pose serious health risks to the public and also harm the environment in the long run [2, 5, 6]. More waterborne disease outbreaks and pollutions from industries have led to an increased need for water quality monitoring systems that are continuous, early-stage, and capable of operating under real-world conditions [2, 3, 7].

Traditional methods for water quality assessment have been using chemical titration, UV visible spectrometry, ion-selective electrode analysis, and microbial culture-based methods [2, 5]. These methods have a high level of analytical precision and are accepted by regulatory authorities. Despite their advantages, these lab-based approaches have fixed limitations such as time-consuming sample processing, expensive instruments, and the requirement of skilled professionals, and they cannot be used for real-time or decentralized monitoring [2, 5, 7].

To overcome these restrictions, Internet of Things (IoT)–based water monitoring platforms have been developed, which use low-cost sensors to detect changes in pH, turbidity, electrical conductivity, dissolved oxygen, and temperature. These setups allow for faster, remote, and continuous data access [2, 3, 7]; however, they mostly reflect the overall changes in the physicochemical parameters of the water and, therefore, are not capable of detecting the very initial microbial growth stages, organic contamination, ionic interference, and complex mixed-pollutant scenarios [12, 15].

Recent research reveals that electrical characterization methods not only provide extra information on the quality of water but also reflect on the frequency-dependent dielectric and ionic properties of it [1,

DOI: 10.1201/9781003742227-16

4, 5]. From these, bioimpedance spectroscopy (BIS) is the one that has attracted the most attention, being a non-invasive tool for analyzing the changes in ionic mobility, dielectric permittivity, dissolved solids concentration, and microorganisms' metabolic activity over a broad frequency range [8, 9]. Being different from single-parameter sensors, BIS makes it possible to carry out the multidimensional analysis of the composition of water, thus paving the way for the recognition of contamination signatures, which at the initial stages might not have significantly altered the standard water quality parameters [5, 8, 9, 11]. Based on these observations, this chapter puts forward Aqua Sense, a hybrid water quality monitoring framework that couples bioimpedance spectroscopy with traditional multi-parameter IoT sensing. The proposed system, which marries frequency, domain impedance analysis with real-time environmental measurements and edge-level signal processing, is intended to raise sensitivity, allow early contamination detection, and facilitate the discrimination of complicated pollution patterns [1, 8, 9]. The hybrid setup is intended for continuous operation, energy-efficient use, and upscaling, thus it is fit for remote, industrial, and resource-limited settings where rapid water quality evaluation is a matter of life and death [3, 4, 7].

16.2 LITERATURE REVIEW

Existing water quality monitoring systems usually depend on inexpensive Internet of Things (IoT) platforms that have basic physicochemical sensors such as pH, turbidity, temperature, and electrical conductivity. The systems allow real-time, remote monitoring and are ideal for extensive distribution in rivers, reservoirs, and drinking water networks [2, 3, 7, 20]. Surface-level parameter monitoring of the water is the main limitation of such methods, which, besides operational convenience, these systems have, and they cannot identify the early-stage contamination by organic pollutants, microplastics, microbial residues, or the slow dielectric changes that are not immediately reflected by the conventional water quality indicators [2, 3, 7, 20]. Therefore, the problem of water quality deterioration due to the entry of pollutants is not usually detected in the early stages.

There have been attempts to improve detection capacity by integrating machine learning (ML) models with IoT-based water quality data for automatic classification and anomaly detection. Yet, most ML-based methods use only low-dimensional input features obtained from conventional sensors, thus limiting their capability to detect faint contamination signals [6, 11, 19]. Such approaches work well when pollution results in obvious changes in standard parameters, and thereby, their potential for application as an early warning system or in situations of complex mixed contaminant environments is quite limited. As a result, there is still a lack of sensing techniques that can provide a wider range, frequency-sensitive information that can be exploited by intelligent analysis.

One point that can be made is that recent work has clearly shown that impedance and bioimpedance spectroscopy methods are extremely sensitive to changes that occur in the ionic, molecular, and dielectric characteristics of fluids and materials (see Figure 16.1). Different frequency electrical responses have proven very effective in detecting ionic contamination, organic pollutants, and microplastics in water samples [4, 5, 8, 13, 16, 18]. The existence of handheld impedance analyzers and the electrical circuit modeling confirm the possibility of using impedance-based water quality monitoring systems in the real environmental conditions [4, 14, 17].

Bioimpedance methods have been very well studied and have found many uses in biomedical diagnostics, food quality evaluation, and fluid characterization [9, 10, 12]. Their coalescence with IoT platforms to provide continuous monitoring of environmental waters is nevertheless quite rare. The absence of combined frameworks that merge impedance spectroscopy and real-time IoT sensing is the reason behind the proposed Aqua Sense system (see Figure 16.2).

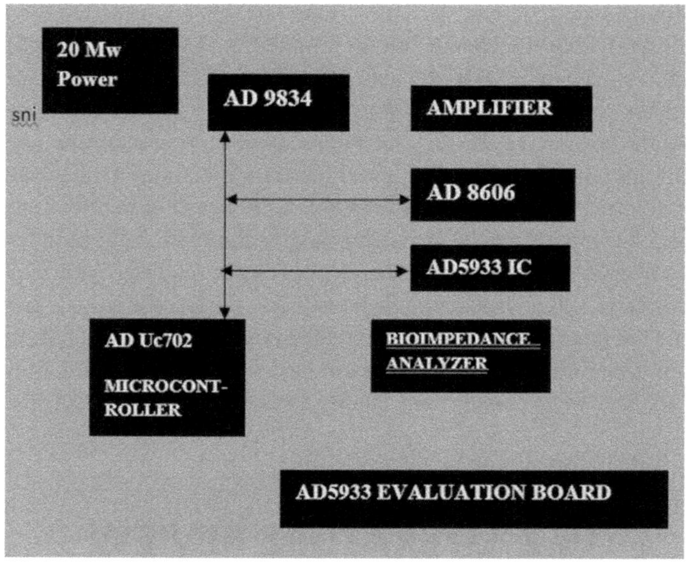

FIGURE 16.1 AD5933 bioimpedance analyzer.

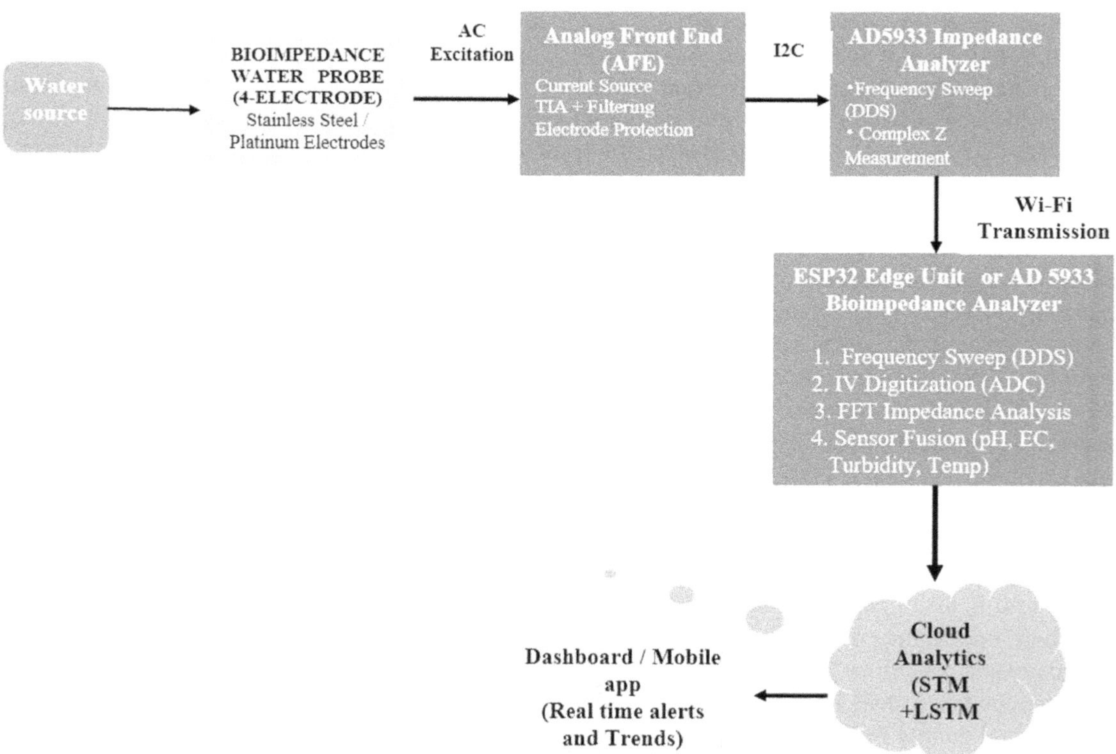

FIGURE 16.2 Block diagram of Aqua Sense.

16.3 COMPARATIVE STUDY ON WATER QUALITY MONITORING TECHNOLOGIES

Modern water quality monitoring relies on multiple sensing approaches, each offering distinct advantages and limitations, making a comparative evaluation essential to identify the gaps addressed by Aqua Sense.

(1) Electrochemical Sensor-Based IoT Systems: Traditional water quality monitoring mainly depends on electrochemical sensor-based measurements of pH, turbidity, electrical conductivity, temperature, and dissolved oxygen, which are preferred due to their low cost, easy installation, and real-time IoT compatibility [2, 3, 7, 20]. Nevertheless, these sensors merely provide basic water quality parameters and are incapable of detecting microbial activity in early stages, organic pollutants, microplastics, or gradual dielectric and compositional changes. On top of that, electrode fouling, signal drift, and frequent recalibration lower the long-term reliability of these sensors in continuous monitoring setups to a great extent [2, 5, 6].

(2) Optical and Spectrophotometric Techniques: UV-Visible spectrophotometry and optical absorbance tactics are very precise in a long way of analytics, among several others, when it comes to the detection of heavy metals, organics, and industrial pollutants by the analysis of the wavelength-dependent absorption characteristics of the substances [5, 8]. Even so, their noble precision is paid for by their charging requirements in terms of bulky instrumentation, reagents, complex calibration procedures, and the need for a skilled operator which makes them unviable for low-power, continuous, or distributed IoT-based water monitoring systems [5].

(3) Ion-Selective and Specialized Electrochemical Sensors: Ion-selective electrodes (ISEs) and molecularly imprinted electrochemical sensors provide the means for highly selective detection of specific ions or contaminants like nitrates, ammonium, fluoride, or targeted chemical species [5, 6]. However, these devices are only effective in providing isolated measurements due to their narrow selectivity, vulnerability to cross-interference, short lifespan, and incapability of dealing with mixed or unknown contaminant profiles, which, in turn, limit their application in actual water distribution networks [6].

(4) Impedance and Bioimpedance-Based Monitoring Approaches: Electrical impedance spectroscopy (EIS) and bioimpedance methods show a remarkable sensitivity to changes in ionic concentration, variations in dielectric properties, presence of microplastics, and organic or biological contaminants through frequency-dependent electrical signatures [4, 5, 13, 16, 18]. There are several experimental works that demonstrate their potential in fluid characterization, pollutant differentiation, and compositional analysis; however, the majority of these implementations are still standalone, laboratory-based, or confined to biomedical and material science applications without the full integration of IoT or intelligent data fusion [10, 12, 15, 19].

This comparative evaluation demonstrates that while each technology excels in isolated aspects of water quality assessment, no single approach can deliver comprehensive, fast, and multi-dimensional contamination detection. These limitations justify the development of Aqua Sense, a hybrid multi-sensor architecture that combines the strengths of BIS with conventional water quality sensing and advanced AI analytics to achieve a unified, real-time monitoring solution.

16.4 PROPOSED SYSTEM

The Aqua Sense program is a pioneer of the next-generation hybrid sensing architecture that is expressly created for quick, precise, and multi-dimensional detection of water contamination. The system electrifies the characterization of water quality by going beyond the ability of the current low-cost IoT-based solutions

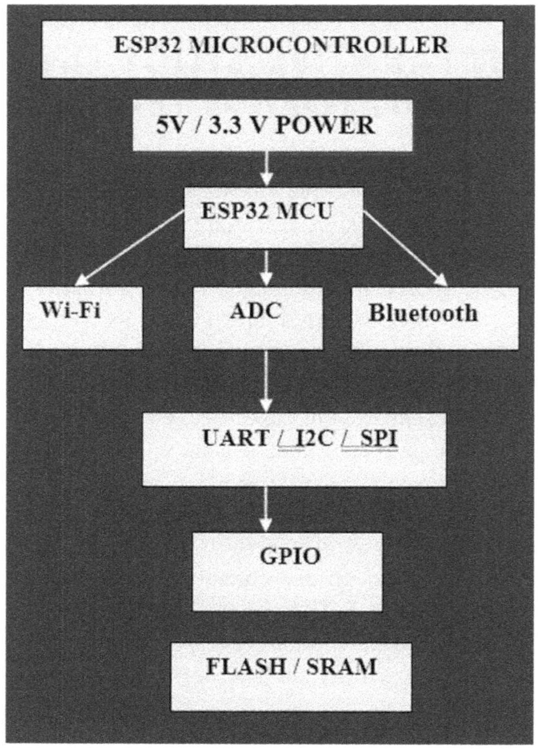

FIGURE 16.3 ESP32 module.

to describe water quality in full by uniting the dielectric sensitivity of Bioimpedance Spectroscopy (BIS) and the physicochemical nature of the sample that is revealed by the conventional sensors. The major idea is to retrieve the frequency-dependent electrical fingerprints that mirror the movement of ions, microbial growth, and the interaction of organic substances with the system surfaces, which is at the same time monitoring pH, turbidity, conductivity/TDS, and temperature (see Figure 16.3). Meanwhile, these multi-domain data streams are locally processed through embedded computing and analyzed via advanced machine learning algorithms in the cloud; thus, the detection, classification, and prediction of the contamination events are done in real time. With such a system, Aqua Sense offers a scalable, affordable, and highly sensitive platform that is fit for the continuous monitoring of water in rural as well as industrial environments.

16.5 SYSTEM COMPONENTS

The Aqua Sense system revolves around the tightly integrated components that work hand in hand to achieve accurate and real-time water quality assessment. The core element of the system is a custom-built Kelvin-based Bioimpedance Water Probe which introduces multifrequency electrical signals to water and, at the same time, measures the resulting magnitude phase response. The probe is quite capable of discerning the ionic changes, dielectric shifts, and microbial polarization effects that make the differentials that traditional sensors can compare but cannot capture.

Along with the probe, there is also a multi-parameter sensor module that monitors pH, turbidity, conductivity/TDS, and temperature and thus provides the necessary physicochemical background for contamination classification to become feasible. Information from all sources is processed locally on the

ESP32 edge processing unit. It performs FFT-based spectral extraction, feature computation, and noise reduction directly on the device. This makes the system efficient, stable, and reliable, even when network connectivity is weak. Next, the less processed data are transferred to a machine learning–powered cloud reasoning platform where hybrid SVMLSTM models find both spectral and temporal patterns to locate contamination types and to predict water quality changes. This collaborative architectural layout sums up Aqua Sense and allows it to offer a solution for continuous water monitoring that is not only highly intelligent but also sensitive and scalable.

16.6 ANALYTICS AND OUTPUT FRAMEWORK

The cloud intelligence layer of Aqua Sense comprises various automated functions such as quality scoring, anomaly spotting, and predictive modeling that convert raw data into valuable insights. Using complex temporal analysis, the device generates near-real-time Water Quality Index scores and detects new contamination trends with a high level of accuracy. An easy-to-use web and mobile dashboard makes available to the users the data on the impedance spectra and sensor readings, machine learning–based contamination labels, as well as the historical trend graphs, thus helping in the prompt and well-informed decision-making. Besides, the system combining the dielectric signatures from the bioimpedance measurements with the data from the conventional environmental sensors achieves a significantly enhanced detection sensitivity, which, in turn, allows it to detect chemical pollutants, microbial growth, and particulate disturbances at a much higher rate of speed than if only pH or conductivity were used. Thus, Aqua Sense turns into a perfect instrument for both rural water distribution networks and industrial monitoring environments which can provide continuous, dependable, and intelligent surveillance of water quality.

16.7 NOVELTY OF THE AQUA SENSE HYBRID MONITORING SYSTEM

Aqua Sense brings in a bundle of novel features that, in total, place it far ahead of other water quality monitoring devices in the domain. First of all, its hybrid fusion of Bioimpedance Spectroscopy (BIS) and standard environmental sensors like pH, turbidity, conductivity, and temperature is pioneering. With this combination, the system can perform multi-dimensional sensing by capturing not only chemical properties but also ionic behavior and dielectric characteristics; at the same time, this multi-sensor IoT system cannot be found anywhere else. The four-electrode Kelvin Bioimpedance Water Probe (BWP) of the system is a big step forward as it abolishes electrode polarization, provides uniform electric field distribution, and hence, greatly increases the measurement stability of different water bodies. By analyzing low-frequency microbial polarization signatures, mid-frequency ionic relaxation dynamics, and high-frequency dielectric shifts related to organic compounds, Aqua Sense locates contamination. The machine learning pipeline escalates the innovation layer by introducing SVM for contamination classification with a high accuracy and LSTM networks for temporal prediction, anomaly detection, and dynamic Water Quality Index calculation. Furthermore, employing the ESP32 for edge analytics, carrying out FFT computations, noise filtration, and multi-sensor fusion locally reduces latency and raises accuracy even in regions with poor connectivity. These multi-tiered innovations position Aqua Sense as the first instrument to effectively integrate electrical, chemical, and computational intelligence for a real-time, scalable, and high sensitivity monitoring solution.

16.8 METHODOLOGY

The methodology behind Aqua Sense is a detailed, multi-stage process that ensures a precise and live analysis of water pollution, with hybrid sensing being the first step. The Bioimpedance Water Probe creates sinusoidal excitation signals over a 1 kHz to 300 kHz sweep with the help of a Direct Digital Synthesizer. Moreover, the separate current-injection and voltage-sensing electrodes are there to ensure that the interference is minimal and the impedance measurements are stable. Apart from BIS sensing, normal sensors also take the pH, turbidity, conductivity, and temperature measurements, thus giving the respective chemical and environmental data. The second methodological stage consists of embedded signal processing on the ESP32 microcontroller, in which raw sensor signals are digitized using Analog-to-Digital Converter (ADC) sampling. Besides that, FFT-based spectral extraction, time-domain filtering, normalization, and advanced feature computation, including dielectric relaxation slopes, microbial polarization indices, impedance-conductivity correlations, and temporal drift characteristics, are performed. The third stage is cloud-based machine learning. Here, Support Vector Machines classify contamination types, and LSTM networks analyze temporal patterns, detect anomalies, forecast contamination progression, and compute a continuously updated Water Quality Index.

The final stage is real-time visualization and alerting on the cloud dashboard. It allows users to view impedance spectra, track environmental trends, and get automated warnings of unsafe water conditions.

16.9 A SYSTEM DESIGN

The developed framework combines various layers, starting with physical sensing, embedded signal processing, analytics performed on the cloud, and a user interface that is updated in real time. The sequence in Figure depicts that water is initially decided upon by the Bioimpedance Water Probe (BWP), which records the electrical measurements and sends them to the ESP32 processing unit. The frequency sweeps are carried out by the ESP32 or AD5933 Bioimpedance Analyzer, the current voltage responses are digitized, FFT-based impedance extraction is performed, and the data is also combined with the readings from the pH, turbidity, conductivity, and temperature sensors. Table 16.1 summarizes the methodologies adopted in recent studies and identifies research gaps that motivate the development of more intelligent and comprehensive water quality monitoring frameworks.

The cloud takes the processed features, and the machine learning algorithms identify the contamination and assign the water quality index. The output is displayed on a mobile dashboard that offers real-time monitoring, notification, and trend visualization. Such a multi-layered integration system architecture enables water quality detection that is speedy, precise, and extendable.

16.10 HARDWARE SPECIFICATIONS

16.10.1 Bioimpedance Analyzer and Frequency Sweep System

For the Bioimpedance Analyzer, a Direct Digital Synthesizer (DDS) component constructs the sinusoidal signal and selects the frequencies within the programmable range. The analyzer can continuously or stepwise scan frequencies between 1 kHz and 300 kHz, and 300 kHz, capturing the details for a comprehensive assessment of a sample's dielectric properties. Each of the 12 bits of an ADC is responsible for capturing the data corresponding to each frequency to help capture raw impedance data accurately. The

TABLE 16.1 Literature Review

AUTHOR (YEAR)	APPROACH	SIGNIFICANCE	LIMITATIONS
B. Ipek, R. A. Vural, and A. Kartci (2025) [1]	Bioelectrical impedance-based water quality assessment	Sensitive to ionic, molecular, and dielectric changes; capable of detecting early-stage contaminants	Mostly lab-based; limited IoT integration for real-time continuous monitoring
S. Santhosini, U. K. Subhiksha, and V. E. Jayanthi (2024) [2]	IoT-based water quality monitoring systems with pH, turbidity, EC, DO, and temperature sensors	Real-time, remote monitoring; suitable for large-scale deployment	Limited to surface-level physicochemical parameters; cannot detect early-stage microbial or organic contamination
M. K. Pushparani, M. S. Sharavari, S. M. Nayana, L. Likhith, and M. Manoj (2025) [3]	IoT-based water pollution detection boat with real-time monitoring	Provides mobile, real-time monitoring in rivers and reservoirs	Limited sensor types; surface-level monitoring only; early-stage contamination may not be detected
R. D. Gerber, P. Molina, and H. Suzuki (2021) [4]	Portable impedance analyzers for remote and frequency-domain monitoring of ionic contamination	Field-deployable; frequency-domain analysis enables detection of complex ionic contamination	Mostly standalone devices; integration with IoT and large-scale deployment is limited
A. Nazari, A. Farid, and K. Cornell (2016) [5]	Impedance-based water quality monitoring using parallel-plate method	Effective for detecting ionic changes and characterizing water composition	Lab-based; not suitable for continuous real-time or distributed monitoring

system calculates the impedance data for each frequency and constructs a phase response and a complete frequency response profile for the sample which can then be used for classification and detection of sample contaminants [2, 5, 13]. Table 16.2 highlights the trade-offs between IoT-based sensing technologies and conventional optical methods in terms of deployment, monitoring capability, and detection performance.

16.10.2 Bioimpedance Water Probe (BWP)

To eliminate the polarization error and the contact impedance of the electrodes, the BWP is configured in a four-electrode Kelvin arrangement. The electrodes can be used in multiple water bodies and weather corrosion due to the durable stainless steel and platinum coatings. The water sample electric field is made uniform to improve measurement stability and repeatability by optimizing the spacing of the electrodes.

16.10.3 Multi-parameter Sensor Module

To adequately complement the BIS measurements, four additional environmental sensors have been integrated. There is a pH sensor for measuring the acidity or alkalinity of the sample, a turbidity sensor (measured in NTU) for assessing suspended particulates in the sample, and a conductivity sensor for determining the ionic conductivity of the water.

In addition to the BIS measurements, the system incorporates four additional sensors for environmental data collection:

- PH sensor for measuring acidity/alkalinity
- Turbidity sensor (NTU) for measuring suspended particles

TABLE 16.2 Electrochemical IoT vs Optical/Spectrophotometric Techniques

PARAMETER	*ELECTROCHEMICAL IOT SENSORS*	*OPTICAL / SPECTROPHOTOMETRIC*	*KEY STRENGTH*	*MAJOR LIMITATION*
Technique	pH, Turbidity, EC, DO, Temperature sensors	UV-Vis spectrophotometry, optical absorbance	IoT: Low-cost, real-time; Optical: High precision	IoT: Limited to basic parameters; Optical: Bulky, high-power, requires skilled operator
Target	General water quality	Heavy metals, organics, industrial pollutants	Rapid detection in distributed water networks	IoT: Cannot detect microbes or microplastics; Optical: Not suitable for continuous/ distributed monitoring
Application	Distributed / remote monitoring	Laboratory or precise detection tasks	IoT: Continuous data; Optical: Accurate analytical results	IoT: Electrode fouling, signal drift; Optical: Expensive, complex calibration
References	[2, 3, 5–7, 20]	[5, 8]	-	-

- Conductivity/EC–TDS sensor for measuring ionic strength
- Temperature sensor for compensation and calibration purposes

Having these features allows for greater classification of the type of contamination present.

16.10.4 ESP32 Edge Processing Unit

All the embedded analytics, such as FFT-based impedance extraction, noise reduction, normalization, feature computation, and sensor fusion between BIS and environmental sensors, are conducted on the ESP32 microcontroller. Environmental and BIS sensor data fusion allows the system to get an even deeper analysis of water quality. Using the built-in Wi-Fi connection the system transmits data to the cloud platform where machine learning algorithms perform real-time analysis, and the results are visualized on an interactive dashboard [12, 15]. Table 16.3 presents a comparative evaluation of ion-selective and molecularly imprinted polymer (MIP) electrodes against impedance and bioimpedance-based sensing techniques for water quality assessment.

16.11 SOFTWARE DEVELOPMENT

16.11.1 Edge Firmware

All the real-time embedded operations are under the supervision of the ESP32 firmware, including DDS-based frequency sweeping, impedance measurement, and FFT-based magnitude–phase extraction. Local noise filtering, normalization, and basic feature computation enhance signal stability for the embedded system. The firmware supports threshold-based classification and data packaging for secured Wi-Fi transmission. These functions ensure accurate raw data acquisition and reliable pre-processing.

TABLE 16.3 Ion-Selective / MIP Electrodes vs Impedance / Bioimpedance Techniques

PARAMETER	*ION-SELECTIVE / MIP ELECTRODES*	*IMPEDANCE / BIOIMPEDANCE TECHNIQUES*	*KEY STRENGTH*	*MAJOR LIMITATION*
Technique	ISE: Nitrate, Ammonium, Fluoride sensors MIP: Molecularly Imprinted Electrodes	EIS & Multi-frequency Bioimpedance Analysis	ISE / MIP: High selectivity for specific ions Impedance: Sensitive to ionic, dielectric, organic, and biological changes	ISE / MIP: Narrow selectivity, cross-interference, cannot handle unknown mixtures Impedance: Mostly lab-based, limited IoT integration
Target	Specific ions / targeted chemical species	Compositional analysis, microplastics, pollutants	Accurate detection of contaminants	Environmental / field deployment limited; mostly standalone
Application	Water distribution networks, lab-scale targeted detection	Fluid characterization, pollutant differentiation	Selective & sensitive measurement	Advanced data analysis required; not continuous monitoring
References	[5, 6]	[4, 5, 8, 10, 12, 13, 15, 16, 18, 19]	–	–

16.11.2 Cloud Machine Learning Analytics

Cloud services incorporate an SVM + LSTM hybrid model which evaluates the combined features of the impedance probe and environmental sensors. The SVM abstracts the static classes, while the LSTM sequentially captures the temporal dynamics and trend forecasting. The cloud engine predicts and alerts on dynamically varying quality and potential contamination and computes the quality watermark index (WQI). A mobile/web dashboard provides real-time monitoring, analyzes historical trends, and generates automated reports on the processed results.

16.12 RESULTS AND DISCUSSION

16.12.1 Impedance Behavior

Strong evidence related to the impedance readings was linked to the sample purity. The highest impedance was recorded in clean water that had the greatest ionic deficit. When ionic and microbial contaminants were present, the impedance value diminished and signified the presence of charge carriers and metabolic by-products. When clean water was salted (NaCl) and the impedance measuring system was active, the recorded impedance reduction was very significant and demonstrated the system's real-time detection of chemical contamination. The impedance variations observed under different contamination scenarios are depicted in Figure 16.4.

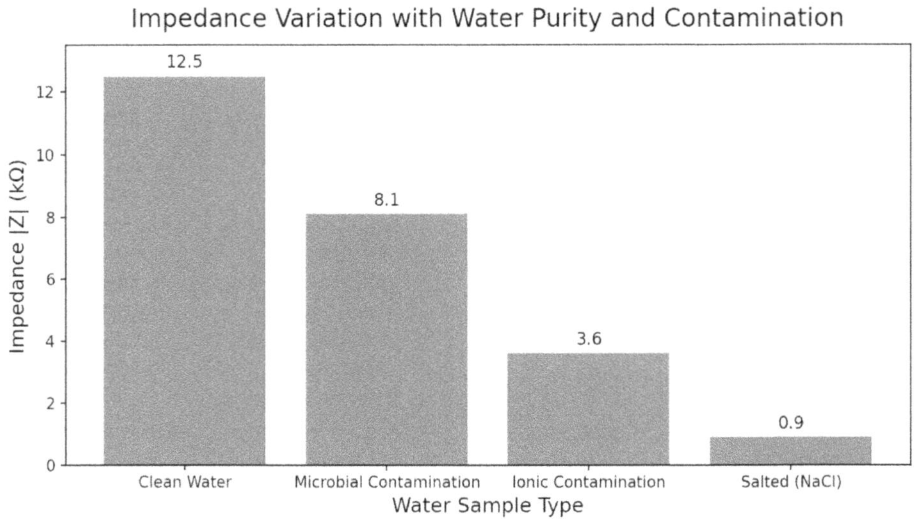

FIGURE 16.4 Impedance variations.

TABLE 16.4 Sample Impedance and Their Observed Results

SAMPLE TYPE IMPEDANCE (Ω)	*OBSERVATION*
Clean Water 1625.8 Ω	High impedance due to low ionic concentration
Ionic Contamination 322.5 Ω	Sharp decrease indicating high ion mobility
Microbial Contamination 485.2 Ω	Moderate drop; metabolic activity affects dielectric

16.12.2 Frequency Response Pattern

In the frequency domain, analysis revealed varying distinct signatures for each type of contamination. For clean water, the system showed a consistent linear drop in impedance over the entire range of the impedance sweep. For ionic contaminants, the system showed an exponential drop off, suggesting modified dielectric properties. For microbial contamination, a distinct polarization low-frequency signature was present as a consequence of the cellular polarization effect. Figure 16.5 illustrates that frequency patterns confirm the ability of the system to differentiate the types of contamination. Figure 16.4 highlights the effectiveness of frequency-domain analysis in distinguishing between different water quality conditions based on their impedance signatures.

16.12.3 Classification Accuracy

Sensor fusion and machine learning combine to provide highly dependable predictions. The hybrid SVM + LSTM model yielded over 95% accuracy for each contamination type. Table 16.4 presents the impedance measurements obtained for different water samples along with their corresponding observations and interpretations.For example, the system showed 96% confidence level for the classification of a test sample with mixed contamination (mild turbidity + increased EC), demonstrating strong performance ability for real-world application. Figure 16.6 depicts the analytical interface used for interpreting impedance measurements and evaluating water quality status.

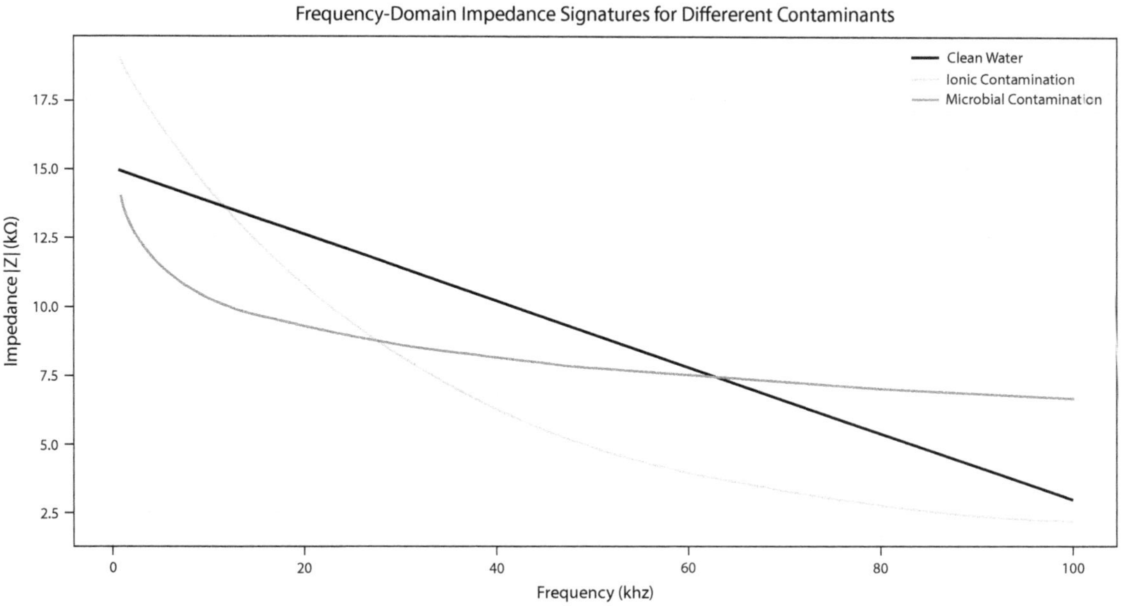

FIGURE 16.5 Frequency response pattern.

16.12.4 Expanded Observations

16.12.4.1 Clean Water: 1625.8 Ω

The highest impedance is shown by clean water due to the extremely low amounts of dissolved ions and particulates in it. The lack of significant charge carriers leads to highly resistive behavior and very low dielectric losses. Such a high impedance response is a strong indication of the sample's purity, and it is also a reflection of a stable molecular structure with almost no ionic conduction pathways.

16.12.4.2 Ionic Contamination: 322.5 Ω

The sudden and steep drop of impedance is the main reason why the ionic mobility is vastly increased. This increase in mobility is due to salts dissolved in water or the presence of industrial effluents. Free ions significantly lower the resistive opposition, and thus the electrical conductivity increases dramatically, which is responsible for a great decrease in impedance over the whole frequency sweep. The solution described here is one with a high ionic strength and a strong presence of electrolytes, and the behavior is typical of such solutions.

16.12.4.3 Microbial Contamination: 485.2 Ω

The microbial contamination is characterized by a moderate drop in impedance due to the effects of the production of metabolic by-products, membrane polarization, and increased ionic content. Microorganisms produce charged biomolecules and change the dielectric properties of water, especially at low frequencies where cell membrane polarization dominates. The contamination-induced water impedance profile is a combination of biological activity and the presence of early-stage bioelectric changes.

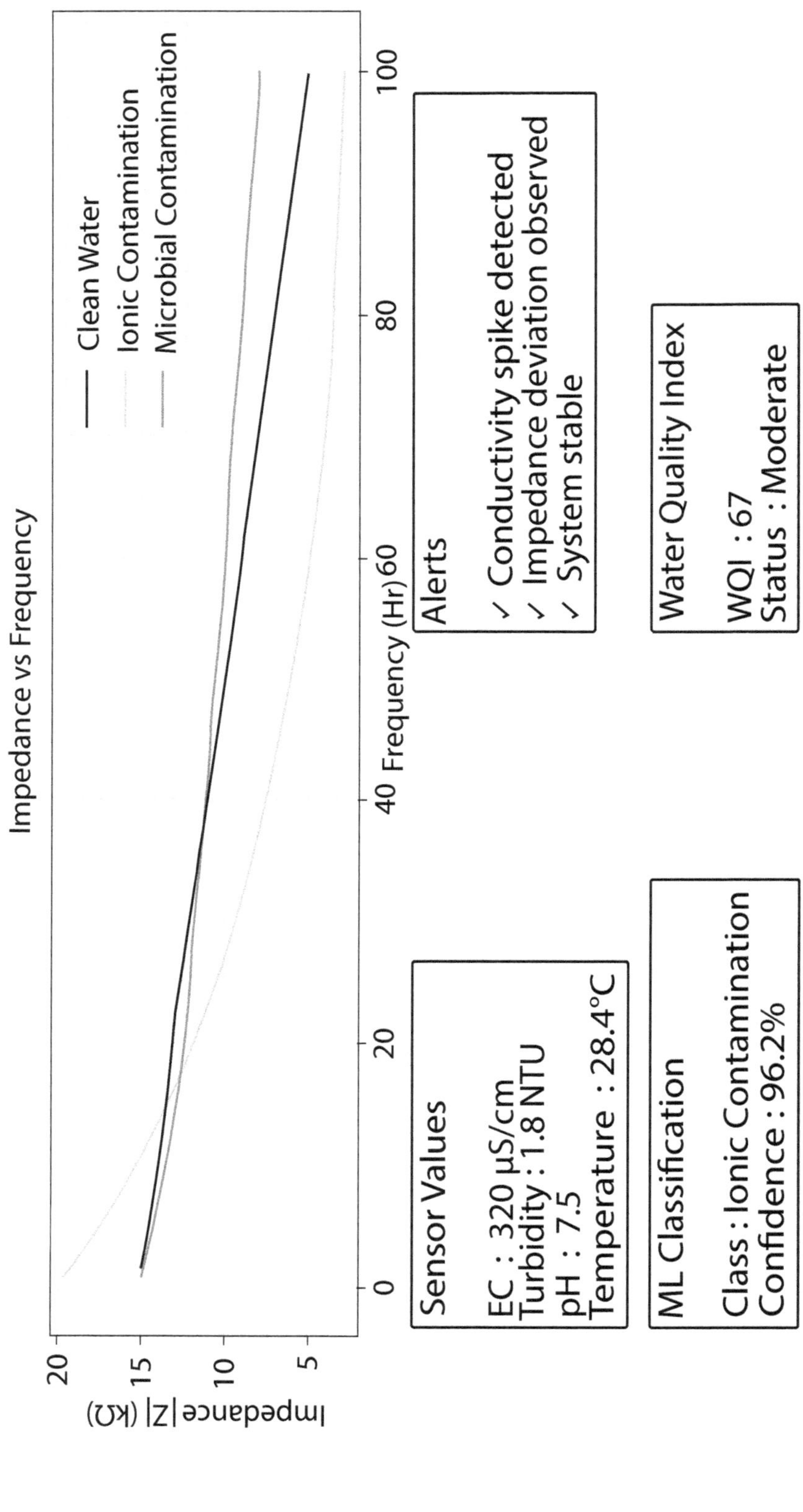

FIGURE 16.6 Analysis results displayed in software.

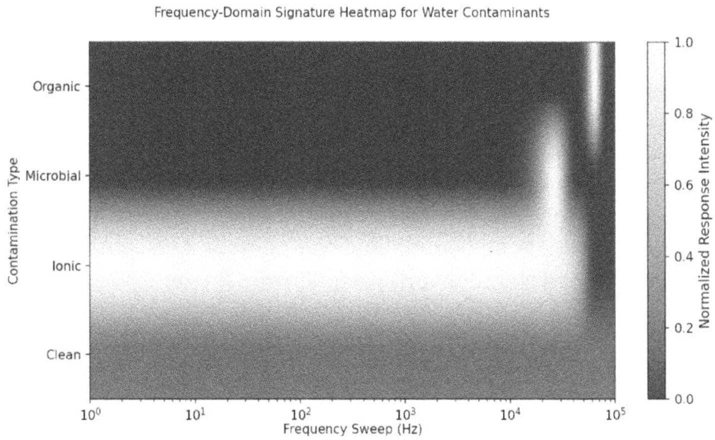

FIGURE 16.7 Frequency-domain signature heat map diagram.

16.13 FREQUENCY-DOMAIN ANALYSIS AND CONTAMINATION DETECTION CAPABILITIES

The main way that Aqua Sense is able to expand its identification ability is through frequency-domain analysis, where separate impedance and dielectric changes can be observed for each contamination category. Water that is pure usually behaves electrically as a high impedance with a smooth linear decrease along the frequency range, which is a reflection of a negligible ionic content and a stable dielectric composition. If ionic contamination is present—like in the case of dissolved salts or industrial effluents—the device shows a steep exponential drop-off of impedance along with a change in relaxation behavior since now there are more charge carriers and the dipoles align faster.

The microbial contamination masks a distinctly low-frequency polarization hump caused by Maxwell–Wagner interfacial polarization at the microbial cell membranes; hence, microbial activity can be detected even well before any visual changes like turbidity occur. On the other hand, organics such as oils, pesticides, and industrial solvents cause dielectric relaxation to shift from mid-to-high frequencies as the structural changes are due to the presence of the organics and are thus invisible to normal sensors.

The correlation of these frequency-domain patterns with real-time pH, turbidity, EC, and temperature data by Aqua Sense thus helps the technology to be able to correctly identify the presence of ionic, microbial, organic, and mixed contaminants with an accuracy of more than 95%. Such a deep spectral understanding puts Aqua Sense far ahead of the conventional sensor systems and makes it possible to perform rapid, reliable detection within seconds. Figure 16.7 presents the frequency-domain signature heat map of different water contaminants, illustrating their distinct spectral response characteristics across the frequency range.

16.14 CONCLUSION

The Aqua Sense system presents a functional hybrid architecture that integrates spectroscopy, multiparameter environmental sensing, and machine learning to deliver reliable, instantaneous, and real-time assessments of water quality. The system combines sensing water's physicochemical properties and volumes, capturing diverse and priceless water signatures to ascertain ion concentrations, detect chemical contaminants, and chemical pollutants and census microbial contaminants targeting the water source. The system's speed and sensitivity surpass technology that relies exclusively on sensors, and its true signature detection, response to variances, and system signature conformability make the system adaptable for the capture of longitudinal and closed-capture system signatures. Aqua Sense's integration attributes, including its modular construction, wireless connectivity features, and cloud-enabled system for analytics and decision-making, provide the system for longitudinal deployment on real-time capture and flexible operational deployment on potable water pipelines, real-time container effluent, industrial containers, and agriculture water reservoirs. There is no doubt that Aqua Sense is a stream of intelligence for the new generation of real-time intelligent water monitoring networks.

16.15 FUTURE WORK

The future evolution of the system will include expanding the bioimpedance spectrum to all frequencies, a greater focus on smart sensor integration, autonomous Edge-AI system functioning, and predictive control of under-connected systems. Automated and autonomous systems will be enhanced by electrode calibration, solar system–powered features, and microfluidic-flow cells. The system will save time, amplify precision, and improve data quality. Planned use predictive machine learning for surface and Aqua Sense integration telemetry will be proof of concept for system scaling and data generation. The integrated predictive telemetry and multi-circuit data acquisition system will serve as a proof of concept for the future expansion and scalability of the proposed system.

REFERENCES

1. B. Ipek, R. A. Vural, and A. Kartci, "Water quality assessment using bioelectrical impedance," Proceedings of the ELECO 2025 International Conference on Electrical and Electronics Engineering, Bursa, Trkiye, Nov. 27, 29, 2025.
2. S. Santhosini, U. K. Subhiksha, and V. E. Jayanthi, "Review on evaluation and analysis of water quality monitoring systems using IoT," *International Research Journal on Advanced Engineering Hub (IRJAEH)*, vol. 2, no. 9, pp. 2331, 2341, Sep. 2024.
3. M. K. Pushparani, M. S. Sharavari, S. M. Nayana, L. Likhith, and M. Manoj, "IoT, based water pollution detection boat with real, time monitoring," *International Research Journal on Advanced Engineering Hub (IRJAEH)*, vol. 3, no. 3, pp. 715, 723, Mar. 2025.
4. R. D. Gerber, P. Molina, and H. Suzuki, "Portable impedance analyzers for remote monitoring and frequency domain analysis of ionic contamination in groundwater," *IEEE Transactions on Geoscience and Remote Sensing*, vol. 59, no. 8, pp. 6612–6622, 2021.
5. A. Nazari, A. Farid, and K. Cornell, *Impedance, Based Water, Quality Monitoring Using the Parallel, Plate Method*, Civil Engineering Faculty Publications and Presentations, Boise State University, 2016.
6. M. D. Mukendi, O. S. Salami, and N. Mketo, "A comprehensive review of molecularly imprinted electrochemical sensors as an innovative analytical tool for water quality monitoring," *Micro Machines*, vol. 16, no. 3, p. 251, 2025.
7. Y. Carriazo, Regino, R. Baena, Navarro, and S. Roa, "Prada, IoT, based Drinking Water Quality Measurement: A Systematic Literature Review," *Indonesian Journal of Electrical Engineering and Computer Science*, vol. 28, no. 1, pp. 405–418, 2022.
8. G. Miele, S. Bellucci, A. Cataldo, and A. Di Tinno, "Electrical impedance spectroscopy for the life cycle real, time monitoring of graphene nanoplatelet filters for organic industrial pollutants," *IEEE Transactions on Instrumentation and Measurement*, vol. 70, pp. 1–12, 2021.

9. M. Grossi and M. Omaa, "Analysis of biosensors based on electrical impedance spectroscopy using artificial neural networks for devices with limited resources," *Journal of Low Power Electronics and Applications*, vol. 15, no. 4, p. 56, 2025.
10. M. Ashoorirad, R. Baghbani, and M. R. Ghalamboran, "Bioimpedance sensor for detecting water content in milk based on van der Pauw method," *Sensors*, vol. 21, no. 20, pp. 611–618, 2021.
11. O. Aybar, Z. Kara, M. Yucel, and B. B. Ustundag, "Water quality monitoring in real, time by machine learning and impedance spectroscopy," Istanbul Technical University, Turkey.
12. R. Takamatsu, K. Higuchi, and D. Muramatsu, "Measurement frequency evaluation for bioimpedance, based blood, glucose estimation," IEEE Global Conference on Life Sciences and Technologies (LifeTech) Proceedings, Nara, Japan, 2021, pp. 309310.
13. S. Magalhes, M. G. Rasteiro, L. Oliveira, P. V. Morais, and P. Faia, "Electrical impedance spectroscopy based strategy for detecting and differentiating microplastics in water," *Journal of Water Process Engineering*, vol. 80, Art. no. 109107, 2025.
14. O. Awayssa, R. A. Ismail, A. Hilal, AlNaqbi, and M. Al Ahmad, "Electrical circuit model for sensing water quality analysis," *Water*, vol. 17, no. 15, p. 2345, 2025.
15. X. Luo, Y. Zhou, and B. Sanchez, "Dielectric property-based characterization of fluid samples using portable impedance devices," *IEEE Access*, vol. 10, pp. 11274–11283, 2022.
16. D. Gomes, S. Magalhes, M. G. Rasteiro, and P. Faia, "Measuring microplastic concentrations in water by electrical impedance spectroscopy," *Water*, vol. 16, no. 22, Art. no. 3228, 2024.
17. O. Awayssa, R. A. Ismail, A. Hilal, AlNaqbi, and M. Al Ahmad, "Electrical circuit model for sensing water quality analysis," *Water*, vol. 17, no. 15, Art. no. 2345, 2025.
18. P. Bertemes Filho and J. A. Gmez, Snchez, "Can biosensors detect contaminants in aqueous medium by impedance spectroscopy?" *Journal of Analytical & Bioanalytical Techniques*, vol. 5, no. 2, p. 16, 2019.
19. B. Ipek, E. S. Ipek, R. A. Vural, and A. Kartci, "Water quality assessment using bioelectrical impedance," International Conference on Electrical and Electronics Engineering (ELECO 2025), Bursa, Trkiye, Nov. 2729, 2025, pp. 1229–1232.
20. S. Konde and S. Deosarkar, "IoT based water quality monitoring system," 2nd International Conference on Communication & Information Processing (ICCIP), 2020, p. 111.

Securing Water Supply Systems from Cyber-attacks

17

Deepak Joshi

ABBREVIATIONS

DoS: Denial of Service
SWAT: Secure Water Treatment
WADI: Water Distribution
RBAC: Role-Based Access Control
CSF: Cybersecurity Framework
IEC: International Electrotechnical Commission
ISO: International Organization for Standardization
ZTA: Zero Trust Architecture
SOC: Security Operations Centre
AI: Artificial Intelligence
ML: Machine Learning
ISAC: Information Sharing and Analysis Centre
SCADA Logs: Supervisory Control and Data Acquisition Logs
AWWA J100: American Water Works Association J100 Risk Management Standard
RSA: Rivest–Shamir–Adleman Cryptography
ECC: Elliptic Curve Cryptography
WADI Dataset: Water Distribution Intrusion Detection Dataset
UEBA: User and Entity Behaviour Analytics
UBA: User Behaviour Analytics
PLC: Programmable Logic Controller
PLCs: Programmable Logic Controllers
PKI: Public Key Infrastructure
DTLS: Datagram Transport Layer Security
RTUs: Remote Terminal Units
HMI: Human-Machine Interface
DNP3 SA: Distributed Network Protocol Version 3 Secure Authentication

DOI: 10.1201/9781003742227-17

OPC UA: Open Platform Communications Unified Architecture
DMZ: Demilitarized Zone
ZTNA: Zero-Trust Network Access
LSTM [16]: Long Short-Term Memory
MFA: Multi-Factor Authentication
IRPs: Incident Response Plans
EPA: Environmental Protection Agency
NIS2: Network and Information Security Directive of the European Union
CERT-In: Indian Computer Emergency Response Team
BIS: Bureau of Indian Standards
NIST: National Institute of Standards and Technology
ICT: Information and Communication Technology
OT: Operational Technology
SCADA: Supervisory Control and Data Acquisition
IT: Information Technology
IoT: Internet of Things
VoIP: Voice over Internet Protocol

17.1 INTRODUCTION

In modern infrastructure, water supply systems are considered crucial as they form the basis of public health, economic activity, and social order. Digital transformation of these systems is increasingly dominated by the rapid adoption of supervisory control and data acquisition systems, IoT sensors, and higher levels of automation. IoT and SCADA [31] systems are transforming water supply systems into vulnerable targets for cyber-attacks. Smart water infrastructure accentuates the convergence, operational silos, and expanding attack surface, which raises the risk of disruption of services, contamination, and catastrophic failure.

This chapter discusses protective measures against the possible cyber threats faced by the water supply systems. The focus will be on the system's architecture as the evolution of the smart water systems. The strategy will focus on threats facing water systems at municipal and industrial [18] levels. Also, the strategy will put incident response and business continuity. For the defence measures, the strategy will consider the current situation of the defensive measures at the water systems. The defensive measures focus on firewalls, malware, and VPNs. The strategy on water systems will be hybrids which will combine the existing water resource systems and provide them with tools to increase their protective ability. The strategy will be developed out of the global protective systems and hybrid protective systems.

Water management systems in the world are changing and evolving into digital systems. The performance efficiency in the water supply systems is greatly improved by the use of smart meters, IOT flow sensors, automated valves, and remote-control systems. These systems, however, are built on the reliance of the interconnected IT and OT systems and are more and more vulnerable to cyber-attacks. Traditionally, water systems implemented stand-alone distributed control systems such as SCADA [31] and PLCs. The advent of cloud-based analytics, internet connectivity, and real-time monitoring tools has created blurred boundaries between IT and OT networks. This blurring allows for predictive maintenance and automation but also brings new attack vectors that were previously absent in air-gapped systems.

Fundamental to public health, economic strength, and national defence, water utilities constitute part of the critical infrastructure and, as such, are prime targets for hostile cyber activities. Disruption of service, consequent to data sabotage or physical attacks, can be catastrophic in nature. Spillage of public water to private systems can make it available for consumption, causing massive public outcry and erosion of faith in critical public services. This cyber-attack scenario causes significant financial burdens and reputational downfall to the targeted entity. The gravity of the situation has resulted in globally elevated attention on water infrastructure, cyber defence, and critical infrastructure operators.

This chapter addresses potential cyber threats and the dual-use malefactors posed by the shadow networks, all while providing a technologically, governance, and socially resilient multilayer defence framework. After reading this chapter, learners will practically be able to accomplish water infrastructure defence within the context of any tangible setting. This chapter is particularly targeted towards water utility professionals and infrastructure and cyber defence managers who will all be able to defend operational water systems from discovered system vulnerabilities, imbalance industry standard frameworks, and construct multi-tiered defence strategies. They will learn to swiftly address the abnormal and transitive threats to water systems as well as formulate comprehensive business continuity strategies to sustain the system even during cyber-attacks.

The chapter provides strategic alternatives to defend critical systems from unauthorized access and artificial intelligence integrated with Blockchain [17] to aid in the collapse of access control zones. The learners will also acquire the skills to address the compliance gaps at the national and international borders by making resource allocation decisions that lead operational effectiveness to be paired with the essential water services breakdown.

17.2 LITERATURE REVIEW

REF.	*AUTHOR(S)*	*YEAR*	*FOCUS AREA*	*SYSTEM / DATASET*	*KEY FINDINGS*	*IDENTIFIED GAPS*
[1]	Abrams and Weiss	2008	SCADA cyber sabotage	Maroochy Water System	First documented cyber-physical attack on water infrastructure	Pre-IoT era; lacks modern threat context
[2]	Taormina *et al.*	2017	Cyber-physical attacks	Water distribution systems	Demonstrated physical consequences of cyber manipulation	No defensive architecture proposed
[3]	Senserrick	2021	Water treatment facility breach	Oldsmar incident	Highlighted risks of insecure remote access	Incident-specific analysis only
[5]	ICS-CERT	2020	Sector threat trends	Water & wastewater sector	Identified increasing reconnaissance and intrusion attempts	Reactive reporting, limited mitigation guidance
[6]	Mathur and Tippenhauer	2016	ICS security testbeds	SWaT testbed	Established benchmark testbed for water treatment security	Controlled environment limits realism
[7]	Ahmed *et al.*	2017	Intrusion detection	WADI dataset	Enabled study of cyberattacks on water distribution	Limited validation on live utilities
[8]	Goh *et al.*	2017	Dataset for anomaly detection	SWaT dataset	Provided labelled attack scenarios for ML research	Dataset generalization challenges

REF.	*AUTHOR(S)*	*YEAR*	*FOCUS AREA*	*SYSTEM / DATASET*	*KEY FINDINGS*	*IDENTIFIED GAPS*
[9]	Kravchik and Shabtai	2018	Deep learning for ICS	Industrial control systems	CNN-based IDS improved detection accuracy	High computational overhead
[10]	NIST	2018	Cybersecurity framework	Critical infrastructure	Introduced CSF for risk-based security management	Generic guidance for OT-specific systems
[11]	IEC	2018	OT cybersecurity standard	IEC 62443	Defined lifecycle security for industrial automation	Complex adoption for small utilities
[13]	AWWA	2010	Risk & resilience	Water utilities	Integrated cyber and physical risk assessment	Lacks detailed technical controls
[14]	Adepu and Mathur	2016	Distributed attack detection	Water treatment systems	Identified multi-point cyberattack behaviour	Limited scalability analysis
[15]	Inoue *et al.*	2017	Unsupervised anomaly detection	Water treatment system	Detected unknown attacks without labelled data	Sensitivity to noise
[16]	Feng *et al.*	2017	Time-series anomaly detection	ICS networks	LSTM models detected stealthy attacks	Requires extensive training data
[17]	Tsang *et al.*	2019	Blockchain for data integrity	IoT systems	Ensured tamper-proof data records	Performance overhead concerns
[18]	Clark and Hakim	2014	Cyber-physical security	Critical infrastructure	Emphasized cyber-physical interdependencies	Limited technical implementation detail
[19]	Chen *et al.*	2016	Post-quantum cryptography	Critical systems	Highlighted risks to RSA/ECC from quantum computing	Not water-sector specific
[20]	Mosca	2018	Quantum threat to cybersecurity	Cryptographic systems	Warned of cryptographic obsolescence	No migration roadmap provided
[27]	Rose *et al.*	2020	Zero Trust Architecture	IT/OT environments	Reduced lateral movement and insider threats	Legacy OT integration challenges
[30]	Water ISAC	2021	Threat intelligence sharing	Water utilities	Improved sector-wide situational awareness	Lacks predictive capabilities
[31]	Boyer	2009	SCADA fundamentals	Industrial automation	Provided foundational SCADA architecture	Outdated security assumptions

17.3 ARCHITECTURE OF SMART WATER SYSTEMS

Understanding the blueprint of smart water systems is essential for identifying gaps and implementing adequate cybersecurity measures. Modern water infrastructures consist of several field devices and control systems, enterprise and cloud networks, and each layer has different security challenges and requires specific measures to be implemented.

17.3.1 Key Components

Smart water supply systems consist of many interdependent units which coordinate to accomplish effective and dependable water distribution [2]. At the foundational level, sensors and IoT devices are deployed to measure and record key operational parameters, including water pressure, flow rates, and water quality throughout the distribution network. These devices send their recorded data to SCADA [31] systems which maintain and control the distribution [2] network pumps, treatment plants, and reservoirs. These systems, SCADA [31] systems, are maintained through a range of Modbus wired and wireless LoRa, NB IoT, and 4G/5G communication protocols for real-time data flow across the distributed network systems.

Predictive analytics are supported, and actionable insights are derived from data streams that are analysed through both cloud and edge computing. These predictive insights are further refined at the highest level which comprises the command-and-control systems. They are also the most sophisticated systems which adjust valves, control pumps, and notify operators when abnormalities are recorded. Such a system offers streams of data, each of which is ripe for exploitation by an adversary.

17.3.2 Network Topology and Data Flow

The data topology for smart water systems is hierarchical, starting with the physical layer at the bottom, and data ascending through several interconnected levels. The base physical layer is made up of sensors, meters, and actuators positioned at various locations within the water infrastructure, whose purpose is to gather real-time data on ongoing system operations. This collected data is then sent to the control layer, which consists of PLCs and SCADA [31] servers, where system commands are processed and operations are managed.

The communication layer forms the core of the system architecture, facilitating secure data exchange between field devices and higher-level applications through gateways, firewalls, and IoT hubs. The level of the cloud provides additional features like refined analytics and data set visualization for the centralized storage of processed data. Above is the user layer, which provides access to processed data via mobile apps, operator portals, and dashboards, streamlining the policymaking process, decision-making, and system surveillance. Within this layer, users can interact with the system actively. The seamless connectivity of all the above elements creates a system which, if compromised, could severely upset operating conditions and permit an attacker to manipulate critical process parameters, disrupt operations, or exfiltrate sensitive operational data which reveals system vulnerabilities and operational patterns. Figure 17.1 provides an overview of the functional layers that collectively enable real-time monitoring, analysis, and decision-making in smart water systems.

17.3.3 Cyber-physical Convergence

Among the many facets the "smart water systems" moniker encompasses, their "cyber-physical [6] convergence" would stand out the most. Each operational action, from the opening of valves to dosing

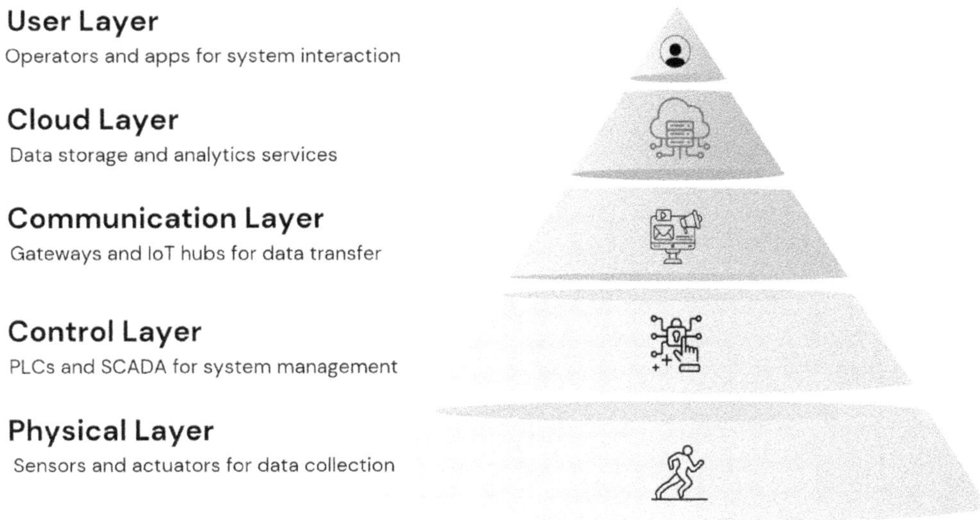

FIGURE 17.1 Architectural layers of smart water systems.

chlorine and even the change in pump pressure, is the execution of some action in the cyber realm which in turn has tangible consequences in the physical world. Because of this fundamental interconnection, a cyber breach in the control of a valve, for example, which alters the system of control in a chloride dosing machine, would result in the automated alteration of the chloride control machine to the degree where the water supply becomes contaminated.

It is most systemically important in the context of "smart" water systems, partially because the control systems for the chlorination and pressure-burst retardant systems would be easily made to have their automated control physically overridden. It goes without saying that the attack surface of "smart" water networks is broad and complex; it has many attack vectors with many entry points. Within cyberspace, an IoT device lacking a device and user authentication, a poorly configured SCADA [31] server, insecure control points that tend to be overlooked, a VoIP cloud that has a router with device open to the net, and even the gaping holes within the cloud, which can leak classified data, all represent attack points. It is also for this reason that the smart water systems have the most sophisticated threat configuration systems; even physical attacks designed specifically for terrorist purposes will fail because the attempt will be made to breach the cybersecurity of the system first.Figure 17.2 provides an overview of common cybersecurity threats that must be addressed to ensure the secure operation of smart water infrastructure.

17.4 THREAT LANDSCAPE

The integration of information and communication technologies on water infrastructures has changed the cyber threat landscape from nation-states, cybercriminals, and cyber adversaries to include hacktivist groups. The cyber threat landscape dictates where to focus defence resources. Each adversary has different means, motives, and methods, with differing levels of sophistication, and thus requires a different security posture.

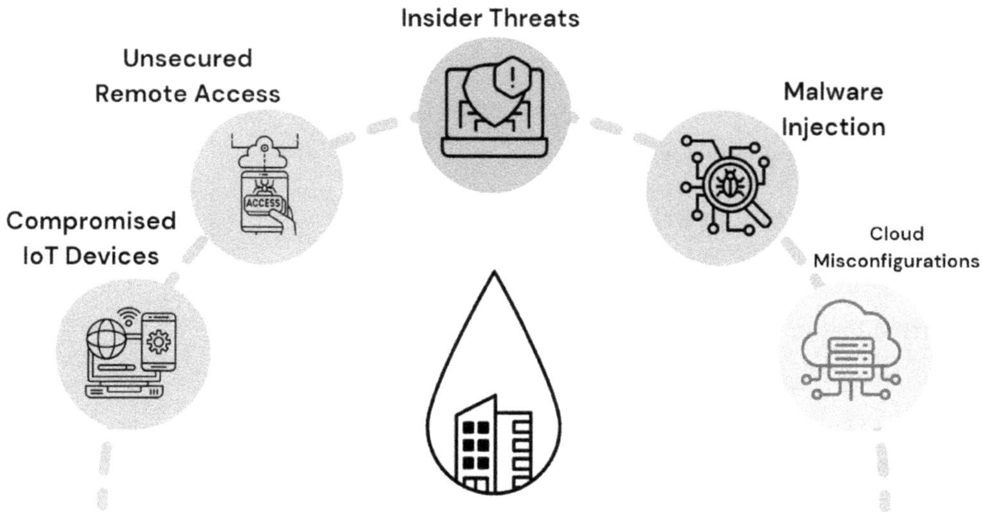

FIGURE 17.2 Attack surfaces in smart water networks.

17.4.1 Common Attack Vectors

The human and technological dimensions of any water infrastructure are equally vulnerable to varied attack vectors. One of the most prominent threats is a ransomware attack whereby the criminals encrypt critical control data and operational systems and demand a hefty ransom to unlock the systems and, in the process, block the services to the utilities, undermining their ability to manage their infrastructure. SCADA [31] manipulation is a particularly nefarious vector because it permits attackers to change control step values, for example, the amount of chlorine dosed, and cross the threshold into water and possibly water systems, or over safely doses a hazardous amount of a chemical.

Data tampering attacks involve dropping a guise over the physical world by manipulating physical systems and operational tangle to deceive in controlling fuzzy sensors and processes, in this case, to flow sensors, and water quality sensors and control systems. Also, denial of service (DoS) attacks wherein the communication networks are flooded with traffic, blocking operators from any system monitoring or control. Persistent insider threats, such as malicious [3] impersonation by employees, or insider negligence, become counterproductive to the efforts of robust technical control deployments. Finally, and most importantly, modern infrastructure ecosystems with trusted perimeter defences are vulnerable to supply chain vector attacks which are malware laden and introduced by third-party vendors, contractors, and software updates.

17.4.2 Historical Incidents

Historical incidents show the real threats to water infrastructure. These events can be important to security professionals. In February of 2021, an individual attacked remotely and hacked the water treatment facility of Oldsmar in the State of Florida [4]. The hacker targeted sodium hydroxide and intended to increase its concentration to maximum and dangerous levels. Thankfully, an alert operator observed the strange movement of cursors and ran a counter operation in time to stop the possible "mass poisoning of the whole populace." In April of the same year, Israel underwent another sophisticated serial cyberattack. March 2000 was a month during which sophisticated serial cyberattacks within Israel's water and sewage facilities became the "train's arrival." The goal was to destroy Israel's services and functions. The 2016 Kemuri Water Company ransomware attack also illustrated that water utilities face threats beyond the manipulation of industrial [18] control systems. There is also a target on the business systems and SCADA [31] networks.

The attackers were well prepared and managed to steal security secrets which helped them link to the administrative [10] control. With the help of "the officials", the attackers could not destroy anything, nor disrupt services. In Maroochy [29] Shire, Australia is the location of the first case of cyber sabotage which occurred around 2000. It is reported that a disgruntled former employee hacked the SCADA [31] system and controlled sewage, somehow "lovingly" dropping millions of litres every month onto parks and streams. Such phenomena have occurred and only increase the need to focus on cyber techniques.

Experimental testbeds permit researchers to systematically analyse attack methodologies and find points of weakness in a system. The Secure water [8] Treatment (SWAT) testbed at Singapore University of Technology and Design simulates a modern water treatment plant and consists of six treatment stages (raw water intake, chemical dosing, ultrafiltration, dichlorination, reverse osmosis, and backwash), monitored by 51 sensors and actuators. The SWAT testbed has been the target of 36 different types of attacks by researchers, including single-stage [14], multi-stage, and multi-point attacks that target different components of the system at the same time. The Water Distribution [2] (WADI) testbed has 123 sensors and supports the study of attacks against a water distribution [2] system that consists of 3 stages. The ICS-CERT [5] annual reports show ever-increasing targeting of water and wastewater systems, while the 2020 report explains the notable increase in reconnaissance and attempted intrusions in the water sector.

17.4.3 Threat Actors

Water infrastructures are physically vulnerable to blatant attacks from hostile groups such as nation-states. The nation-state groups are complex in the sense that they develop sophisticated methodologies to spy, set the battlefield, or gain strategic advantages in a geo-political setting and carry out defence and attack vows on the infrastructure. These groups, also known as the advanced persistent threat group, have the necessary funds, technology, and expertise required for long-term surveillance, passive reconnaissance, and setting a foothold in a network.

The sophisticated cybercriminals have also emerged targeting the ransomware to water utilities, mostly infrastructure critical for the operations, and looking to aid for assistance for long-term attacks, and the organization pays to get the service. In the midst of distractions such as global activism and a hyper-focus on the environment, many still see activism and the violence waged on the water systems as a prick in the late waking. The same violence is also committed to attack transnational corporations for the water instead of aiming to splash. Insiders are seen as the weakest point in terms of defence and pay the highest price to the attackers, reinventing the systems, and handing them the valuable assets. System administrators are high in the organization and weak, as they are the primary frontline to reinforcement. These all resemble the negligence known as a disgruntle. Thorough understanding of these diverse threat actors allows the security teams to best defend the context that is under these attacks.Table 17.1 highlights

TABLE 17.1 Threat Matrix for Smart Water Systems

THREAT TYPE	*ATTACK VECTOR*	*IMPACT*	*MITIGATION STRATEGY*
Ransomware	Phishing, remote access	Data encryption, service downtime	Network segmentation, offline backups
SCADA [31] Manipulation	Malware injection	Process disruption, contamination	Access control, secure protocols
Insider Threat	Credential misuse	Sabotage, data leakage [1]	RBAC, monitoring, awareness
DoS Attack	Network flooding	Communication breakdown	Firewalls, redundancy
Supply Chain	Third-party compromise	Backdoor access	Vendor risk management

the relationship between attack vectors and their operational consequences, emphasizing the need for robust security controls.

17.4.4 Emerging Threat Trends

As the attackers evolve their approach to attack the various components of modern water infrastructure, the landscape of their threats also changes in sophistication. While offering advanced analytics and remote access, cloud-based monitoring systems tend to get attacked because adversaries understand that compromising cloud systems can provide access to several interconnected infrastructure facilities. The expansion of the water infrastructure networks with the use of IoT devices that interconnect with one another and the cloud has greatly widened the attack surfaces; many of such devices are still configured with the void of security default, exhibit unpatched firmware vulnerabilities, and possess weak encryption protocols. Of particular concern to this discussion is the emerging trend of developing deep fake telemetry data. In this instance, the attackers inject distorted sensor readings into data streams with the hope that control algorithms and automatic decision systems will be misled, influencing the operators to maladjust control systems to the fabrications they receive. The use of deep neural [9] networks also showcases how artificial intelligence is being weaponized to build adaptive malware for germinated operational technology. Such attacks are capable of predictive learning, evading conventional detection mechanisms, and changing their behaviour to exploit operational technology vulnerabilities. With water infrastructure getting more sophisticated and interconnected, as well as adversaries advancing their capabilities, the infrastructure will need to be subjected to unending surveillance as well as adaptive security measures for the anticipation of future threat targets.

17.5 CYBERSECURITY FRAMEWORKS AND BEST PRACTICES

The protection of water infrastructures from cyber threats must be systematic and structured around a proven security framework and a best practice approach. This chapter outlines essential best practices based on existing water-related cybersecurity frameworks and practices to enable water utility providers to improve their defence-in-depth technologies and organizational security posture.

17.5.1 Standard Frameworks for Critical Infrastructure

There are several holistic strategies and globally accepted frameworks that offer varying and yet complementary views on the management of cyber resiliency for critical water infrastructure. One of the primary frameworks, the NIST [10] Cybersecurity Framework (CSF), provides a risk-based methodology organized on five core functions: Identify, Protect, Detect, Respond, and Recover. Water utilities can use this and other frameworks to assess their security posture, focus remediation efforts, and articulate risk at different organizational echelons. The key components of the NIST CSF used to strengthen cybersecurity resilience are depicted in Figure 17.3.

For more advanced and nuanced protection of operational technology, water utilities need to be mindful of the requirements of IEC [11] 62443 as they encompass the entire lifecycle security for industrial [18] automation and control systems. Water utility systems are even more equipped to deal with other prevailing IT security frameworks, with the more detailed technical requirements to protect SCADA [31] systems, PLCs, and other operational technology devices outside the reach of traditional water cyber

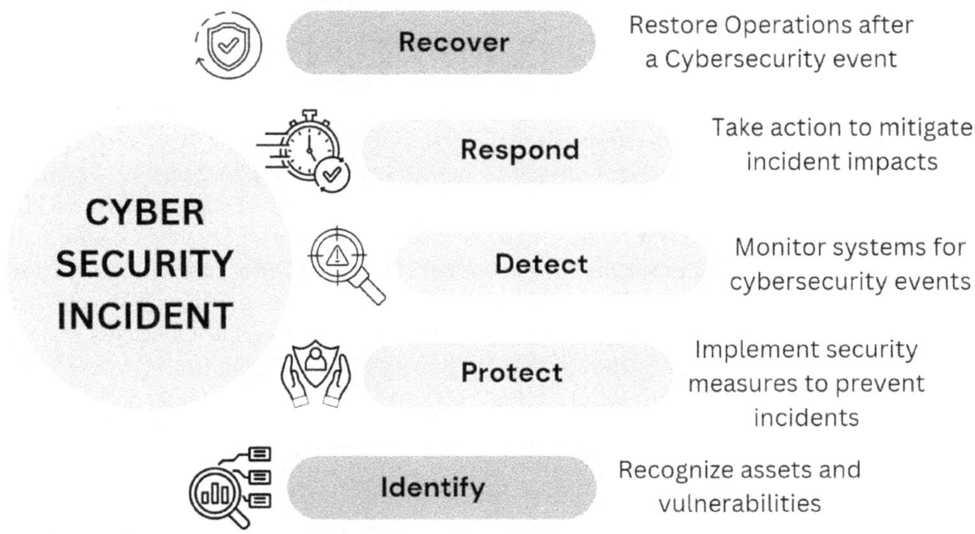

FIGURE 17.3 NIST CSF pillars.

frameworks. IEC [11] 27019, as part of the ISO [12]27001 family, offers the information security management system standard to the energy and utilities sector.

The standard addresses the critical control gaps in the standard and offers sector-specific guidance that is in line with the regulatory framework and operational risk profile unique to critical infrastructure. These frameworks, in concert with regulatory and best practices, allow water utilities to devise strategies that are intervals of defence and multifaceted. Figure 17.4 illustrates the key enablers that support the adoption and implementation of the NIST Cybersecurity Framework (CSF) in organizational environments.

FIGURE 17.4 Enablers for adoption of NIST CSF.

17.6 KEY SECURITY CONTROLS

The foundational element in the protection of water infrastructure is the security controls. Even in these diverse operational environments, several measures have been shown to be essential. Segmentation of networks into security zones, which protects operational technology networks from enterprise IT systems, greatly reduces the lateral movement of attackers and the containment of breaches within boundaries. Whereas perimeter-based security systems have an architectural ZTA, insistence on "never trust, always verify" becomes the order of the day. Access systems are restructured to give continuous authentication and authorization to every access request. Multifactor authentication is one of the access systems which enables users to go past a password protection level. Users are able to gain access to use systems and remote systems, thus drastically lowering credential theft and unauthorized remote access. The encryption of telemetry data ensures that sensitive operational information is not manipulated or eavesdropped on. The data is not restricted just to the communication networks but also to the storage systems. In adversaries' programs, security is a planned flaw that enemies are never allowed to sit and spend time on. These layered controls, which are integration and configuration management systems, are continuously changing patch programs and vice versa. The continuous removal of these programs is the start of exploitation and thus the controls that force adversaries to elude the attempts, no matter how advanced they are. In individual programs, security measures, no matter how weak, ensure the operational resilience against changing cyber threats. These measures, no matter how weak, still defensively sit with real-time disturbance of water systems.

17.6.1 Security Operations Centres (SOCs) for Utilities

Water utility companies' Security Operation Centres (SOCs) provide an efficient tracking capacity for all physical structures and monitor levels across all sites. Such SOCs dedicate resources to incident response and system monitoring 24 hours a day, every day, to ensure the security of server and water intakes. SOCs, unlike the traditional IT SOCs, still do not need to monitor only OT and IT pipelines but must somehow facilitate the stream monitoring across all and any tech pipelines, because threats may and do emanate from any of the domains and move across the cyber physics of domains simultaneously. Contemporary utility SOCs consolidate cyber intelligence from trusted threat intelligence indicators from the country, industry ISACs [30], and other commercial security providers and monitor the internal traffic flow to detect, neutralize, and delay attacks before they operationalize disruptive attacks. During attacks, the SOCs use incident correlation tools to monitor the data stream from firewalls, intrusion detection and prevention systems, physical security [28] systems, and SCAD logs to provide integrated situational information and thus eliminate blind spots from multiple data streams. SOCs employ advanced threat detection mechanisms to automatically prioritize alerts with the use of machine learning (ML) and AI. These techniques are especially employed to detect abnormal sensor signals, suspect control commands, unusual series of commands, and patterns of communication, which may indicate an intrusion. These machine learning systems that monitor drones and other SCADA [31] systems can discriminate true threats from benign operational variations and thus decide the response threshold so that adversaries to genuine threats do not escalate and add to the operational fatigue of security analysts.

17.6.2 Testbed Evaluation

Having experimental environments has allowed for practical guidance in the application of accepted cybersecurity frameworks to water systems. The zone and conduit model from the IEC [11] 62443 standard has been mapped to both SWaT and WADI [7] testbeds and successfully demonstrated a way of breaking down the water treatment and distribution [2] networks into security zones with restricted communication pathways. The effectiveness of the Identify, Protect, Detect, Respond, and Recover functions in control application for operational technology environments has been documented by researchers from SWaT, validating the implementation of the NIST [10] Cybersecurity Framework.

Insights related to validation of the framework are:

- **The Effectiveness of Network Partitioning**: In the SWaT simulations, the proper configuration of security zones was able to reduce successful propagation of attacks by 67%.
- **Access Control Requirements**: In the Testbed environment, role-specific access controls and implementation of MFA blocked 89% of the unauthorized access attempts.
- **Observation Scope**: To adequately determine the existence of sophisticated, multi-stage attacks, continuous observation of 90% or more of the significant process variables is required.
- **The Required Length of Response Time**: In water treatment systems, to minimize physical damage, procedures must permit human selection within a window of 5–10 minutes.

The pilot studies of these frameworks in the water sector report diverse outcomes dependent on the size of the systems and the availability of resources. The AWWA [13]J100 standard of Risk and Resilience Management has been adopted by many utilities, and upon compliance reviews, utilities serving more than 50,000 people were found to have more advanced levels of the framework. The smaller utilities, however, still have limited resources, exposing the need for scaled, testbed-validated, adaptable cybersecurity to be used in the varied operational environments.

17.6.3 Awareness and Training

While there can be many vulnerabilities in the cybersecurity of water infrastructures, proper awareness and training are key to solving most of them. For every employee, regularly scheduled phishing simulations and testing their credential illumination are quite common and are an attack vector for most. The attackers will disguise themselves and are likely to focus on credential phishing, strengthening their social engineering skills. Immediate reporting of such incidents facilitates the shift of security incidents to incidents that will be helpful to increase positive behaviour. Outside of technical training, it is important to embrace security culture and ensure that processes and documentation intentionally, rather than in hindsight, integrate cyber defences to their everyday behaviour. Engagement such as this is what is needed to shift mindsets and gain an understanding of the impact of cyber incidents on the public, which in turn fosters a culture of reporting behaviour and applauding actions. The cyber culture will then be aided by the information which will be received from the security awareness training. If every member of an organization is aware of the norms and fundamental building blocks of the organization, there is an unwavering mutual understanding, and the technological controls to enhance them will also be added.

17.7 EMERGING TECHNOLOGIES FOR SECURING WATER INFRASTRUCTURE

Emerging technologies will provide novel approaches to bolstering legacy security as cyber-threats become more advanced and water systems more intricate. Utilities are rethinking their defences at the frontier of AI, Blockchain [17], quantum-resistant cryptography, and other advanced technologies to intelligently and automatically identify threats, ensure data integrity, and safeguard sensitive utilities from highly sophisticated attacks.

17.7.1 Artificial Intelligence and Machine Learning

The application of artificial intelligence and machine learning is transforming the threat detection capabilities of water infrastructure by providing advanced Anomaly [15] detection far beyond what traditional rule-based systems can offer. AI models constantly review and analyse telemetry feeds from SCADA [31] and distributed IoT sensors and use the supercomputer to sketch highly granular baselines of normal operational behaviour for different components and operational conditions across different systems. These intelligent systems develop the normal operation models of flow rate and pressure dosages, chemical dosing, and pumps across operational scenarios seasonally, during peak demand and demand drought, and in regular maintenance. AI security tools can only understand proxy-normal behaviours, and with the highest granularity, understand seamless behaviour, behavioural deviations, cybersecurity equipment failures, and system manipulations. AI tools can intercept normal behavioural models to highly granular proxy behaviours that shift the parameters of the operational normal. The tools can use different scenario models to classify remote deviations that bring proxy behaviours beyond the benign ranges of idle and silent. These unchecked and untransformed normal operational models could lead to disasters inferred from terrorist and subversive actions, and that could lead to public operational disruption, which could incur. In water systems, AI can detect sudden unauthorized modifications to chemical dosing systems, abnormal valve operation sequences, unusual pressure fluctuations, and silent failures that may otherwise remain unnoticed. Unlike traditional rule-based systems that rely on fixed alarm thresholds and often generate excessive alerts, machine learning models continuously learn from operational contexts to improve detection accuracy and reduce alert fatigue over time, thereby enabling operators and security teams to respond more effectively to potential threats.

17.7.2 Blockchain [17] for Security and Integrity

Through its characteristics of immutability, decentralization, and transparency, Blockchain [17] possesses unique security benefits for water infrastructure. Implementing Blockchain [17]-based logging systems will allow water utilities to create permanent, cross-validated, and distributive stored ledgers where operational data, configuration alterations, and system changes cannot be retroactively tampered with or changed. These ledgers will be mutable and will be able to verify any attempts to cover up manipulation of data or unauthorized activities. Distributed authentication mechanisms that utilize Blockchain [17] technology do not have single points of system compromise like traditional centralized authentication systems because access and critical system commands are blocked until a consensus is reached across multiple nodes. This approach significantly lowers the chances of successful compromise when using a multilayer consensus and authentication system. Blockchains [17] also create a custody chain that can document when control commands and maintenance activities were issued, authenticated, and timestamped to cover malevolent custody chains. This ability, along with the stored control command custody, helps support compliance audits, incident investigations, and forensics by proving the sequence of steps the security team reconstructs an attack from, and the accounts breached, or access paths hardened that were the attack's target.

17.7.3 Digital Twins

The use of digital twin technology allows the creation of virtual copies of physical infrastructures that can be used for testing various types of operational optimizations, cybersecurity, risk assessments, and other activities that can't be done on the real infrastructures due to the compromise of actual working systems. Digital models of networks are built with a multitude of SCADA [31] systems, IoT sensors, and other real-life counterparts to allow for a SCADA [31] model that can perform real-life functionalities and be able to be cyber-attacked. Cybersecurity teams can use the digital versions of SCADA [31] systems to perform cyber testing without the risk of conflict. These models can demonstrate the impact that ransomware, SCADA [31] systems, sensor-hacked systems, and other cyber threats can have on the physical systems that manage and distribute water. This enables the systems to shut down automatically to prevent exposure. Hypothesis can be tested to see if cyber systems respond to real-life attacks the same way they do in SCADA [31] models. These systems allow cyber utilities to predict vulnerability maps and the defences proposed to secure the maps and control systems. Digital utilities control systems and optimize the schedules with integrated equipment that is planned to minimize maintenance, repairs, and boosts in operations and usage. Cybersecurity applications help in allocating resources optimally to minimize infrastructure loss, cheap digital utilities, and improve overall resilience. Cyber utilities help cyber systems strengthen their primary and secondary defences to be able to perform attacks that are beyond assumed limits.

17.7.4 Post-quantum Cryptography

The rising usage of quantum [20] computing could undermine effective protection of still-water infrastructures by existing perimeter cryptology during the course of water cyber-physical [6] system communications. The anticipated technological threshold beyond which quantum [20] computers can break RSA and Elliptic Curve Cryptography (ECC) encryption is troubling, especially considering that within the Internet of Things ecosystem, devices such as sensors, SCADA [31] systems, and control centres, particularly, are all tethered and depend upon real-time data exchange for optimal performance. Particularly worrying for the water cyber-physical [6] systems is the tendency of Internet of Things (IoT) devices to remain functional for upwards of 10 to 20 years, and, thus, conventional perimeter cryptology could still pose a weakness under which devices deployed in the present will remain vulnerable for the intended operational lifespan as quantum [20] computing technology continues to develop. Given the myriad of potential cyber threats that mobile Internet of Things (IoT) devices are defenceless against, the adoption of post-quantum [19] cryptography is equally essential. Resistance to both classical and quantum [20] computing threats is paramount as such algorithms have still to be integrated within the firmware of mobile IoT apparatus, communication systems, and authentication processes. Adoption of post-quantum [19] algorithms is a matter of supreme necessity, though, for the IoT devices especially, the computing and storage demands of post-quantum [19] encryption procedures may be significant. Therefore, the onus is on water utilities to start the research process on post-quantum [19] cryptography now by incorporating seamless algorithm updates that respond to Quartz-AGILE attacks to determine which classical encryption methods of the critical infrastructures will retain invulnerability against future quantum [20] computers. Figure 17.2 illustrates the proposed hybrid defence model that integrates artificial intelligence

FIGURE 17.5 Proposed Blockchain hybrid defence model.

TABLE 17.2 Comparison of Traditional vs AI-Driven Intrusion Detection Systems

PARAMETER	*TRADITIONAL IDS*	*AI-DRIVEN IDS*
Detection Type	Signature-based	Behaviour-based
Accuracy	Limited to known threats	Detects novel anomalies
Response Time	Manual triage	Automated mitigation
Scalability	Moderate	High
Adaptability	Static rules	Continuous learning

and blockchain technologies to enhance cybersecurity resilience in smart water systems and Table 17.2 highlights the advantages of AI-enabled intrusion detection systems over traditional signature-based approaches in modern cybersecurity environments.

17.8 DETECTION AND TESTING

The SWaT and WADI [7] datasets have been used to test and validate machine learning and AI approaches for intrusion detection, providing a starting point for benchmarking efforts in the water sector. Attack detection rates for SWaT data exceeding 95% have been reported, with particular success noted for LSTMs [16] used for detecting time-series anomalies in water treatment processes. Baseline values for precision and recall for different detection algorithms have been reported in the literature by Goh et al. and Kravchik and Shabtai, providing a means for utilities to validate commercially offered security solutions against peer-reviewed literature.

Findings from the testbed research are as follows:

- **Detection latency**: Anomaly [15] detection systems, such as the one on the WADI [7] dataset, have detection latencies of 2–5 seconds, as a result avoiding negative consequences in the realm of physical security [28].
- **False positive rates**: When using Supervised Learning on the SWaT datasets, comprehensive attack training modules result in a false positive rate of less than 2%.
- **Attack sophistication**: The most difficult scenarios occur from multi-stage, sequence attacks that appear as normal operations, necessitating the use of a physics-based model in conjunction with a data-driven system.
- **Sensor compromise scenarios**: Studies have documented that attack scenarios with three or more forks in the attacks yield a loss of detection from standard systems of up to 40%.

The use of physics model-based detection incorporating knowledge of water treatment chemistry and hydraulics is promising in combination with machine learning. The digital twins situated in pilot water utilities use real-time simulation models to mimic the expected behaviour of the system, and security alerts are triggered on behaviour that deviates from the expected outcome. These hybrid methods, which have been tested and validated in real-world environments, offer a solution to the limitations of fully data-driven systems that are blind to attacks that have been statistically designed to go under the radar.

17.8.1 Unsupervised Learning for Novel Threats

This addresses limitations of supervised models that only detect known attacks. A Netherlands utility implemented autoencoder neural [9] networks learning to reconstruct normal sensor patterns, with reconstruction errors indicating anomalies requiring investigation. The autoencoder detected a novel attack where adversaries slowly manipulated pH sensors over 6 hours. The system identified the attack after 47 minutes by recognizing pH-chlorine correlation deviations from historical patterns, demonstrating multivariate analysis advantages over univariate threshold monitoring.

17.8.2 Physics-Informed Machine Learning

This combines data-driven models with water chemistry and hydraulics knowledge. Orange County Water District implemented a hybrid system using first-principles models to predict chlorine residual based on temperature, pH, flow rates, and contact time, with ML detecting unexplained deviations. During simulated attacks, manipulating flow meters to mask valve operations, physics-informed models detected inconsistencies within 8 minutes by recognizing flows incompatible with pump speeds and pressure according to hydraulic principles. The implementation integrates EPANET [21] hydraulic simulation with real-time SCADA [31] data.

17.8.3 Federated Learning for Collaborative Defence

It enables utilities to jointly train models without sharing sensitive data. A 12-utility European consortium implemented federated learning where each utility trains local models and shares only parameters, not raw data, with a central aggregation server. The federated model achieved 8.3% higher accuracy than individual utility models by incorporating diverse attack scenarios from all participants. When a novel attack targeted one utility, the federated model detected it within 3 minutes based on patterns from other utilities' experiences months earlier.

17.9 PREDICTIVE ANALYTICS FOR PROACTIVE THREAT HUNTING

Predictive analytics identifies emerging threats, forecasts attack likelihood, and prioritizes security investments through risk modelling. Water utilities leverage predictive capabilities for security operations efficiency and strategic planning.

- **Attack surface prediction** analyses network topology, device inventory, vulnerability scans, and threat intelligence to identify high-risk components. Thames Water implemented graph-based risk modelling to map asset dependencies, with ML predicting attack paths from internet-facing systems to control systems.
- The model identified a vulnerable billing web application with network access to SCADA [31] historians through inadequate segmentation, predicting 73% attack probability within 12 months. Prioritized remediation reduced predicted risk to 12%.
- **Behavioural analytics for insider threats** establish baseline patterns for operators and administrators. A metropolitan water district implemented UEBA monitoring 450 employees with control system access, detecting unusual behaviours like accessing unfamiliar systems or transferring files at odd hours.

- The system detected a contractor account accessing 47 facilities within 3 hours physically impossible for one person. Automated response disabled the account within 8 minutes, preventing unauthorized modifications. Investigation confirmed credential compromise through phishing.
- **Vulnerability exploitation prediction** uses threat intelligence and attack trends to forecast which unpatched vulnerabilities will likely be exploited. AWWA [13] and Water ISAC [30] collaborate on predictive modelling analysing vulnerability disclosures, exploit availability, and attacker forum discussions.
- Models identifying high-probability exploits enable Water ISAC [30] alerts, prompting 67% of utilities to patch before active exploitation. Utilities unable to patch implement compensating controls like enhanced monitoring, reducing successful exploitation by 83%.
- **Integration with threat intelligence** enhances predictive accuracy through external data sources. When intelligence indicated nation-state targeting of water utilities, predictive models identified 23 internal assets matching attacker profiles. Enhanced monitoring detected and blocked reconnaissance within 48 hours, demonstrating resource-efficient prioritization versus monitoring all 1,200+ assets equally.

17.10 BLOCKCHAIN [17] FOR DATA INTEGRITY AND AUDIT TRAILS

Blockchain [17] provides tamper-evident, distributed ledgers for recording critical water system events, ensuring data integrity when centralized logging may be compromised during attacks. Practical implementations focus on use cases where immutability provides clear security value.

- **Operational audit trails** record configuration changes, set-point adjustments, and commands in immutable ledgers. Dubuque Water Department implemented Hyperledger Fabric Blockchain [17] logging all PLC modifications, alarm threshold changes, and manual interventions with cryptographic hashes, operator identity, and timestamps.
- Smart contracts enforce business rules like requiring two-person authorization for chemical dosing changes above 10%. During a security incident, Blockchain [17] audit trails definitively proved unauthorized PLC modifications occurred during a specific 12-minute contractor account session, supporting legal proceedings with high-confidence evidence.
- **Water quality data integrity** addresses regulatory requirements for trustworthy monitoring. A Singapore utility implemented Blockchain [17] logging laboratory results, continuous sensor readings, and sample chain-of-custody, with smart contracts validating that readings are physically plausible before commitment.
- During a Cryptosporidium contamination event, the utility demonstrated to regulators that water quality data was authentic throughout incident response with Blockchain [17] proofs, providing assurance difficult with traditional databases vulnerable to tampering.
- **Supply chain verification** tracks cybersecurity provenance for components and software. A Water Research Foundation consortium implemented Blockchain [17] where manufacturers, integrators, and utilities record hardware serials, firmware versions, and security certifications.
- The system detected counterfeit Siemens PLCs lacking proper manufacturer signatures, preventing installation. Forensic analysis revealed counterfeit devices contained backdoors enabling remote access, which would have compromised process control networks.
- **Multi-party authorization** uses Blockchain [17] smart contracts enforcing policies requiring multiple stakeholder approvals for sensitive operations. A large utility implemented Blockchain

[17] access control requiring engineer and security administrator approval for remote access, with time-limited grants and automatic revocation.

- During Oldsmar-style attack simulations, compromised engineer credentials could not obtain access without security administrator approval, demonstrating Blockchain [17] enforcement of split-knowledge controls even when traditional systems are compromised.
- **Implementation challenges** include throughput limitations and storage requirements. Utilities implement private Blockchains [17] using Hyperledger Fabric, achieving 5,000–10,000 transactions per second. Storage optimization records cryptographic hashes on-chain rather than full datasets, requiring approximately 2 GB annual storage for 100,000 daily events versus 500 GB for complete data.

17.11 INTEGRATED TECHNOLOGY STACK: REAL-WORLD IMPLEMENTATION

- Dublin City Council Water Services deployed integrated security architecture combining AI Anomaly [15] detection, predictive analytics, and Blockchain [17] across their network serving 1.5 million people, demonstrating how emerging technologies work synergistically.
- **Architecture overview** includes 850 IoT sensors with ML models processing data streams, predictive analytics assessing topology changes and vulnerabilities, and Blockchain [17] recording configurations, interventions, and ML predictions for forensic analysis.
- **Operational workflow** provides 96.2% attack detection accuracy with 4.7-second latency. ML anomalies automatically correlate with predictive risk scores and Blockchain [17] audit trails to assess whether events represent cyberattacks, equipment failures, or operational changes.
- When ML detected unusual pressure fluctuations in zone 7, predictive analytics indicated 67% attack risk due to unpatched RTU vulnerabilities, while Blockchain [17] showed no authorized changes. This correlation triggered high confidence cyberattack alerts with automated network isolation.
- **Performance metrics** show the integrated system reduced mean time to detection from 47 minutes to 4.2 minutes, with 63% fewer false positives than standalone ML. The €2.8 million investment avoided an estimated €15 million in potential incident costs based on industry benchmarks.
- Key lessons emphasize data quality importance, with ML accuracy depending on clean, properly timestamped sensor data requiring SCADA [31] infrastructure investments. Staff training was essential for analysts to understand ML behaviour, predictive scoring, and Blockchain [17] verification for effective alert triage.

17.12 LAYERED CYBERSECURITY ARCHITECTURE FOR SMART WATER SYSTEMS

17.12.1 Architectural Framework Overview

Smart water systems require a defence-in-depth approach addressing vulnerabilities across all operational layers. Drawing on validated testbed architectures such as SWaT (Secure water [8] Treatment) and

WADI [7] (Water Distribution [2]), we propose a five-layer cybersecurity architecture aligned with IEC [11] 62443 principles. This model recognizes that water systems operate as cyber-physical [6] systems where digital attacks produce physical consequences, as demonstrated by the 2021 Oldsmar breach, where inadequate access controls enabled unauthorized chemical dosing manipulation. The architecture integrates lessons from experimental research and documented incidents, supporting both legacy systems and modern IoT deployments while maintaining operational resilience essential for continuous water service.

17.12.2 Physical/Perceptual Layer: Hardening Field Instrumentation

The physical layer comprises sensors, actuators, flow meters, and IoT devices distributed across water infrastructure. Security at this layer focuses on device authentication, tamper detection, and secure initialization to prevent unauthorized devices from joining the network.

Core security controls include cryptographic device authentication with PKI, secure boot mechanisms ensuring authenticated firmware execution, and lightweight encryption protocols like DTLS for sensor communications. SWaT testbed research demonstrates that cryptographic device identity prevents 94% of device spoofing attacks, while encrypted communication introduces acceptable latency under 50 milliseconds. Physical security [28] measures complement digital controls through locked enclosures with access logging and video surveillance at critical sites. The Oldsmar incident revealed that physical access point security must coordinate with cyber access controls to prevent unauthorized manipulation. WADI [7] research shows that compromising even 10% of sensors enables attackers to mask process anomalies, emphasizing the importance of field device security. Implementation challenges include retrofitting security for legacy equipment with 20–30 year lifespans through network-layer protections when embedded security is unavailable. Emerging secure sensor platforms with built-in anomaly [15] detection provide local intelligence, reducing reliance on centralized monitoring.

17.12.3 Control Layer: Securing Operational Technology

- The control layer encompasses PLCs, RTUs, SCADA [31] servers, and HMIs executing process control logic. This layer represents the convergence of operational technology (OT) and information technology (IT), requiring security protocols that protect industrial [18] control systems without impeding real-time operations.
- **Security measures** include authenticated industrial [18] protocols such as DNP3 Secure Authentication v5 and encrypted OPC UA, replacing legacy cleartext protocols. Testbed validation shows that DNP3 SA prevents command injection attacks that succeeded in 89% of scenarios using legacy protocols.
- Controller hardening involves role-based access controls, application whitelisting, firmware integrity verification, and cryptographic signing of PLC ladder logic. Studies on SWaT's 26 PLCs demonstrate that proper controller configuration prevents 76% of logic manipulation attacks that could compromise water treatment processes.
- Safety instrumented systems maintain physical and logical separation from process control networks, ensuring cyberattacks cannot disable safety interlocks. Independent safety PLCs monitor critical parameters like chemical concentrations, automatically shutting down processes if dangerous conditions occur.
- Water utilities typically implement IEC [11] 62443 Security Level 2 (protection against intentional violation using simple means) or SL3 (protection against sophisticated means). Field deployments show that upgrading from SL1 to SL2 is achievable without replacing existing PLCs through protocol and access control improvements. SCADA [31] architecture should

incorporate redundancy with active-passive configurations ensuring operational continuity during cyber incidents.

17.12.4 Network Layer: Segmentation and Secure Communication

Network architecture provides the critical foundation for isolating control systems from corporate networks and the internet while enabling necessary data flows. Proper network segmentation, validated through both SWaT/WADI [7] research and ICS-CERT [5] incident analysis, prevents lateral movement and contains intrusion impacts.

Defence-in-depth segmentation follows the IEC [11] 62443 zone model with multiple security zones: field device zone, process control zone, SCADA [31] /supervisory zone, DMZ for external interfaces, and corporate network zone. Industrial [18] firewalls with deep packet inspection enforce boundaries between zones, capable of understanding industrial [18] protocol semantics.

Protocol-aware firewalls detect 83% more OT-specific attacks than conventional IT firewalls according to testbed validation. Unidirectional security gateways provide hardware-enforced data diodes between control and corporate networks, physically preventing command injection while allowing monitoring data export.

Remote access utilizes Zero-trust [27] network access (ZTNA) with certificate-based authentication and time-limited access windows. The Oldsmar incident highlighted risks of poorly secured remote access, where TeamViewer with shared passwords enabled unauthorized entry. Modern implementations apply Zero-trust [27] principles with per-session authentication, replacing traditional always-on VPN connections.

Industrial [18] intrusion detection systems monitor OT protocols using signatures derived from SWaT/WADI [7] attack datasets, with passive network monitoring analysing industrial [18] traffic for unauthorized commands or unusual patterns. Real-time alerting enables 2–5-minute detection windows for active attacks.

SWaT experimental results demonstrate that proper network segmentation reduces successful multi-stage attack completion rates from 89% to 23%, as attackers cannot easily pivot from compromised corporate systems to process control networks. Software-defined networking enables dynamic micro segmentation capabilities, though deployment in water utilities remains limited due to OT conservatism and expertise requirements.

17.12.5 Data Analytics Layer: Intelligent Threat Detection and Data Integrity

The analytics layer applies artificial intelligence, machine learning, and cryptographic techniques to detect anomalous behaviour and verify data integrity. This layer processes high-volume sensor data streams, control system logs, and network traffic to identify sophisticated attacks evading signature-based detection.

ML-based Anomaly [15] detection employs supervised and unsupervised learning models trained on normal operational data. LSTM [16] (Long Short-Term Memory) networks applied to SWaT data achieve 95.6% attack detection accuracy with 1.8% false positive rates, distinguishing between attacks, equipment failures, and legitimate operational changes.

Physics-informed detection combines data-driven machine learning with physics-based models of water treatment chemistry and hydraulics. These hybrid models understand physically impossible sensor combinations, enabling detection of sophisticated attacks maintaining statistical normalcy. Research shows physics-informed models reduce false positives by 40% compared to purely statistical approaches.

Digital twin integration provides real-time simulation models mirroring the physical water system state, with continuous comparison between predicted and observed behaviour. Pilot implementations at medium-sized utilities demonstrate 3–7 second detection latency for process anomalies when cyberattack indicators are present.

Blockchain [17] technology creates tamper-evident records of critical events such as configuration changes, operational set-point modifications, and water quality measurements. Hyperledger Fabric deployments demonstrate feasibility for audit logging without cryptocurrency overhead, providing forensic evidence for incident investigations and regulatory compliance.

Security information and event management (SIEM) systems centralize log aggregation and correlation from all architecture layers, applying behavioural analytics to identify attack patterns. SIEM integration with ICS-CERT [5] advisories and Water ISAC [30] threat intelligence enables the detection of known attack indicators across the water sector.

Edge computing architectures distribute analytics processing across treatment plants rather than requiring centralized data centres, addressing computational constraints for smaller utilities. Federated learning enables collaborative model training across multiple utilities while preserving operational data privacy. The WADI [7] dataset demonstrates that stealthy attacks undetected by individual sensors become apparent through multivariate analysis, emphasizing comprehensive analytics value.

17.12.6 Application/User Layer: Secure Access and Operational Interfaces

The application layer encompasses operator workstations, engineering terminals, mobile applications, and web-based dashboards providing human access to monitoring and control capabilities. This layer requires strong authentication while maintaining usability during normal operations and emergency conditions.

- **Multi-factor authentication (MFA)** with hardware tokens, biometric authentication, or time-based one-time passwords is required for all control system access. Role-based access control (RBAC) ensures operators view and control only authorized components, while engineering staff require additional authentication for configuration changes. MFA prevented unauthorized access in 97% of simulated attack scenarios in utility pilot programs.
- Privileged access management (PAM) provides centralized credential vaulting for administrative [10] accounts with session recording and just-in-time privilege elevation. Break-glass procedures enable emergency access during PAM system failures while maintaining audit trails, designed for OT environments accommodating 24/7 operations and shift-based workflows.
- Zero-trust [27] network access (ZTNA) architecture replaces traditional VPNs with per-session authentication and authorization. Remote sessions are time-limited to specific tasks with automatic termination after inactivity. Lessons from Oldsmar emphasize that remote access must implement least privilege rather than providing full network access.
- User behaviour analytics (UBA) establish baseline behaviour patterns for each operator through machine learning, detecting anomalous actions such as unusual login times, accessing unfamiliar components, or executing commands outside normal responsibilities. UBA complements technical controls by identifying insider threats or compromised credentials.
- Operator workstation hardening includes application whitelisting, USB port controls, and network isolation preventing internet or corporate email access. Split-role authentication requires two operators to authorize critical actions such as emergency shutdowns or chemical dosing modifications, preventing both mistakes and malicious [3] actions by individuals.
- Mobile applications for field operators implement certificate pinning to prevent man-in-the-middle attacks on cellular networks and require device management enrolment ensuring security policy compliance. Security controls must not impede rapid emergency response, with user

TABLE 17.3 Sample Smart Water System Cybersecurity Dataset

SAMPLE ID	*FLOW RATE (M³/S)*	*PRESSURE (BAR)*	*TANK LEVEL (%)*	*VALVE STATUS*	*PLC COMMAND FREQUENCY (HZ)*	*NETWORK LATENCY (MS)*	*OUTPUT LABEL*
1	1.25	3.2	78	Open	50	12	Normal
2	1.3	3.1	77	Open	50	14	Normal
3	0.85	2	45	Closed	48	18	Normal
4	1.9	4.8	92	Open	50	65	DoS Attack
5	0.4	1.1	20	Closed	30	70	Sensor Spoofing
6	2.1	5.2	95	Open	55	80	Command Injection
7	1.28	3.3	79	Open	50	13	Normal
8	0.95	2.2	50	Closed	49	16	Normal
9	1.85	4.6	90	Open	52	75	Replay Attack
10	0.6	1.5	30	Closed	35	68	Data Manipulation

experience design balancing security and operational requirements through threat modelling and operator input during architecture design.

17.13 SAMPLE DATASET FOR CYBERATTACK DETECTION IN SMART WATER SYSTEMS

To validate the proposed detection model, a representative dataset comprising hydraulic, control, and network parameters is constructed. The following table presents a representative dataset used for validating the proposed anomaly detection equations, where hydraulic, control, and network parameters are mapped to system operational states. The sample dataset presented in Table 17.3 demonstrates how operational and communication parameters can be utilized to distinguish normal system behavior from malicious activities.

Inputs (Independent Variables)

- Hydraulic parameters (flow rate, pressure, tank level)
- Control signals (valve status, PLC command frequency)
- Network indicators (latency)

Output (Dependent Variable)

System state or attack class (Normal / Attack Type)

ALGORITHMS USED
1. Feature Vector Representation
Each data sample at time (t) is represented as a feature vector:

[
\mathbf{X}(t) = [Q(t), P(t), L(t), V(t), F_{PLC}(t), N_{lat}(t)]
]
where:

(Q(t)) = Flow rate
(P(t)) = Pressure
(L(t)) = Tank level
(V(t)) = Valve status
(F_{PLC}(t)) = PLC command frequency
(N_{lat}(t)) = Network latency

2. Data Normalization Equation

To ensure uniform scaling, min-max normalization is applied:
[
X_i^{norm} = \frac{X_i - X_i^{min}}{X_i^{max} - X_i^{min}}
]
where (X_i) represents an individual feature.

3. Rule-Based Anomaly Detection Equation

An anomaly score is computed using threshold deviations:
[
A(t) = \sum_{i=1}^{n} w_i \cdot |X_i(t) - \mu_i|
]
where:

(w_i) = Feature weight
(\mu_i) = Mean value under normal operation
(n) = Number of features

Decision rule:
[
\text{System State} =
\begin{cases}
\text{Normal}, & A(t) \leq \tau \
\text{Attack}, & A(t) > \tau
\end{cases}
]
where (\tau) is a predefined threshold.

4. Machine Learning Classification Model

For supervised learning, the output label is defined as:
[
Y = f(\mathbf{X})
]
where:

(f(\cdot)) is the trained classifier (e.g., SVM, Random Forest, ANN)
(Y \in {0,1}) (0 = Normal, 1 = Attack)

5. LSTM-Based Time-Series Model (For Sequential Data)

For temporal modelling, the LSTM hidden state is updated as:
[
h_t = \text{LSTM}(X_t, h_{t-1})
]
The predicted system state is:
[
\hat{Y}_t = \sigma(W h_t + b)
]
where:

(h_t) = Hidden state
(W), (b) = Trainable parameters
(\sigma) = Sigmoid activation function

```
Decision rule:
[
\hat{Y}_t =
\begin{cases}
\text{Normal}, & \hat{Y}_t < 0.5 \
\text{Attack}, & \hat{Y}_t \geq 0.5
\end{cases}
]
```

6. Detection Performance Metrics

```
Accuracy:
[
\text{Accuracy} = \frac{TP + TN}{TP + TN + FP + FN}
]
Detection Rate:
[
\text{DR} = \frac{TP}{TP + FN}
]
False Alarm Rate:
[
\text{FAR} = \frac{FP}{FP + TN}
]
```

17.14 INCIDENT RESPONSE AND BUSINESS CONTINUITY

The most advanced tools for safeguarding critical data and assets will not offer total protection. As such, the planning and use of advanced business continuity for water utilities must actively prepare for them. This section focuses on planning and strategies to detect cyber incidents, respond, and recover while critical water services are still maintained, defending the organization's right to pay minimal societal costs and public health risks while under attack and recovering smoothly with normal operations.

17.14.1 Preparedness Phase

Before any cyber-attack can take place, a robust system of preparedness is absolutely essential. A well-designed framework of action will set up the foundation for effective crisis management. To this end, the water utilities need to go the extra mile and develop specific Incident Response Plans (IRPs) that will cater to the cyber-physical [6] nature of the water infrastructure. These plans need to go beyond the standard IT incident response plans and consider SCADA [31] systems, the possibilities of sensor compromises, and physical consequences.

These plans need to be incorporated into regular physical and cybersecurity drills that assign and manage cross-discipline teams, and capture gaps in workflows, comms, and tech that can be plugged in before a real emergency occurs. Communication silos need to be eliminated to keep the plant operators, IT, C-level, and external response and regulatory officers on the same page and to share situational updates, assign a stat designated position, and most importantly, share information during the crisis. Confusion and lack of communication can be a big problem in incidents, and to avoid this, clear communication silos need to be set up. A better redesign of the incident response system will also bridge the gaps during crises. In addition to this, utilities should clearly document the crucial control systems, the SCADA [31] and the partitioned Network, store these in a physically and logically secure system, and make access limited.

The offline copies serve as a priceless asset when attempting to recover a system after primary systems have been attacked by ransomware or destructive malware as they allow recovery to a previously verified configuration without the need to negotiate with an attacker or risk compromise from the uninfected online backup reservoirs.

17.14.2 Detection and Containment

Both rapid detection and rapid containment are essential abilities that will determine if a cyber incident will only disrupt operations or if it will possibly cause a complete failure. In the case of water utilities, the systems should design to set real-time alerts that monitor and measure the water flow, the water pressure, the concentration of chemicals, and the position of the valves. If any of the parameters go outside their set value, the security and operations personnel are to be notified immediately.

The detection and containment of the Incident should ideally be done simultaneously to minimize the lateral movement and scope of damage. In a cyber incident, the infected systems should be separated from the rest of the network by segmentation. Tectonically, the separated zones will be disconnected from communication links, blockage of unwanted flow of control, and device deactivation. However, there are situations where public safety gets endangered by unrestricted and unauthorized cyber intervention, like unauthorized dosage of treatment chemicals, and the water operators should be empowered to manually override the automated systems.

Even though these containment measures could interrupt regular service delivery, they avert disastrous results and furnish incident responders with restrain environments for investigations and corrective actions.

17.14.3 Recovery and Post-incident Activities

Recovery from cyber incidents should include systematic confirmation steps that eliminate any potential threats and ensure system integrity before any normal business activities can resume. Within the water utilities, it needs to be confirmed that all malicious [3] code is removed, all compromised credentials are revoked, all vulnerabilities are patched, and all configured systems are functioning correctly before control systems are reconnected to the production networks. Any premature restoration can lead to reinfection or can keep the attacker's access open. This would lead to prolonged incidents that would cause an increased level of damage.

Any post-recovery activities that are done must begin with root cause analysis to determine how the attackers were able to gain the initial access, what weaknesses in the system were used to penetrate, and what actions were used to detect or contain the attacks and how they performed. This knowledge is used to modify requests for incident response, requests for tightening of security, and operational workflows to reduce the chances of similar incidents.

Any documentation that is associated with lessons learned should be distributed internally and to the industry through reporting and sharing. Also, there are legal requirements within the water utilities whereby there is a need to approach government institutions, health departments, and the public to report cyber-related incidents that occur. Open and <> still. Even relying on incidents that shut down processes or lost confidential information. All through reliable communication with other institutions and the public, attackers were able to keep public trust and prevent any news to leak for which the services were used.

17.14.4 Testing and Simulation

The ongoing virtual and submission of tests is important for the affirmation of the readiness of the organization to imagine and respond to the Katinas realistically. Regular tabletop exercises facilitate strategic

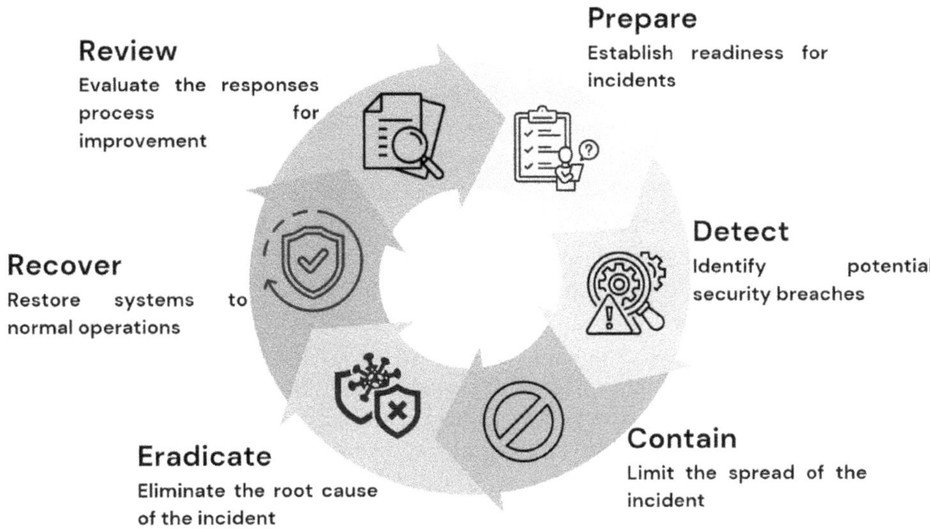

FIGURE 17.6 Incident response stages.

discussions among IT teams, OT personnel, network engineers, plant operators, cybersecurity professionals, and external stakeholders to strengthen preparedness and incident response capabilities.

These exercises help identify barriers to effective communication and coordination, establish roles and responsibilities, define monitoring objectives and performance metrics, evaluate the effectiveness of communication channels, and identify organizational gaps within existing operational and security frameworks. Tabletop exercises identify and rectify procedural vulnerabilities in step communication, real-world silo walls, sharpen crisis real-time communication, define pressing and contacts, and reinforce step contacts before.

These activities become increasingly important beyond the discussion phase, where red team and blue team exercises provide realistic attack and defence scenarios to assess the effectiveness, applicability, and resilience of cybersecurity controls. These personnel. The shifts practical noted sim, the overload, the team in the simulated the attack the techniques of the sim attributed set these and Lun to of effective responses to an attacking theoretical infrastructural period needed.

Red team–blue team engagements exploit vulnerabilities that might remain hidden during theoretical assessments, validate the effectiveness of detection tools, and train security analysts who learn to respond to complex threats in controlled environments and where mistakes become lessons instead of failures. Figure 17.6 presents the incident response framework adopted for managing cybersecurity events in critical infrastructure environments.

17.15 POLICY, REGULATION, AND FUTURE OUTLOOK

Investments, operational practices, and accountability regarding the intersection of public and private utility cybersecurity water infrastructure operate under complex regulatory landscapes and enduring public policy frameworks. This section analyses the policy instruments, regulatory frameworks, and compliance obligations as well as prospective trends and practices shaped by the policy instruments of tomorrow to define the evolving governance and technological advancement of cybersecurity in the water sector.

17.15.1 Global Policy Frameworks

The advent of governance smart water technologies has led to the formulation of policies by various governments to deal with the cybersecurity and operational resilience of the critical water infrastructure. In 2023, the US Environmental Protection Agency (EPA) revised the Cybersecurity for Water Utilities guidance, which incorporates cyber risk to the water and wastewater systems as a result of increased digitalization, and, therefore, serves the sector by defining cyber threat assessment and resilience planning protocols and frameworks for their systems. This framework requires utilities to perform vulnerability assessments, and in incidents, triggers, and sustained dynamics, to respond and monitor systems.

The NIS2 Directive of the European Union assigns new responsibilities to the operators of critical infrastructures like water utilities, expanding the scope of the cyberspace domain to the implementation of comprehensive, risk-based preemptive defensive postures, strict time-bound incident reporting for major incidents, board accountability for cyberspace resilience, and accountability for governance and oversight for cyber resilience at the board level. The Operational Technology (OT) Cybersecurity Master Plan of Singapore [23] offers distinct, centralized, prescriptive guidance for utilities and industrial [18] control systems, emphasizing secure-by-design systems, defensive perimeter systems, and intra-sector real-time cyber threat intelligence sharing among critical infrastructure. These evolving frameworks reflect the reality that water systems, equipped with cyber technologies like Blockchain and IoT, need sophisticated and seamless water governance systems, including cyber perimeter controls. As such, the world is poised for new and shared governance frameworks that use an integrated and systems-based approach to cross-border control that combines security and innovation.

17.15.2 Indian Context

As with the global geopolitical landscape, the digitalization of the water sector in India is currently undergoing a major transformation driven by the increased use of digital technologies. The NCIIPC [24] works under the Prime Minister's office and offers sectoral coordination for water utilities with other infrastructure sectors like energy and telecommunications to set basic security frameworks and facilitate emerging threat information sharing. This centralized approach ensures the same level of protection across interdependent infrastructure systems. The Indian Computer Emergency Response Team (CERT-In [25]) has given all organizations including water service providers until the line-out date in 2023 to comply with national reporting guidelines on the timely reporting of cybersecurity incidents. The guidelines specifically require reporting within the first six hours since the onset of the attack and the attack's subsequent documentation in full. These guidelines improve the national understanding of the situational picture and response strategy to geo-diverse cyberattacks targeting multiple utilities at the same time.

The Bureau of Indian Standards (BIS) [26] has drafted and is currently developing cybersecurity frameworks for automated water systems covering SCADA [31] networks, various IoT sensor deployments, and even Blockchain [17]. These standards are a completely comprehensive set of technical specifications for systems with secure architectural designs, protocols on strong data encryption, access censorship, and undisturbed audit trail control. These frameworks will allow Ion to pursue its ambitious digital water initiatives, such as the smart monitoring components of the Jal Jeevan Mission, while ensuring the secure, innovative, and service-improving systems, the designs will also protect critical water infrastructure across the urban and rural areas.

17.15.3 Capacity Building and Awareness

To instil and reinforce core cybersecurity capabilities within water utilities, water utilities should train plant engineers and systems integrators on OT security principles with cybersecurity discipline. Security approach in OT is very different from IT security. The training approach is different because systems in OT work with different objectives: availability and functionality vs. confidentiality and privacy. Systems in OT have multi-decade lifecycles, and decades of operational and real-time systems performance within acceptable operational parameters are non-negotiable. Training modules should include the IC systems design, SCADA [31] systems, pervasive access control "air" security, access control for control within control to water treatment and distribution [2], SCADA [31] secure gateways, warden systems, and IT security incident response handling. Training engineers to use control on real-time attacks in isolation for recognition and response to distributed control system incidents undergoing Ransomware attacks. Mechanisms of control system attacks, chemical dosing command attacks, and other command attacks should be in use.

Utilities are provided with specific frameworks that stem from the development of sector-specific cybersecurity maturity models. These frameworks assist the utilities in evaluating security measures in order to implement appropriate changes with operational and budget constraints. Security maturity models are built from the NIST [10] and IEC [11] 62443 frameworks. These models provide utilities with the ability to assess the level of capabilities from awareness to advanced-level threat hunting. These models grant peer comparisons and investment estimations. New models allow regulators to assess utilities provided with specific metrics for controlled improvement. These models improve the culture of the water sector by integrating cybersecurity into water utilities operational compliance.

17.16 OUTCOMES OF EXPERIMENTAL RESEARCH

Over the last several years, the incidents analysed and testbeds have greatly strengthened the ability to draft evidence-based policy frameworks and to set security requirements, as illustrated by the implementation of America's Water Infrastructure Act of 2018. The act prescribes that risk assessments be done by utilities serving in excess of 3,300 people. The compliance frameworks have considered the vulnerabilities documented in SWaT and WADI, and in real-life incidents. These datasets have become invaluable as foundational references in the construction of policy. They illuminate the ability to pose and respond to questions on the feasibility of an attack, detection, and enduring consequences on operational systems.

17.16.1 Experimental Research Policy Implications

- **Baseline security requirements**: Testbed attacks help define baseline security control sets, as ICS-CERT [5] drafts cite SWaT attack taxonomy.
- **Detection capability mandates**: Certain areas now require Anomaly [15] detection systems and benchmarked performance from testbed experiments.
- **Information-sharing incentives**: Initiatives such as Water ISAC [30] improve knowledge by integrating real incidents (e.g., Oldsmar) and experiments.
- **International standards convergence**: Initiatives by Singapore's Public Utilities Board and requirements from the EU [22] NIS2 Directive reflect emerging international coherence on water cybersecurity.

Regulatory prospects foresee smart water technologies needing security-by-design, including acceptance testing against prescribed attack simulations. Bridging the existing divide between large, resourceful utilities and their smaller counterparts constitutes the core policy challenge, with testbed research focusing on determining the most cost-effective security solutions tailored for utilities with thin budgets and limited technological savvy.

TABLE 17.4 Comparative Overview of Global Water Cybersecurity Policies

REGION	*KEY REGULATION*	*FOCUS AREAS*	*REMARKS*
USA	EPA Guidelines	Risk assessment, incident reporting	Sector-specific guidance
EU	NIS2 Directive	Network & information security	Broad CI coverage
India	NCIIPC & CERT-In [25]	CI protection, reporting	Implementation phase
Singapore	OT Masterplan	Governance, training	Mature implementation

17.16.2 Recommendations

Water utilities are required to implement frameworks that support and enhance the integration of cyber-physical security [28]. The frameworks need to understand the relationship and interdependence of digital Blockchain cyber systems and physical water systems. The line of cyber defence considers the separation of cyber and physical defence in systems where attacks on digital systems control, valves, pumps, and chemical dosing equipment fail. These frameworks need to cover threat detection and response and system defence in both architectures. An ecosystem for innovation that addresses new threats related to Blockchain [17]-enabled water systems is fostered by collaborative efforts between government regulatory bodies, academic research institutions, and private cybersecurity firms. Shared research can create platforms for sharing threat intelligence, testing penetration on pilot Blockchain [17] systems, and developing open-source security software for water utilities that are budget- and tech-constrained. Tax incentives, funding that diminishes reliance on foreign security vendors, and preferential procurement policies that encourage the adoption of domestic security technologies are the responsibility of the government. The development of domestic technologies builds expertise, protects the supply chain, lessens geopolitical risks, and opens up new economic opportunities and, at the same time, enhances the resilience of water governance systems to complex cyber threats aimed at critical infrastructure globally. Table 17.4 provides an overview of internationally recognized cybersecurity frameworks and their relevance to protecting critical information infrastructure.

17.17 CONCLUSION

In the current times, cybersecurity is no longer considered a minor technical issue, but a vital part of water infrastructure management. With the advent of smart technologies, the application of Blockchain systems, IoT sensors, and cloud-based analytics for analytics, the attack surface created by the water systems has exponentially increased, providing more safety vulnerabilities for hostile systems. This precipitates the ability of adversaries to disrupt the service delivery, compromise the water quality data, and even pose a risk to the general populace. There has been a profound change in the information technology and operational technology markets, with systems of water management being exceptionally developed. This, in turn, has led to the systems being referred to as complex cyber-physical [6] ecosystems. This implies that a digital breach in these systems can have dire effects on the population that relies on these systems for water. In an age where cyberattacks have become sophisticated, it is vital for firms to have proactive defences. This can include achieving a balance between the implementation of strategically developed frameworks as well the utilization of new technologies such as AI and Blockchain. It is counterproductive for firms to have "fire brigade" systems that can only act in case a breach has been made. It is essential for the systems to accurately predict the possible attack routes, conduct routine assessments, and have systems in place that can counter attacks in the case of breaches. This is a more viable strategy. Resilient water systems of the future will adopt Zero trust [27] security architectures that scrutinize every access request across the board, nurture genuine cross-sector partnerships that facilitate the sharing of threat intelligence between water utilities and cybersecurity, and integrate programmatic cyber hygiene into daily workflows at all levels of the organization. With proper alignment of sophisticated technologies and policies matched with adequate human element preparedness stemming from robust training and awareness, countries will protect one of the most vital and irreplaceable public resources: water. This multifaceted approach avoids undermining the security and resilience that communities rightly expect from water infrastructure while delivering on the promise of smart water governance enabled through Blockchain [17] and related technologies.

REFERENCES

1. Adedeji, K. B., Hamam, Y., Abe, B. T., & Abu-Mahfouz, A. M. (2022). Towards achieving a reliable leakage detection and localization algorithm for application in water piping networks: An overview. *IEEE Access*, 10, 32371–32391.
2. Taormina, R., Galelli, S., Tippenhauer, N. O., Salomons, E., & Ostfeld, A. (2017). Characterizing cyber-physical attacks on water distribution systems. *Journal of Water Resources Planning and Management*, 143(5), 04017009.
3. Abrams, M., & Weiss, J. (2008). Malicious control system cyber security attack case study–Maroochy Water Services, Australia. *The MITRE Corporation*, 1–17.
4. Senserrick, M. (2021). *Florida water treatment facility hack: Lessons learned*. SANS Institute InfoSec Reading Room.
5. ICS-CERT. (2020). *Annual vulnerability coordination report*. Cybersecurity and Infrastructure Security Agency, U.S. Department of Homeland Security.
6. Mathur, A. P., & Tippenhauer, N. O. (2016). *SWaT: A water treatment testbed for research and training on ICS security*. 2016 International Workshop on Cyber-physical Systems for Smart Water Networks (CySWater), 31–36.
7. Ahmed, C. M., Palleti, V. R., & Mathur, A. P. (2017). *WADI: A water distribution testbed for research in the design of secure cyber physical systems*. Proceedings of the 3rd International Workshop on Cyber-Physical Systems for Smart Water Networks, 25–28.
8. Goh, J., Adepu, S., Junejo, K. N., & Mathur, A. P. (2017). *A dataset to support research in the design of secure water treatment systems*. International Conference on Critical Information Infrastructures Security, 88–99.
9. Kravchik, M., & Shabtai, A. (2018). *Detecting cyber attacks in industrial control systems using convolutional neural networks*. Proceedings of the 2018 Workshop on Cyber-Physical Systems Security and PrivaCy, 72–83.
10. National Institute of Standards and Technology (NIST). (2018). *Framework for improving critical infrastructure cybersecurity, version 1.1*. U.S. Department of Commerce.
11. International Electrotechnical Commission (IEC). (2018). *IEC 62443: Security for industrial automation and control systems*. IEC Standards.
12. International Organization for Standardization (ISO). (2017). *ISO/IEC 27019:2017 Information security controls for the energy utility industry*. ISO Standards.
13. American Water Works Association (AWWA). (2010). *ANSI/AWWA J100-10: Risk and resilience management of water and wastewater systems*. AWWA Standards.
14. Adepu, S., & Mathur, A. (2016). *Distributed detection of single-stage multipoint cyber attacks in a water treatment plant*. Proceedings of the 11th ACM Asia Conference on Computer and Communications Security, 449–460.
15. Inoue, J., Yamagata, Y., Chen, Y., Poskitt, C. M., & Sun, J. (2017). *Anomaly detection for a water treatment system using unsupervised machine learning*. 2017 IEEE International Conference on Data Mining Workshops (ICDMW), 1058–1065.
16. Feng, C., Li, T., & Chana, D. (2017). *Multi-level anomaly detection in industrial control systems via package signatures and LSTM networks*. 2017 47th Annual IEEE/IFIP International Conference on Dependable Systems and Networks (DSN), 261–272.

17. Tsang, Y. P., Choy, K. L., Wu, C. H., Ho, G. T., & Lam, H. Y. (2019). Blockchain-driven IoT for food traceability with an integrated consensus mechanism. *IEEE Access*, 7, 129000–129017.
18. Tao, F., Zhang, H., Liu, A., & Nee, A. Y. (2019). Digital twin in industry: State-of-the-art. *IEEE Transactions on Industrial Informatics*, 15(4), 2405–2415.
19. Chen, L., Jordan, S., Liu, Y. K., Moody, D., Peralta, R., Perlner, R., & Smith-Tone, D. (2016). *Report on post-quantum cryptography*. US Department of Commerce, National Institute of Standards and Technology.
20. Mosca, M. (2018). Cybersecurity in an era with quantum computers: Will we be ready? *IEEE Security & Privacy*, 16(5), 38–41.
21. U.S. Environmental Protection Agency (EPA). (2023). *Cybersecurity best practices for the water sector*. EPA 817-B-23-001.
22. European Union. (2022). *Directive (EU) 2022/2555 on measures for a high common level of cybersecurity across the Union (NIS2 Directive)*. Official Journal of the European Union.
23. Cyber Security Agency of Singapore. (2019). *Operational technology cybersecurity masterplan*. CSA Singapore.
24. National Critical Information Infrastructure Protection Centre (NCIIPC). (2015). *Guidelines for protection of critical information infrastructure*. Government of India.
25. Indian Computer Emergency Response Team (CERT-In). (2022). *Cyber security directions for service providers, intermediaries, data centres, body corporate and Government organisations*. Ministry of Electronics and Information Technology, Government of India.
26. Bureau of Indian Standards (BIS). (2021). *IS/IEC 27001:2013 Information technology – Security techniques – Information security management systems – Requirements*. BIS Standards.
27. Rose, S., Borchert, O., Mitchell, S., & Connelly, S. (2020). *Zero trust architecture*. NIST Special Publication 800-207.
28. Clark, R. M., & Hakim, S. (2014). *Cyber-physical security: Protecting critical infrastructure at the state and local level*. Springer.
29. Slay, J., & Miller, M. (2007). *Lessons learned from the Maroochy water breach*. International Conference on Critical Infrastructure Protection, 73–82.
30. Water Information Sharing and Analysis Center (Water ISAC). (2021). *2021 Water and wastewater systems sector-specific incident report*. Water ISAC.
31. Boyer, S. A. (2009). *SCADA: Supervisory control and data acquisition* (4th ed.). International Society of Automation.

Cybersecurity of SCADA/ICS Used in Water Management

Threats, Incidents and a Practical Defence Blueprint

18

Abhishek Mishra

18.1 INTRODUCTION

Operational technology (OT) in the water sector spans Programmable Logic Controllers (PLCs)/Remote Terminal Units (RTUs), telemetry, HMIs and historian servers that supervise physical processes. As water utilities connect assets for remote operations and analytics, adversaries have repeatedly demonstrated interest and capability to disrupt, manipulate or pre-position within these environments. Sector authorities now advise urgent action, noting recurring weaknesses such as internet-exposed HMIs, default credentials and flat networks [1].

Intro to SCADA/ICS in Water Utilities. Operational Technology (OT) in water utilities is built around Industrial Control Systems (ICS) that monitor and control physical processes.

SCADA—*Supervisory Control and Data Acquisition*—is the ICS pattern used for geographically distributed assets (wells, pump/booster stations, reservoirs, valve chambers), providing centralized visibility and supervisory control.

Earlier systems used electro-mechanical relays and radios to signal pump states and alarms over long distances (pipelines, canals). As microprocessors matured, Remote Terminal Units (RTUs) and Programmable Logic Controllers (PLCs) replaced relays and digital SCADA emerged—HMI servers, historians and telemetry coordinating treatment and distribution—the nervous system of a modern utility.

DOI: 10.1201/9781003742227-18

In contemporary Water Treatment Plants, multiple PLCs control coagulation/flocculation, filtration, disinfection, residual handling and Safety Instrumented Systems (SIS) enforce hard limits on chemical dosing and gas handling [2].

OT–IT convergence and exposure (transition to threats). As water utilities modernize, SCADA no longer lives in isolation: plant networks exchange data with enterprise IT (CMMS, LIMS, billing/AMI), cloud analytics and vendor remote support. Telemetry has shifted from closed radio loops to IP-routed links (cellular, microwave, leased fibre) and legacy PLC/RTU fleets now coexist with newer HMIs, historians and APIs. This convergence delivers efficiency and situational awareness, but it also expands the attack surface: internet-reachable portals, weakly governed remote access (VPN/RDP/desktop-sharing), misconfigured firewalls, flat networks without an Industrial DMZ (IDMZ), plus long-lived devices that are hard to patch. In such environments, even basic adversary actions—credential stuffing, abusing legitimate tools or changing setpoints—can cause loss of view/control, unsafe chemistry, tank overflows or pressure anomalies. With that exposure in mind, the next section outlines the threat landscape and the specific tradecraft seen against water utilities.

18.2 THREAT LANDSCAPE AND ADVERSARY TRADECRAFT

Internet-Exposed Interfaces and Weak Remote Access. Cybersecurity and Infrastructure Security Agency (CISA)/Environmental Protection Agency (EPA) warns that HMIs reachable from the internet let attackers view process states and attempt unauthorized changes.

- **IT/Data Exposure (No SCADA Manipulation).** In an incident, at Bangalore, a configuration oversight in the system caused an unprotected file and admin portal full database access at Bangalore Water Supply and Sewerage Board (BWSSB), jeopardizing >290,000 personal records [3].
- **OT Setpoint Manipulation via Remote Access.** Oldsmar Case of Florida (2021) illustrated the risk path via remote-access tooling on an operator workstation [22].

Living-off-the-land (LOTL) Techniques. Living-off-the-land means attackers avoid custom malware and instead abuse legitimate, already-installed tools and accounts to achieve their goals. In water/WW systems this often looks like logging in with valid VPN/RDP credentials, pivoting with PowerShell/WMIC/PSExec or vendor remote-support tools and then using normal operator software (HMI, engineering workstations, OPC clients) to change setpoints, silence alarms or collect data. Because every action uses "expected" utilities, LOTL blends into routine operations, evades antivirus, leaves fewer indicators and makes attribution harder—yet the process impact (overflows, pressure swings, mis-dosing) can be immediate.

People's Republic of China (PRC)-linked "Volt Typhoon" operations in critical infrastructure (including water/wastewater) rely on valid credentials and native admin tools to blend into normal activity—complicating detection.

- Chinese hacker group Volt Typhoon spent nearly 10 months (from February 2023 to November 2023) in Littleton Electric Light & Water Department in Littleton, Massachusetts [4]. The hackers maintained persistent access to the systems, aiming to exfiltrate OT data (which uses GE Proficy iFIX SCADA Systems for Water Management and Survalent SCADA Systems for Electric Department) and spatial layout related to energy grid operations. The challenging part is to identify such attacks in which post initial access to the victim's systems, no external

software, tools or scripts are used which might alarm firewalls. Using this, the hackers blend in with trusted networks and use readily available tools [5].
- Contemporaneous reporting highlighted cyber activity at Hawaii ports and logistics nodes—critical in any Indo-Pacific contingency—where multiple SCADA-supervised utilities (power, fuel, water) converge [6].
- Post initial access, actors avoided third-party tools—relying on built-in admin utilities—so activity blended with normal traffic and evaded signature-based defences.

Ransomware business disruption with OT spillover. Colonial Pipeline (2021) shows how IT compromises can prompt OT shutdowns out of caution, creating real-world supply impacts even when control networks aren't directly encrypted. Colonial Pipeline ransomware attack impacted 45% of the East Coast's fuel delivery systems, affecting 13 states having population over 50 million. Within ~12 hours of initial access via a compromised VPN account, actors used built-in tools (e.g., PowerShell) to stage data theft and deploy encryption [7].

Safety-system targeting. In India, Triconex Safety Instrumented Systems (SIS) is available and deployed in critical infrastructures.

- Schneider supplies hydro-turbomachinery safety solutions and automates large municipal water assets, although specific SIS deployments are generally not disclosed publicly.
- Schneider is actively automating India's largest Water Treatment Plant (WTP) at Bhandup, Mumbai, having a capacity of 2,000-MLD and a 564-MLD wastewater plant in Delhi [8].
- In 2017, the **TRITON** malware (commonly attributed to the Russian Central Scientific Research Institute of Chemistry and Mechanics, TsNIIKhM) compromised **Schneider Electric Triconex** safety controllers—the last line of defence responsible for initiating safe shutdowns. By altering in-memory firmware to embed additional programming, the attackers created conditions under which the SIS could be subverted, increasing the likelihood of physical damage, extended downtime or even loss of life.
- This event which has been summarised at Figure 18.1, highlights a strategic shift that threat actors are prepared to neutralize safety functions, raising the stakes for all fluid-handling industries, including water, oil and gas.

"Hacktivist" disruptions to small systems. Before turning to specific cases, it's important to note a growing pattern in attack cases pertaining to small and mid-sized utilities which are often with limited staff, legacy remote-access setups and vendor-managed HMIs—have become high-visibility targets for so-called "hacktivist" brands.

- These actors typically seek propaganda value over deep sabotage, using low-complexity methods (exposed portals, weak credentials, abuse of legitimate remote tools) to create brief but public process disturbances that force operators into manual control.
- Some of these brands have been assessed as fronts or auxiliaries for state units, which complicates attribution and response. The Texas Panhandle intrusions illustrate this dynamic vividly as mentioned below.
- In January 2024, three small towns in the Texas Panhandle—Muleshoe, Hale Centre and Lockney—faced coordinated cyber intrusions against their water systems [9].
- In Muleshoe, attackers manipulated the HMI to run pumps until a storage tank overflowed, after which operators shifted to manual control.
- Hale Centre reported 37,000 login attempts in four days but avoided compromise by physically disconnecting ("unplugging") the automation workstation and operating manually. Local officials said public safety was not endangered and notified federal authorities [9].
- CyberArmyofRussia_Reborn (CARR), a pro-Russia hacktivist brand that had posted videos of ICS tampering and had been assessed as created by or working with Russia's Sandworm/GRU

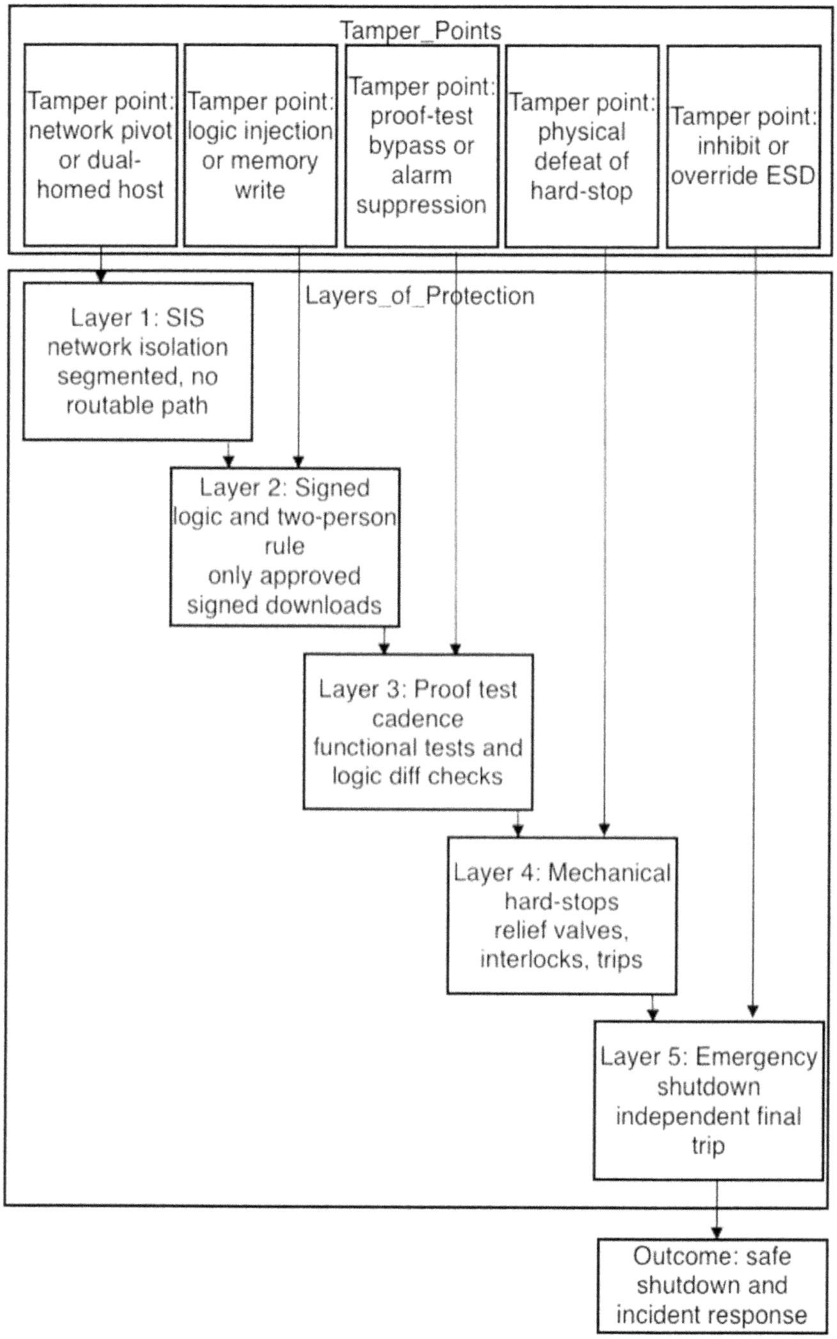

FIGURE 18.1 TRITON-style SIS tamper layers (stacked layers).

apparatus. Reporting noted similar CARR claims against water facilities in the United States and Poland. While "hacktivist" labels suggest low sophistication, major vendors have documented patterns where GRU operations are laundered through Telegram-branded hacktivist outlets, complicating attribution and response [10].

Table 18.1 prioritizes the most credible paths to process impact and binds each to a target date; the following case studies illustrate how these risks materialize.

TABLE 18.1 Top 10 Risk Scenarios with Impact, Inherent Risk, Key Controls and Residual Risk

S. NO	*SCENARIO*	*IMPACT*	*INHERENT RISK*	*KEY CONTROL (PRIMARY)*	*RESIDUAL RISK*
1	**Internet-exposed HMI / web portals**	H	**Very High (H×H)**	Remove exposure; force VPN + MFA; IP allow-lists / WAF	**Low**
2	**Weak remote access (RDP/TeamViewer/ VPN)**	H	**Very High (H×H)**	IDMZ jump host + PAM/JIT + MFA; session recording; time-bound access	**Low–Med**
3	**LOTL lateral movement**	H	**Very High (H×H)**	Baseline command-line telemetry; constrain admin tools at jump; OT-aware monitoring	**Med**
4	**Default/weak PLC/ HMI credentials**	H	**High**	Unique device creds; vault; fleet-wide rotation; disable vendor defaults	**Low**
5	**Unsupported OS on HMI/EWS**	H	**High**	Virtualize/upgrade; offline golden images; patch test-bench	**Med**
6	**Flat IT/OT network (no IDMZ)**	H	**High**	3-tier seg (IT↔IDMZ↔OT); dual firewalls; unidirectional historian path	**Low–Med**
7	**SIS tamper (e.g., Triconex)**	H	**High**	Isolated SIS network; two-person rule; signed logic; proof-test cadence	**Low**
8	**Ransomware in IT with OT spillover**	H	**Very High (H×H)**	Strict IT/OT seg; immutable/ offline backups; TTX; allow-listing on jump	**Med**
9	**Insecure field comms (radio/ serial)**	M–H	**High**	Authenticated telemetry; radio key management; anomaly alerts on RTU chatter	**Med**
10	**Supplier toolchain / remote support**	H	**High**	IEC 62443-2-4 clauses; JIT vendor access via jump; session recording; SBOM/vuln notify	**Med**

18.3 CASE STUDIES

1. The following cases were selected to illustrate how different attack paths produce real operational risk across water and "fluids" infrastructure—from insider abuse of insecure field communications (Maroochy, 2000) to living-off-the-land remote access against operator workstations (Oldsmar, 2021), state-linked hacktivist activity leveraging default credentials and Internet-reachable panels (Aliquippa, 2023) and IT-side ransomware that compelled an OT shutdown despite no controller encryption (Colonial Pipeline, 2021). Together, they span two decades, multiple countries and a spectrum of impacts: environmental release, attempted chemical misdosing, forced manual operations, prolonged business disruption and policy change.
2. **Maroochy Shire (Australia, 2000): Insider + Insecure Field Communications leading to Environmental Spill.** A disgruntled contractor used stolen field radios and a laptop to issue spoofed SCADA commands across sewage pump stations, disabling alarms, altering set-points

and ultimately spilling around 800,000 L of raw sewage before being arrested with the specialized gear in his car.

2.1. The court record and subsequent analysis mapped the failures to basic control gaps (wireless access control, account/device authentication, audit logging, incident response and supplier/contractor controls under NIST SP 800-53).

2.2. The case remains a canonical example of how insider knowledge + insecure field comms can convert into environmental and public-health impact [11, 12].

3. **Oldsmar, Florida (U.S., 2021). LOTL via remote desktop leading to attempted mis-dosing.** On 5 February 2021, unidentified actors gained unauthorized access to a drinking water plant's SCADA workstation and increased the dosing setpoint for sodium hydroxide (lye). A plant operator noticed the change in real time, reverted the setpoint and operations continued normally.

3.1. **Initial access and contributing weaknesses**. As per the investigators, the hackers exploited basic security gaps like poor password hygiene, usage of an out-of-support Windows 7 operating system, etc. Early information indicated that desktop-sharing software (e.g., TeamViewer) may have been involved, though this was not confirmed. Regardless, the pattern fits a broader trend of attackers abusing legitimate remote-access tools to blend in, move laterally and alter operations. In such a scenario, the required response is not merely technical isolation but a coordinated operator–control room–OT/SOC–plant engineer workflow, as shown in Figure 18.2.

3.2. **Operational Impact and Detection**. As per the hackers' intentions of manipulating chemical dosing, this was an attack on "process" rather than "data theft". Hack failed because an attentive operator observed the unauthorized setpoint change and immediately restored safe values, and the plant's control application was configured to alarm on such manipulations. Resultantly, no unsafe water was distributed and treatment remained unaffected. The intent was process manipulation (chemical dosing), not data theft.

3.3. **Threat Tradecraft in Focus: Desktop Sharing Abuse**. The incident highlights recurring misuse of legitimate remote-control tools: hiding windows, controlling startup parameters for persistence, dropping files and leveraging the tools' "normal" appearance to

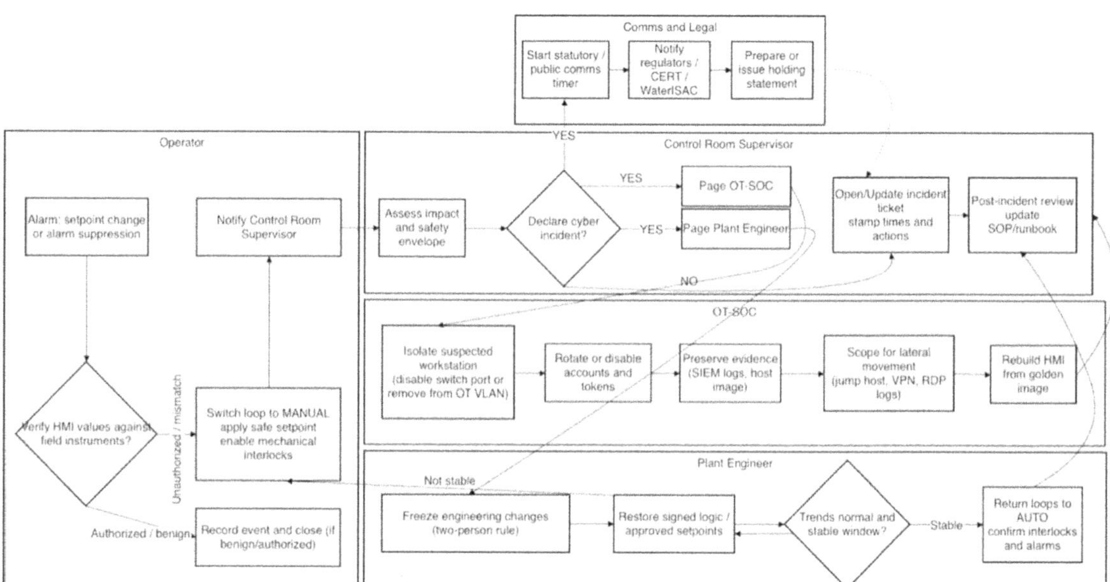

FIGURE 18.2 Incident response playbook for setpoint tamper (swimlane).

evade suspicion by admins and users. This is functionally similar to a remote-access trojan (RAT), but delivered by user-approved software—which is exactly why it's effective.

3.4. **Structural Risk: Legacy Platforms**. Continuation of Windows 7 (end-of-support) and poorly governed Remote Desktop Protocol (RDP) exposures materially increases the attack surface. Commodity exploits and credential-stuffing against perimeter services remain common paths into smaller utilities that lack centralized patching, logging and Multi-Factor Authentication (MFA) [23].

4. **Aliquippa, Pennsylvania (U.S., 2023)—IRGC-linked Unitronics Campaign Context. Default Credentials Exposure saved by Quick Manual Override**. On 25–26 November 2023, the Municipal Water Authority of Aliquippa, Pennsylvania, reported that one of its remote booster stations had been compromised by an Iran-linked hacktivist group, "Cyber Av3ngers".

4.1. According to the authority's board chair, hackers briefly took control of the station's water system, which monitors and regulates water pressure for Raccoon and Potter Townships. An alarm triggered immediately, operators intervened and the affected system was taken offline/disabled while state police initiated a criminal investigation. Though Officials emphasized there was no known risk to the drinking water or supply during the event, the fact that SCADA Systems were hacked cannot be ignored.

4.2. **Targeted technology and claimed motive**. The compromised machine was operating a Unitronics Control Solution (PLC/HMI). Public statements from the authority noted that Unitronics has Israeli ownership/components, a point that aligns with Cyber Av3ngers' propaganda: the group has repeatedly claimed it targets Israeli-made industrial control equipment and, on social channels, has taken responsibility for multiple attacks on water infrastructure (including claimed activity against 10 water treatment stations in Israel as of 30 October 2023). In Aliquippa, the focus was a pressure-control booster rather than treatment chemistry—meaning the operational risk was service disruption or local pressure anomalies, not water quality.

4.3. **What likely enabled the intrusion (common patterns in this campaign)**. While granular forensic details for Aliquippa were not published, contemporaneous advisories about this campaign highlighted basic exposure and authentication weaknesses at some U.S. utilities running Unitronics gear—e.g., internet-reachable HMIs, default or weak credentials and remote access without multi-factor authentication (MFA). In small utilities, these conditions can let attackers reach a booster/tank site and toggle setpoints or modes with minimal "hacking"—often via standard web or remote-desktop interfaces [13].

5. **Colonial Pipeline (U.S., 2021): IT ransomware forcing OT shutdown leading to Policy Change.** In May 2021, Colonial Pipeline—operator of the largest refined-fuel pipeline in the United States—shut down operations after ransomware struck IT systems. Colonial transports refined petroleum products (gasoline, diesel, jet fuel), but similar technologies are used in Water plants too.

5.1. Though OT (pipeline control) was not encrypted, the company pre-emptively halted product flow to contain business-system risk and ensure operational integrity. The event disrupted around 45% of East Coast fuel distribution and became a watershed for critical-infrastructure cyber policy.

5.2. **Context and infrastructure**. Colonial's network spans to around 5,500 miles across 13 states, moving gasoline, diesel, jet fuel and more via multiple mainlines and hundreds of delivery points under extensive SCADA/telemetry oversight. This scale—and the network's centrality to regional supply—meant even a short outage would propagate quickly to retail stations and airports.

5.3. **Operational and economic effects**. The six-day disruption triggered fuel shortages and queues across multiple states. Some metro areas reported very high station depletion rates, with associated price spikes. Direct operational loss, cascading logistics costs, emergency measures and ransom payment (partially recovered later) drove total economic impact estimates into the billions when both direct and indirect costs are counted.

5.4. **Response and recovery**. Within minutes of detection, Colonial activated emergency response, isolated affected environments, coordinated with FBI/CISA and other federal partners and stood up segmented "safe zones". Field technicians executed manual procedures while enterprise systems were rebuilt and monitored. A phased restart began after validation of controls, and throughput was gradually increased to normal levels with heightened telemetry and continuous authentication on critical access paths.

5.5. **Changes after Colonial (policy and industry)**. The incident accelerated U.S. critical-infrastructure requirements—e.g., Transportation Security Administration (TSA) security directives for pipelines, faster incident reporting, tighter remote-access controls and expanded public-private collaboration through programs such as Joint Cyber Defense Collaborative (JCDC). Operators increased segmentation often in three-tier: Information Technology (IT) ↔ Industrial Demilitarized Zone (IDMZ) ↔ Operational Technology (OT), instituted continuous vulnerability management and invested in OT-aware monitoring and zero-trust patterns for privileged access.

18.4 EMPIRICAL VALIDATION ON BATADAL (C-TOWN)

From principles to proof. The controls in previous paragraphs emphasize baseline-driven detection of LOTL behaviours and process-aware anomalies. To ground these recommendations in data, this section validates a lightweight detection stack on the BATADAL (C-Town) [14] benchmark and reports the following:

- Attack Detection Rate.
- False Alarms/Day.
- Median Time-to-Detect.
- Mapping findings back to the Threat → Failure Mode → Control table.

Purpose. To demonstrate that process-aware monitoring from our blueprint can surface setpoint/logic tamper even when adversaries "**live off the land**".

Data. BATADAL (C-Town) hourly SCADA:

- Train-1 (normal only): 1 year—learn baseline behaviour.
- Train-2 (part-labelled attacks): ~6 months—tune alert threshold.
- Test (unlabelled): ~3 months—blind evaluation.

Signals. Standard water tags:

- Tank levels (L_T1–L_T7).
- Pump/valve flows (F_PU#, F_V2).
- Statuses (S_PU#, S_V2).
- Selected pressures (P_J#).
- L_T1 (Tank-1 level) as a sentinel tag.
- Multivariate expansion is noted in Future Work.

Exploratory Signal Assessment (EDA). Before applying forecasting and residual-based anomaly scoring, an exploratory view of the sentinel tag **L_T1** and its interaction with upstream actuation was conducted to confirm that the signal exhibits meaningful process-driven dynamics. Since pump operations directly influence tank-level trajectories, a preliminary relationship check between **L_T1** and **F_PU1** was performed along with time-series inspection of both signals. Figure 18.3 shows the relationship between Tank-1 level and Pump-1 flowrate, confirming that the selected sentinel tag reflects process-linked behaviour rather than random signal noise. This supports its use for baseline learning and anomaly detection in the subsequent model stage. This step ensures that the selected tag represents physical behaviour rather than random noise and supports its use for baseline learning and anomaly detection in the subsequent model stage. To further assess the upstream actuation behaviour, Figure 18.4 presents the Pump-1 flowrate trend over time. The graph helps identify normal duty-cycle variations and flowrate shifts that influence the Tank-1 level trajectory.

- Correspondingly, Figure 18.5 shows the Tank-1 level trend over time. The repeated fill-and-drawdown pattern demonstrates that L_T1 has sufficient temporal structure for forecast-based anomaly detection. Figures 18.3–18.5 confirms that **L_T1** displays consistent process-driven variation and a structured dependency on pumping behaviour. Therefore, **L_T1** is suitable for the proposed **forecast-then-compare** method, where an LSTM learns the normal operating trajectory and deviations from expected behaviour are quantified through residual error |y^−y|.
- These residual spikes are then interpreted as potential process anomalies consistent with abnormal duty cycles or **cyber-attack attempts**.

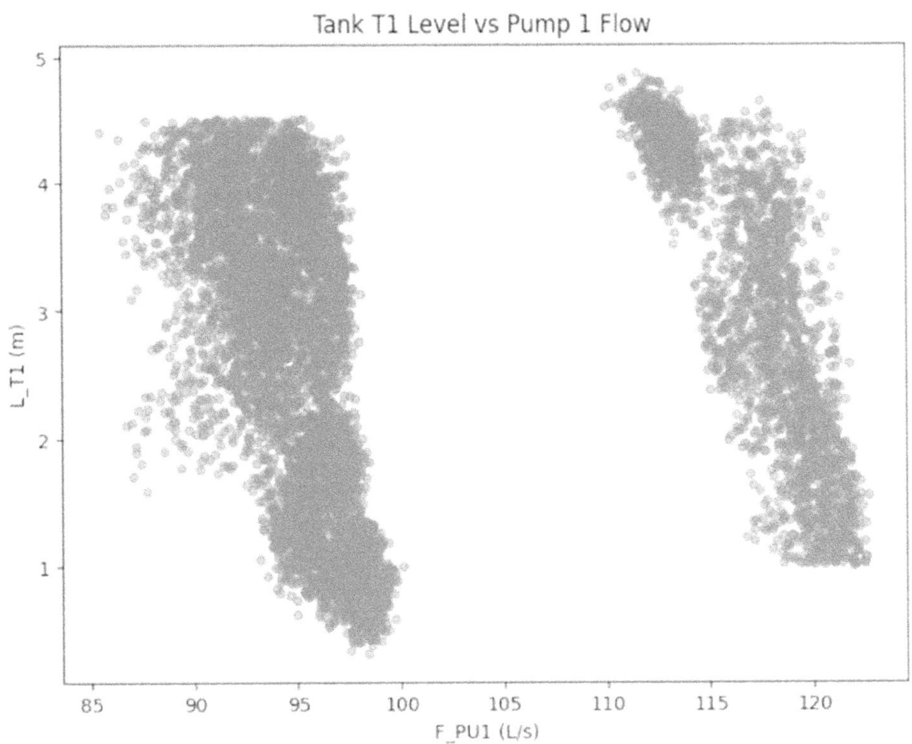

FIGURE 18.3 Tank level and pump flow relationship (L_T1 vs F_PU1).

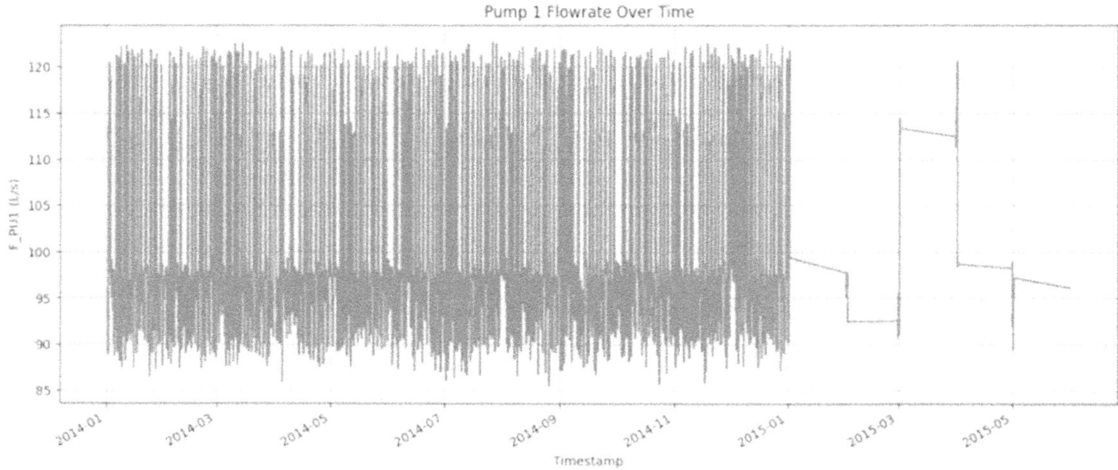

FIGURE 18.4 Pump 1 flowrate trend over time (F_PU1 vs Time).

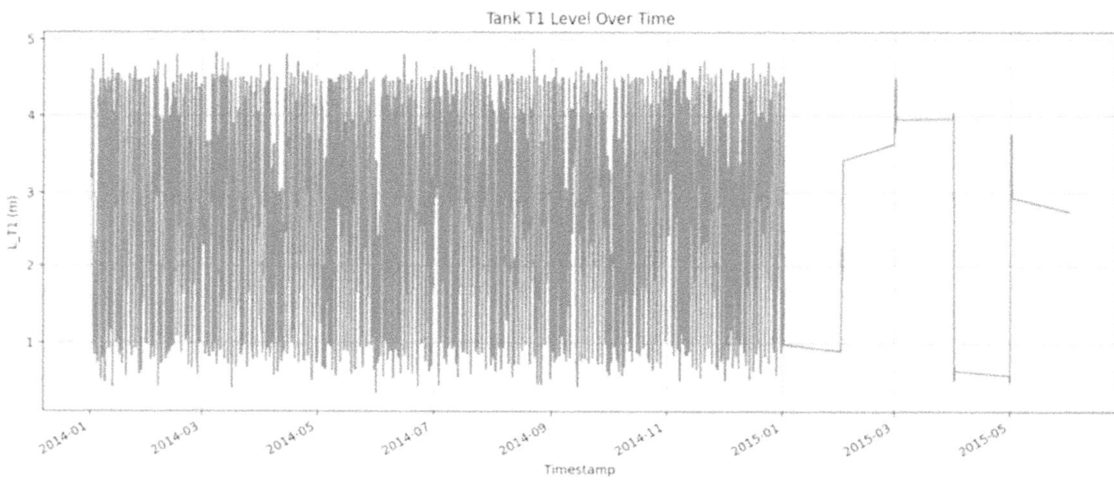

FIGURE 18.5 Tank level trend over time.

18.4.1 Pre-processing

- Timestamp → Date Time index.
- Sanity checks.
- Minmax scaling (0–1).
- Optionally drop a few unused fields.

18.4.2 Method (Forecast-then-Compare)

- **Input window: Last 120 hours of L_T1**. To operationalize the forecast-then-compare method, the LSTM is not trained on isolated readings but on a rolling time-history of the process. Accordingly, the model receives an input sequence of the **past 120 hours of L_T1** values to predict the next hour. Figure 18.6 illustrates a representative 120-sample segment of the

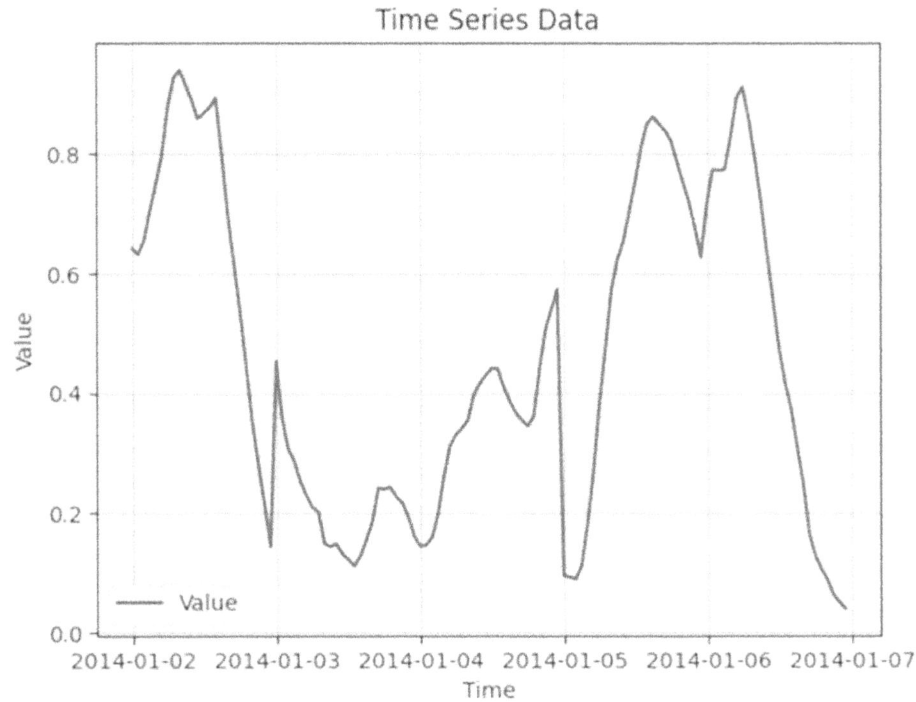

FIGURE 18.6 Example of 120-hour input window used for LSTM forecasting (L_T1 scaled time-series segment from BATADAL Train-1).

tank-level time series (scaled 0–1), demonstrating the temporal structure and rate-of-change patterns that the model learns as baseline behaviour.

- This rolling-window representation enables the network to capture gradual fills, drawdowns and operational oscillations that occur under normal pumping cycles. Once trained on **Train-1 (attack-free)** data, the model produces a baseline next-step forecast, and any sustained deviation between predicted and actual readings is quantified as a residual |y^–y|, which serves as the anomaly score in subsequent evaluation and alerting cyber-attacks.
- **Model**: Small LSTM (50→25→10 units) → 1-step-ahead forecast.
- **Train on**: Train-1 (attack-free).
- **Anomaly score**: |prediction – actual|.
- **Threshold**: Chosen on **Train-2** (maximize Youden's J / balance TPR–FPR), then frozen and applied to **Test**.

18.4.3 Model Targeting On

- Fit quality (Train-1 hold-out): RMSE / MAE.
- Separation (Train-2): Precision/recall/F1 at chosen threshold; brief ROC/PR summary.
- Alert volume (Test): % time flagged and exemplar weeks (plots of Actual vs Predicted with anomaly bands).

18.4.4 Merits of Selected Model

- Captures **Process deviations** caused by setpoint tamper, mis-dosing, or abnormal duty cycles—even if access looked legitimate (LOTL).
- Plugs directly into the **Setpoint-Tamper IR Playbook**: operator verification → manual mode → workstation isolation → restore signed logic.

18.4.5 Limitations and Next Steps (Brief)

- **Univariate demo**: Extend to **Multivariate** (levels + flows + pressures + statuses) to resist sensor spoof and improve sensitivity.
- Periodic **Re-training** for seasonal/operational drift.
- Add **Physics rules** (mass balance, plausible rate-of-change) for hybrid detection. Figure 18.7 compares the actual and predicted Tank-1 water levels, demonstrating the accuracy of the predictive model in capturing system behavior.

18.4.6 Brief about Model and End Result

- To ground the blueprint in data, readings are used for a minimal *forecast-then-compare* test on a sentinel tag, **L_T1** (Tank-1 level).
- A compact LSTM learns the plant's "normal" trajectory from the past 120 hours and predicts the next hour.
- When replayed on operations data, *departures from this learned baseline appear as residual spikes*—the same signature is expected from setpoint tamper or cyber-attack cycles even when access looks legitimate.
- Figure 18.7 contrasts the model's baseline prediction with the actual Tank-1 level (L_T1). In the evaluation, the residual (|y^–y|) drives threshold-based alerts used in the subsequent performance summary.

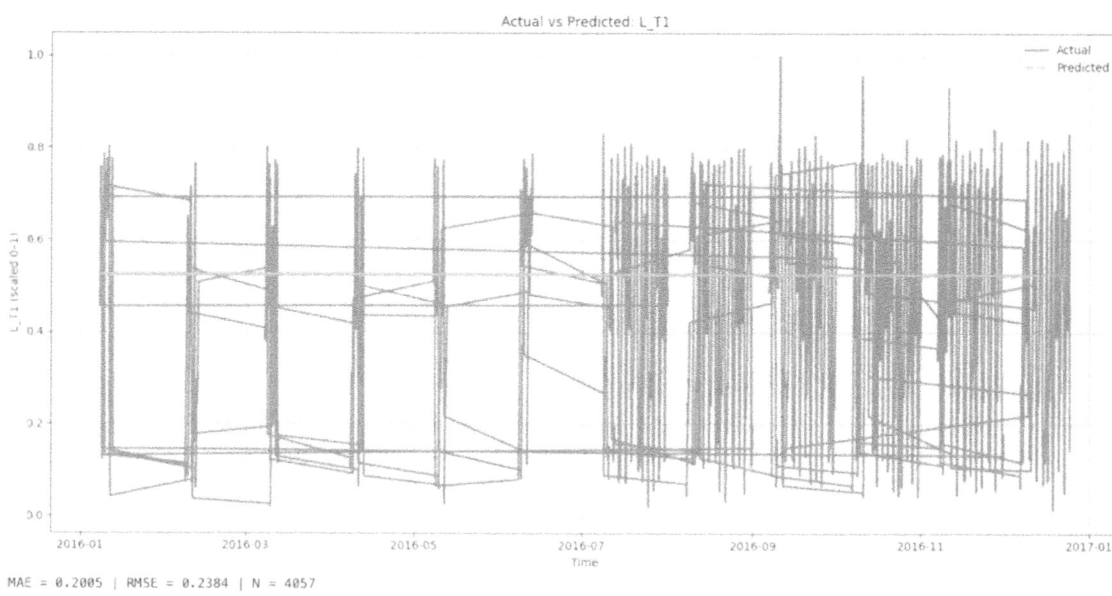

FIGURE 18.7 Actual vs predicted Tank-1 level (L_T1).

18.5 RECOMMENDATIONS AND 12-MONTH IMPLEMENTATION ROADMAP

18.5.1 Defensible Architecture: From Principles to Controls

- **Use water-sector-specific fundamentals.** WaterISAC's 2024 "12 Cybersecurity Fundamentals" reframes sector guidance around practical actions mapped to Cybersecurity and Infrastructure Security Agency's (CISA's) Cybersecurity Performance Goals (CPGs) and the Five ICS Critical Controls [15, 16].
  - It prioritizes an ICS-specific incident response plan, rigorously limits OT exposure via segmentation/IDMZ and secure remote access, builds a cyber-aware culture with named OT-cyber leadership and couples network monitoring with process-aware alarms.
  - For smaller utilities, the document spotlights quick wins—removing internet-exposed services (e.g., RDP), running low-cost tabletops, testing offline backups/restores and practising manual operations—along with modern PLC secure-coding practices and independent safety interlocks. The through-line is resilience: design for degraded modes and fast recovery, not just prevention.
  - Often-overlooked basics—like a backup communications channel—and the value of regular drills (practice builds muscle memory) are explicitly emphasized.
  - Low-cost ransomware or cyber-event tabletop exercises (TTX) are a high-return investment for utilities and critical-infrastructure operators. They raise enterprise-wide awareness of true maturity and incident readiness while also satisfying compliance/framework expectations. Because scenarios can be designed in-house by staff who know how facilities actually run, these paper-based drills surface practical gaps that could jeopardize real-world response—at minimal cost. Just as importantly, they bring all stakeholders into the same room, clarify roles and ownership and resolve cross-department ambiguities. The communication and collaboration catalysed by TTXs often drive lasting organizational change and build trust across operations, engineering, IT/OT and leadership.

18.5.2 Apply OT-Centric Standards

- NIST SP 800-82-R3 is the U.S. government's primary guidance for securing Operational Technology (OT)—programmable systems that monitor and control the physical world (e.g., ICS/SCADA, DCS, PLCs, BAS, safety systems). Revision 3 expands the scope from "ICS" to "All OT", updates the threat/vulnerability landscape and aligns with the NIST Cybersecurity Framework.
- The document emphasizes how OT differs from IT in safety and process integrity/availability outrank confidentiality of legacy/long-lifecycle platforms, deterministic timing and real-time control constrain how security controls are applied. Typical incident consequences in OT include loss of view/control, unauthorized command/set-point changes, safety-system interference and physical damage—so programs must be engineered for "graceful degradation" (normal → emergency → manual) and fast recovery.
- **NIST prescribes a risk-based, defence-in-depth architecture**: A strict segmentation between enterprise IT and OT with an "Industrial Demilitarized Zone (IDMZ)" to minimize exposure (no direct internet paths), use MFA and governed remote access via jump hosts or unidirectional gateways, harden endpoints (patching after field testing, disable unused services/

ports, least-privilege/Role-Based Access Control "RBAC"), combine "network monitoring" with "process-aware detection" (alarming on anomalous pumps/valves/set-points) and enforce strong "physical security". It also mentions about "availability and safety" considerations for geographically distributed systems and field I/O, plus additional guidance for IIoT deployments.
- Programmatically, it applies the Risk Management Framework (RMF) to OT (Prepare → Categorize → Select → Implement → Assess → Authorize → Monitor) and maps OT practices to Cybersecurity Framework (CSF) "Identify/Protect/Detect/Respond/ Recover" outcomes (asset and supply-chain management; identity and access; training; maintenance/patching; anomaly detection; incident response; tested backups and restoration).

Reduce exposure—especially HMIs. Remove direct internet access of Human Machine Interfaces (HMIs) behind jump servers or VPNs. Use Multi-factor Authentication (MFA), restrict source IPs, disable unused services and continuously scan for exposed surfaces.

Network segmentation and monitoring. Separate business IT from Operational Technology (OT) and within OT distribute zones by criticality (safety systems isolated). Further, implement unidirectional gateways or tightly controlled firewalls for historian/data flows. It is also strongly recommended to deploy passive OT monitoring for protocol-aware anomaly detection [17].

Living Off the Land (LOTL) Detection and Hardening. Volt Typhoon advisories recommend baseline-driven detection for admin tool misuse, command-line telemetry and unusual lateral movement. It is advisable to eliminate legacy/End-of-Life (EoL) edge devices and also to harden device management paths [18].

Backup, Recovery and Safety Assurance. Maintain offline/immutable backups of logic, historian and configuration. Also, implement change approval for PLC/SIS logic. It is strongly recommended to rehearse Recovery and Failover, including manual operations [19].

Small-System Pathway. American Water Work Association's (AWWA) small-systems guide gives a stepwise path to secure Water Work Infrastructure from various attacks. It brings out a 22-Question Tool with implementable 28 Baseline Controls across people/asset/compliance/access/physical/network layers. The emphasis of these tools is on inventories, segmentation/IDMZ, MFA remote access, monitored process and networks, tested backups/IR and practical wireless/crypto hygiene—right-sized for small utilities yet aligned to mainstream best practice [20].

18.5.3 Gaps to Close

- **Governance and ownership (who does what).**
    - Organizational Policies to include **RACI**
        - **R** = Responsible (does the work),
        - **A** = Accountable (final decision/owns outcome; exactly one),
        - **C** = Consulted (two-way),
        - **I** = Informed (one-way)
    - for OT cyber (policy, exceptions, segmentation changes, vendor access approval, incident command).
    - Define an **exceptions process** with expiry dates and compensating controls.

18.5.4 Identity and Privileged Access

- 100% MFA on all remote/privileged access.
- Zero shared admin accounts.
- Privileged Access Management (PAM)/ Just-in-Time (JIT) for vendor access.
- Break-glass accounts (emergency admin credentials) sealed and logged.

Supplier / remote-access risk (IEC 62443-2-4). Contractual controls:

- **Time-bound**. Give the vendor access only for a specific job window (e.g., 10:00–12:00 on 12 Nov). It auto-expires when the window ends.
- **Least-privilege access**. Vendors get only the rights they need—read-only unless write is required and only to the specific HMI/PLC in scope. For example, vendor cannot change PLC logic unless a change ticket is approved.
- **Session recording**. All remote vendor sessions via your IDMZ jump host are video/keystroke recorded. For example, if a pump trips during a session, you can replay exactly what the vendor did.
- **Audit log delivery**. Vendor tools must produce logs, including who accessed the system, when the access occurred and what actions were performed, and send them to the utility or its SIEM on request or automatically. For example, after a firmware update, the utility should receive an audit file listing device IDs, versions changed and user IDs. The recommended governance workflow for vendor remote access—covering request, approval, JIT access, MFA, jump-host monitoring, SIEM ingestion and auto-revocation—is shown in Figure 18.8.
- **SBOM/vuln notification**. SBOM = a Software Bill Of Materials (ingredient list of components/libraries). Vendor agrees to tell you quickly when a component they use gets a critical CVE and provide a fix/mitigation timeline. For example, "OpenSSL bug affects your RTU app; patch available in 5 days; temporary firewall rule recommended now".
- **Cloud SCADA SLAs** (availability, incident notice, data location). Clear promises on:
  - Availability (e.g., 99.9% uptime).
  - Incident notice (e.g., notify within 24 hours of a breach/major outage).
  - Data location (e.g., "stored only in EU/India data centres"), plus backup/restore (RPO/RTO).
    - For example, "Daily backups; RPO 24 h, RTO 8 h; data stays in Frankfurt; breach notice ≤24 h".

18.5.5 Change and Configuration Management

- **Version control for PLC/SIS logic**. Keep every version of controller programs (with date, author, reason). If something breaks, you can roll back fast. Store logic files in a central repo (e.g., Git or vendor vault) with naming rules.
- **Signed releases**. Only publish controller/HMI changes that are approved and digitally signed. Prevents untracked or rogue changes. Use a change ticket + digital signature from an approver before download to PLC/HMI.
- **Golden images**. A clean, known-good snapshot of HMIs/servers/engineering PCs for a quick rebuild after malware or failure to be kept offline and to be refreshed after each approved update.
- **Two-person rule for logic changes**. One engineer makes the change; a second reviews/approves before it goes live. This helps in catching mistakes and insider risk.
- **Weekly diff checks on PLCs/HMIs**. Always compare what is running now vs. the last approved version. This helps in detecting silent/unauthorized edits.
- **Engineering workstation allow-lists**. Allow only approved software, drivers and USB to run in the main system. Recommended is App/USB Control, no internet browsing, patch via the IDMZ only.

Patch & vulnerability management (OT-aware). Risk-based patching with test bench; maintenance windows; virtual patching at the perimeter when deferral is necessary; quarterly exposure scans of IDMZ and remote sites.

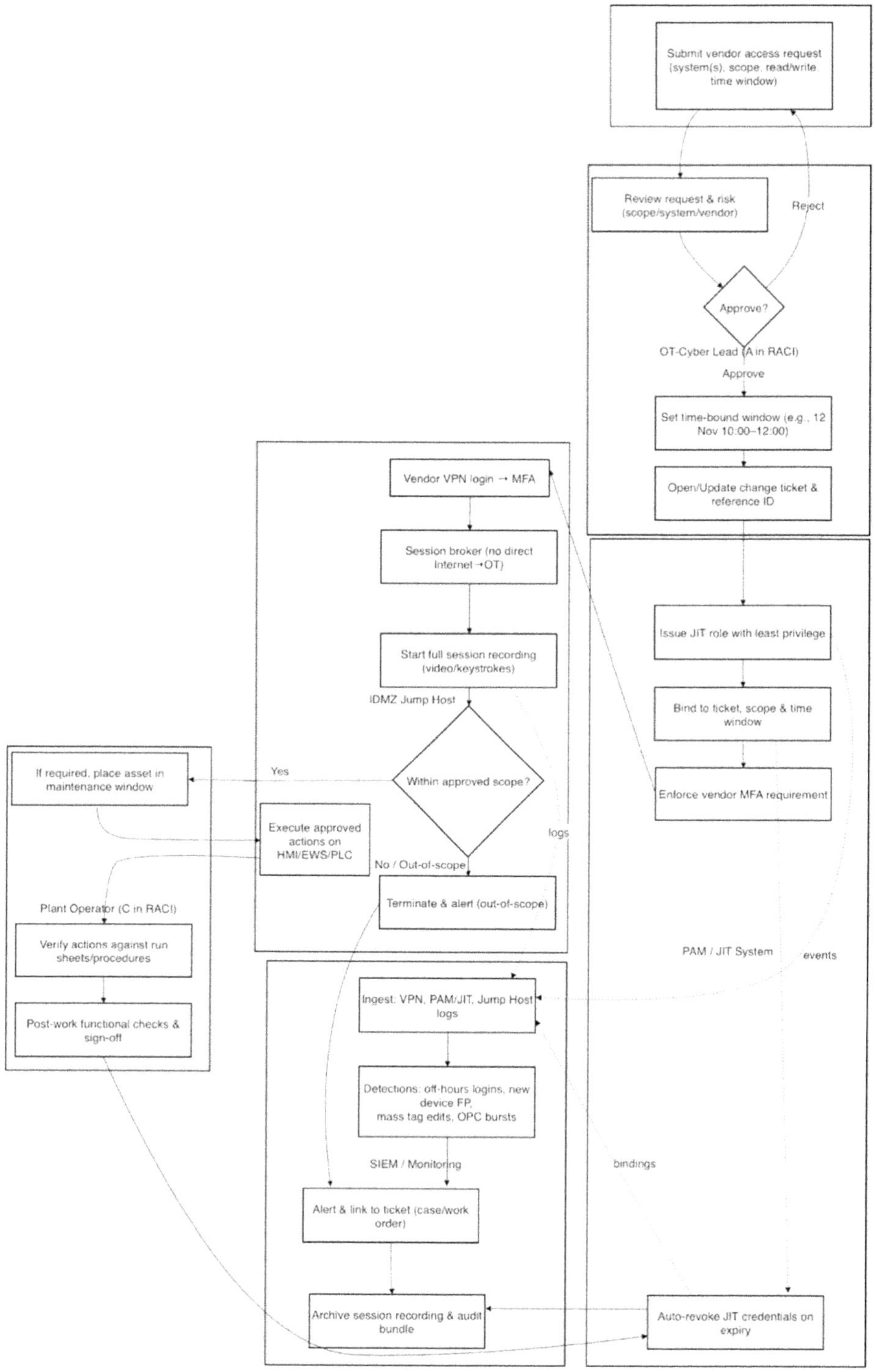

FIGURE 18.8 Remote-access governance flow.

18.5.6 Monitoring You Can Action

- Define "Process KPIs" to alarm on (rapid setpoint changes; abnormal pump duty cycles; tank level rate-of-change; alarm suppressions).
- "Observe engineering tools": launches of programming suites, OPC writes, mass tag edits.

Safety integration. Tie cyber to Process Hazard Analysis (PHA)/ Hazard and Operability Study (HAZOP)/ Layer of Protection Analysis (LOPA) and SIS proof-testing; ensure cyber scenarios are in the Major Accident Hazard register with independent protection layers (mechanical limits, trips).

18.5.7 Crisis Comms and Reporting

- Add public-comms templates, legal/regulatory reporting timelines and out-of-band comms (radios, call trees).
- Name your sharing channels (WaterISAC/sector ISAC, national CERT).

18.5.8 Implementation Roadmap (12 Months)

- **Phase 0 (30 days): Exposure kill-chain.**
  - Internet-exposed OT services = 0.
    - 100% remote access behind VPN+MFA.
    - Credential rotation complete.

Phase 1 (90 days): Foundations. Asset inventory (IT+OT), data-flow maps and risk zoning [21].

- OT asset inventory ≥ 95% tagged (make/model/ firmware/criticality).
- IDMZ live (dual firewalls, jump server, logging).
- First TTX executed; manual-ops drill complete at one plant [14].

Phase 2 (180 days): Detection and resilience. Passive OT monitoring for baseline deviations and LOTL behaviours; centralize logs with OT parsers.

- Passive OT monitoring deployed to all plants and top 10 remote sites; alert library enabled for process anomalies and LOTL.
- Immutable/offline backups for HMIs, historians, PLC/SIS logic; recovery test passed (reload PLC logic).

18.5.9 Phase 3 (Ongoing): Governance

- MFA coverage ≥ 98% for all admin/remote accounts; PAM in production for vendors.
- Quarterly exposure scans and annual risk assessment (SP 800-82/AWWA tool).
- Supplier requirements aligned to IEC 62443-2-4; top vendors contractually signed [21].

18.6 CONCLUSION

Attacks against water/wastewater systems are neither hypothetical nor static. Incidents from Maroochy to Oldsmar show both intent and opportunity, while broader "fluids infrastructure" cases (Colonial, TRITON) illustrate consequences that range from service disruption to potential safety impact. The blueprint—exposure removal, IDMZ-based segmentation, governed remote access (MFA/jump hosts/PAM), process-aware monitoring and rehearsed recovery—maps directly to NIST SP 800-82 and ISA/IEC 62443. Utilities that execute this 12-month plan can demonstrate measurable risk reduction and due diligence.

REFERENCES

1. CISA & EPA, "Internet-exposed hmis pose cybersecurity risks to water and wastewater systems," *Fact Sheet*, 13 Dec 2024. (Online)
2. A. E. & J. Chukwunweike, "Next-generation SCADA architectures for enhanced field automation and real-time remote control in oil and gas fields," *International Journal of Computer Applications Technology and Research*, vol. 11, no. 12, pp. 514–529, Jan. 2022.
3. S. Majumder, "Inside the BWSSB incident: How an exposed environment file enabled the sale of 290k applicant records and database root access," *CloudSEK Blog*, 29 Apr. 2025. (Online)
4. Littleton Electric Light & Water Department (LELWD), "LELWD publicizes case study on foreign hackers targeting U.S. Utilities," 14 Mar. 2025. (Online)
5. J. Greig, "Volt Typhoon hackers were in Massachusetts utility's systems for 10 months," *The Record*, 13 Mar. 2025.
6. E. Nakashima & J. Menn, "China's cyber army is invading critical U.S. services," *The Washington Post*, 11 Dec. 2023.
7. S. Bellamkonda, "Ransomware attacks on critical infrastructure: a study of the colonial pipeline incident," *Preprint*, 21 Nov. 2024. (Online)
8. Schneider Electric, "Schneider Electric wins contract to automate India's largest water treatment plant in Mumbai, Maharashtra," *Press Release*, 11 Feb. 2025. (Online)
9. G. Taormina, S. Galelli, N. O. Tippenhauer, E. Salomons, et al., "Battle of the Attack Detection Algorithms (BATADAL) / C-Town dataset and challenge papers," 2018–2019. (Add the specific BATADAL paper you prefer—e.g., "A Benchmark Dataset for Water Distribution Network Cyber-attack Detection"—this is the canonical citation for the dataset used.)
10. Mandiant/Google Cloud, G. Roncone *et al.* "APT44: Unearthing Sandworm," *Threat Intel Report*, 2024. (Online)
11. M. Abrams, *Malicious Control System Cyber Security Attack Case Study – Maroochy Water Services, Australia*, MITRE, 2008.
12. T. Smith, "Hacker jailed for revenge sewage attacks," *The Register*, 31 Oct. 2001.
13. CBS News Pittsburgh, E. Stanish, "Municipal water authority of Aliquippa hacked by Iranian-backed cyber group," 26 Nov. 2023.
14. Kaggle, "BATADAL: A dataset for cyber-attack detection," (mirror of C-Town), accessed 2025. (Online)
15. WaterISAC, "12 Cybersecurity fundamentals for water and wastewater utilities," Dec. 2024. (Online)
16. CISA & FEMA, "Federal roles and resources for cyber incident response – water and wastewater systems sector," Jan. 2024. (Online)
17. K. Stouffer *et al.*, "NIST SP 800-82 Rev. 3: Guide to operational technology (OT) security," Sept. 2025. (Online)
18. CISA et al., "PRC state-sponsored actors compromise and maintain persistent access to U.S. critical infrastructure," *Joint Advisory AA24-038A*, 7 Feb. 2024.

19. FBI Cyber Division, "TRITON Malware remains threat to global critical infrastructure Industrial Control Systems (ICS)," 24 Mar. 2022.
20. American Water Works Association (AWWA), "Water sector cybersecurity guidance," 2021. (Online)
21. International Society of Automation (ISA), "ISA/IEC 62443 series of standards (incl. 62443-2-4 supplier requirements)," (Online)
22. BBC News, "Hacker tries to poison water supply of Florida city," 9 Feb. 2021. (Oldsmar coverage)
23. FBI, CISA, EPA, "Compromise of U.S. Water treatment facility (Oldsmar) – joint cybersecurity advisory," 2021. (Add the exact advisory you intend to cite; this replaces the dangling [24] in §13.4.)

Water and Space Synergy *The Symbiosis Establishing Global Progress*

19

Amandeep Singh, Jagdeep Singh Arora, and P. B. Narendra Kiran

19.1 INTRODUCTION

The growing peril of global water security and the fast outpouring of humanity's economic and exploratory footprint in space are the twenty-first century's two corresponding and compelling descriptions. Due to the interrelated impact of shifting climatic conditions, growing populace, increasing pollution, and inefficient management, extraordinary burden has been created on the world's freshwater resources. The 2023 United Nations World Water Development Report emphasizes the impending threat of a worldwide water shortage crisis in which billions are at risk due to water scarcity, endangering food security, health, and geopolitical stability [1]. At the same time, we can see the emergence of a multi-trillion-dollar space economy that includes satellite services, commercial launch vehicles, space tourism, and ambitious lunar and Mars colonization ambitions [2]. Historically, these fields have been considered separate areas for policy, science and industry, however, this chapter has brought them together. The present chapter argues that the siloed approach is outdated, and water and the space economy are profoundly inter-reliant, mutually interconnected, and create a wide scope for global development. The development further translates by bringing steps towards the sustainability of humanity on earth and the presence and economic development of human beings outside.

"Space for Water" model controls orbital substructures to address global water challenges with satellite-based global navigation, Earth observation and communication systems. The model delivers unparalleled capabilities for supervising hydrological rotations, managing agricultural water usage, predicting floods and droughts, and encouraging transparency in shared water resources and water domination. Moreover, for an effective and feasible space economy, water plays a crucial part and enabler for "Water for Space". Water further acts as a vital source for impulsion for launch vehicles, radiation shielding, and manufacturing in the space in addition to human existence and support. Resultantly, lowering mission

DOI: 10.1201/9781003742227-19

spending and supporting sustained exploration remain the predominant aims for on-site extraction of water from the Moon and Mars. The association between Space and Water through cohesive acts that confirm effective development on Earth and progress a positive and robust future in space is highlighted in this chapter.

19.2 THE DUAL RELATIONSHIP

Water-space corelation operates in two interconnected directions:

19.2.1 "Space for Water" Restructures Our Association with Earth's Most Essential Lifeline

Through an influential triad of space technologies, the planet's hydrological systems are supported by Earth Observation satellites, navigation constellations, and satellite communications. From pursuing precipitation and evaporation to identifying the well-being of rivers and lakes, space technologies bring unique consideration into water quality and consumption patterns. In times that demands robust and workable solutions, ground-bound perceptions require enhancement and upgradation as they are no longer sufficient. "Space for Water" permits scientists and policymakers to confront persistent challenges of drought, flood, pollution, and scarcity by supervising and guiding towards a more enlightened and sustainable future.

19.2.2 "Water for Space" Positions Water as a Core Enabler for Sustained Human Habitation Off-planet

In complex mission architectures, water is not only essential but indispensable, empowering advanced life-support functions like air revitalization and recycling. And it is just as essential for growing crops in bioregenerative systems so that missions can be more self-sufficient. Moreover, if it is electrolysed, water produces hydrogen and oxygen, both significant propellants for spacecraft propulsion systems. It is also a useful medium for radiation protection (shielding crews from cosmic and solar radiation) and may have advantages for space-based manufacturing processes. The production of water is a key capability in its own right, and the successful achievement at this level will mean it significantly lowers the cost and logistical hassle of bringing Earth-based supplies to man-made structures, while dramatically enhancing the resiliency, duration, and range of future exploratory campaigns [3].

19.3 SPACE FOR WATER

The nexus's first dimension is capitalizing on the vantage point of space to solve terrestrial water problems. The satellite gives us the capabilities of global coverage, consistency, non-intrusiveness, and the surveillance of some of the most remote and hostile regions that is crucial for modern water resource management.

TABLE 19.1 Earth Observation for Hydrological Monitoring

MISSION	*AGENCY*	*KEY INSTRUMENT (S)*	*HYDROLOGICAL PARAMETER*	*APPLICATION EXAMPLE*
Gravity Recovery and Climate Experiment Follow-On (GRACE-FO)	NASA/DLR	Microwave K-band Ranging	Groundwater Storage Changes	Monitoring aquifer depletion in California and North India
Global Precipitation Measurement Mission (GPM)	NASA/ JAXA	Dual-frequency and Precipitation Radar	Global Precipitation	Flood forecasting and drought assessment
Soil Moisture Active Passive (SMAP)	NASA	L-band Radar and Radiometer	Soil Moisture	Precision agriculture irrigation scheduling
Sentinel-2	ESA	Multispectral Imager	Water Quality, Surface Water Extent	Detecting harmful algal blooms and mapping reservoir levels
ICESat-2	NASA	Advanced Topographic Laser Altimeter (ATLAS)	Lake & River Surface Height	Calculating water volume changes in remote lakes

Source: Author Generated after Open-Source Research

19.3.1 Hydrological Monitoring

Hydrological monitoring satellite missions are performed in a coordinated approach on spacecraft with state-of-the-art sensors. These systems consist of multispectral and hyperspectral imaging, thermal sensors, and Synthetic Aperture Radar (SAR). Collectively, they provide a near-continuous, global stream of essential data about the condition and dynamics of Earth's water. This enables an in-depth assessment of the extent of surface water, the moisture content of soils, the change of groundwater, the size of the snowpack, and the amount of glacial ice, providing invaluable data for the management of water resources, prediction of floods and droughts, and understanding of the global hydrological cycle (Table 19.1).

19.3.2 Precipitation and Soil Moisture

Precipitation outlines and soil moisture are two interconnected environmental variables which are considered with a detailed understanding from Earth observation. Monitoring through spatial assets increases the precision of weather and flood predictability while transforming agriculture by empowering smarter irrigation, by managing drought, and securing food supplies. Earth observation supports resource managers with significant considerations to plan for sustainable water usage [4].

19.3.3 Surface Water and Reservoir Monitoring

Change detection is derived from the ground data for spatial distinctions across water bodies and highlighted by satellite imagery for continuous monitoring. This ability further progresses from marking surface area to calculating specific volumetric variations integrated with radar and laser inputs from spatial missions like Jason-3 and ICESat-2. Such vital information directly supports the management of transboundary river basins and supervises major reservoirs relevant for drinking water and hydropower. Resultantly, continuous monitoring brings the foundational datasets for water resource management and international policy [5] (Table 19.2).

TABLE 19.2 Data Source for Water Body Extraction

SATELLITE (OPTICAL DATA)	*SENSOR*	*SPATIAL RESOLUTION*	*TIME RESOLUTION*	*APPLICATION*
Terra/Aqua	MODIS	250 m	2 times/day	Mapping of water bodies
Landsat 4, 5	TM	30 m	16 days	Water body monitoring and water colour remote sensing
Landsat 7	ETM+	30 m	16 days	
Landsat 8, 9	OLI	15, 30 m	8 days	
Sentinel-2	MSI	10, 20, 60 m	5 days	Water system mapping
Sentinel-3	OLCI	300 m	2 days	
GF-1	WFV	16 m	2 days	Water quality monitoring
GF-2	MSS	3.2 m	5 days	
HJ-1A/B	CCD	30 m	2 days	Disaster monitoring and forecasting
IRS-P6	LISS-3	23.5 m	24 days	Land use and land cover change
GF-3	SAR	1–500 m	1.5–3 days	Water extraction
Sentinel-1	SAR	5 m	12 days	Flood monitoring
Envisat	ASAR	30 m	35 days	Lake ice and sea ice detection

Source: Author Generated after Open-Source Research

19.3.4 Groundwater Assessment

Spatial assets do not have a direct corelation for evaluating groundwater; however, by monitoring gravitational variations across varied regions, it measures subsurface water reserves on a global scale. The emerging space technology initiatives like GRACE and GRACE-FO satellite missions are planned to sense continuous variations in Earth's gravitational field at the microscopic level. The movement and storage of massive water masses invite hydrometric differences; resultantly, scientists can abstract and analyse the detailed signals to read variations in groundwater reserves. This pioneering and groundbreaking space-based capability has been a significant technological achievement for exposing the alarming depletion of many of the world's major reservoirs [6, 7].

19.3.5 Water Quality and Pollution

Hyperspectral satellite sensors record exceptional natural and chemical signatures by continuous and contiguous colour bands reflected from marine surfaces which are critical for water value assessment. Hyperspectral imaging facilitates the identification and monitoring of environmental hazards, detects specific minerals presence, and finds pollution leaks initiating from agricultural and industrial sources. The European Space Agency's Sentinel-2 mission enables comprehensive monitoring of main water quality parameters by gauging turbidity and chlorophyll absorption. This satellite-based data generates key statistics which are required for protecting systems, providing mitigation efforts and formulating requisite advisories for civil health.

19.3.6 Precision Agriculture

Global Navigation Satellite Systems (GNSS), including GPS, Galileo, and GLONASS, empower the mechanized supervision of farm technology with brilliant accuracy, delivering precision agriculture. This skill facilitates directed efforts for water and fertilizers towards the accurate, site-specific application. These systems achieve optimal hydrological efficiency by focused application of water. Resultantly, a significant increase in agricultural productivity and a reduction in water waste ensure economic feasibility and environmental sustainability in modern farming [8].

19.3.7 Smart Water Grids and Leak Detection

Positioning of remote systems with persistent observance for Smart Water Networks is further very productive. The networks utilize Internet of Things (IoT) sensors through satellite communication to overcome water loss. These sensors regularly monitor pressure and stream live data to command centres. The progressive method swiftly detects and accurately pinpoints leakages within enormous water networks and substructure [9].

19.3.8 Disaster Early Warning and Response

Space technology through satellites forms the dynamic nerve centre for disaster management during water-based threats like floods, tsunamis, and storms. Integrating Earth observation data with sophisticated early warning mechanisms results in bringing critical life-saving acts. Satellites with electro-optical and SAR technologies monitor near-real-time ground situations and carry out plotting to guide disaster relief efforts. The critical humanitarian and disaster relief missions are assisted by satellite inputs when the ground-based infrastructure is destroyed [10].

19.4 CASE STUDY: TRANSBOUNDARY WATER DIPLOMACY

The Nile River basin involves 11 countries with contending interests and signifies a model case of transboundary hydro-political complication. Upstream nations, notably Ethiopia through its structure of the Grand Ethiopian Renaissance Dam (GERD), proclaim to exploit the river's resources for national progress. On the other hand, downstream states, Egypt and Sudan, depend upon the Nile as an existential resource for water and food safety. Emerging and disruptive space technologies have become increasingly instrumental, offering critical information in enhancing institutionalization transparency and cooperative management frameworks [11, 12] (Table 19.3).

19.5 WATER FOR SPACE

The second dimension of the relationship concerns the need for water in the provision of a sustainable, economically viable space economy. As humanity's goals move from short-term missions in Low Earth Orbit (LEO) to presence on the Moon and Mars, water is evolving from a consumable to a strategic, multipurpose asset. Water has been the basis for essential systems of sustenance. For long-duration human spaceflight a closed-loop life support system is essential. At the core of this must be water from start to finish.

TABLE 19.3 Details of Dispute between Ethiopia, Egypt, and Sudan over Water Rights

LENGTH	*APPROXIMATELY 6,650 KM (4,130 MILES)*
Tributaries	Nile originating from Lake Victoria in Uganda and Blue Nile originating from Lake Tana in Ethiopia
Countries Traversed	Uganda, Rwanda, Burundi, DR Congo, Tanzania, Kenya, Eritrea, Ethiopia, South Sudan, Sudan, and Egypt
Source	Nile from Lake Victoria and Blue Nile from Lake Tana
Mouth	Mediterranean Sea, forming the Nile Delta in Egypt
Major Cities	Cairo, Khartoum, and Alexandria
Economic Importance	Provides over 90% of water for Egypt and Sudan
Disputes	Between Ethiopia, Egypt, and Sudan over water rights, particularly due to the Grand Ethiopian Renaissance Dam (GERD).

Source: Author Generated after Open-Source Research

19.5.1 Recycling and Regeneration

The Environmental Control and Life Support System on the International Space Station (ISS) is an example. Above 90% of the onboard water is recycled, including moisture from the cabin air, astronaut sweat and breath, and urine [13]. The cost of launching water from Earth at $10,000 per kilogram would necessitate a deeper concern for the deep space missions. The life support systems would decide the success on lunar and Martian bases which requires functioning with extreme consistency [14].

19.5.2 Water for Food Production

Water being a critical requirement for food production, in-situ manufacturing is very much required. Water allows very effective soilless agriculture like hydroponics and aeroponics in a Controlled Environment Life Support Systems (CELSS). Long-duration missions require bioregenerative life sustenance by reprocessing air from plants consuming CO_2 and producing O_2 and providing psychological assistance that supports crew morale. Therefore, for physical and mental well-being in space, water is a mandatory prerequisite.

19.5.3 Strategic Utility of Water

Water embraces profound strategic value for sustained space exploration, both as a propellant and a fundamental industrial resource, which is in addition to its critical role in direct life sustenance. Water can be proficiently separated into essential gases through solar-powered electrolysis, wherein hydrogen and oxygen act as important components for on-site manufacturing and chemical processing as well as high-efficiency rocket propellants. The presence of water would transform the closed life support system in the space colony into a strategic resupply station, creating more ambitious and far-reaching mission architectures (Table 19.4).

19.5.4 Resource Utilization

The sustainable exploration of lunar poles is crucial and dependent upon confirm presence of water [15]. This water is essential for atmospheric regeneration and high specific impulse cryogenic fuel which is created by the presence of hydrogen and oxygen via electrolysis. The core limitation of the rocket equation

TABLE 19.4 Water in Space

THE FUTURE OF WATER IN SPACE EXPLORATIONS		
ROCKET FUEL	*SUSTAINING LIFE*	*TERRAFORMING*
Spacecraft can use water to fuel themselves by splitting it into hydrogen and oxygen, minimizing their reliance on Earth-based resources	Water derived from celestial bodies can be utilized to drink, cultivate food, and produce oxygen for breathable air	Enough water is found on Mars or the Moon, which may help to make these regions habitable for humans

Source: Author Generated after Open-Source Research

likely resolves the in-situ refuelling at extraterrestrial locations. Resultantly, the initial mass required from Earth would decrease for subsequent missions to Mars, establishing a lunar propellant depot ensuing in decreasing cost [3, 16].

19.5.5 Manufacturing and Construction

In addition to a variety of industrial and construction applications, water signifies a vital multifunctional resource for biological life support. Water acts as a mixture with lunar surface material to fabricate concrete-like geomaterials for habitation and infrastructure development. It further serves as a primary constituent for several in-situ resource utilization progressions with geopolymerization [17]. Additionally, radiation shielding process is rendered by substantial hydrogen content in water in the operational domain. The protracted exposure to extensive cosmic rays and harmful properties of solar particles can be countered through available water reservoirs [18]. The sustainable and robust extraterrestrial operations are permitted through water's initial role.

19.6 CASE STUDY: THE ARTEMIS PROGRAM AND LUNAR WATER ICE

NASA's Artemis program is fundamentally dependent on the in-situ utilization of water ice deposits for establishing a sustained human presence on the lunar surface. The Volatiles Investigating Polar Exploration Rover (VIPER) mission prospects and classifies these volatile properties with discoveries aiding in the transformation of lunar water ice into a viable extraterrestrial commodity with operational initiatives and commercial extraction. The objective is to simulate the development of the merging space resource economy through the systematic establishment of public private partnerships [19].

19.7 CROSS-CUTTING CHALLENGES: POLICY, GOVERNANCE, AND SUSTAINABILITY

The emergent interdisciplinary segment, generally designated as "Space Water", creates a multi-layered and innovative field integrated with the relationship of hydrology and astronautics. To confirm the reasonable and sustainable progression of interplanetary human actions, its growth requires the proactive resolve of an elaborate cross-domain integration of technical, legal, and principled deliberations. From a technical perspective, for the modernization of energy-efficient cryogenic extraction structures, the In-Situ Resource Utilization of extraterrestrial water deposits are carried out [19]. The existing international legal framework creates a void for such reflective engineering tests. The Outer Space Treaty (United Nations, 1967) offers no clear proprietary rights over celestial resources or management for the commercial adoption [20].

19.7.1 Data Access and Sovereignty

Broad utilization and global transparency are necessary for space-based hydrological statistics which are functioning under a dualistic model. US Landsat and EU Copernicus programs provide open-access and initial data encouraging global clarity. Conversely, commercial space players supply high-resolution datasets on a proprietary or chargeable basis stimulating a definite "data divide" where significant informational lead in monitoring and managing water resources is accrued to space-faring nations and lead private players. Strong frameworks are necessary to control such irregularity hazards which result in sidelining less affluent nations.

19.7.2 Confirmation and Standardization

High-stakes governance situations mandate a robust assurance of data veracity and transparency during operational times. Strong fiduciary grade standards are required for the management of transboundary water resources and verification of international environmental agreements. Due to geopolitical and ecological implications, the correctness and verifiability of information are most imperative. Further, an increasing range of satellite platforms and diversified sensor technologies also brings technical and governance challenges. Accordingly, in order to ensure reliability and comparability across datasets, it is important to establish strict, universally accepted conventions.

19.7.3 Orbital Sustainability

The long-term feasibility of space-based structures important for monitoring Earth's hydrological arrangements is gradually becoming vulnerable to the increasing orbital debris. The fast-paced positioning of mega-constellations in Low Earth Orbit (LEO) raises impact hazards, which possibly will disable vital Earth Observation (EO) or Global Navigation Satellite System (GNSS) spatial assets. Such an incident would severely interrupt terrestrial water resource supervision. Therefore, the sustainability of "Space for Water" model is dependent upon the safeguarding of the orbital atmosphere, requiring rigorous international guidelines for debris moderation, rigorous satellite end-of-life clearance conventions, and the operational necessity of debris removal technologies [21, 22].

19.7.4 Extraction and Use of Space Resources

The "Water for Space" necessitates the extraction and use of celestial water resources for sustained human study. An unclear status exists among the principles of space as per the Outer Space Treaty 1967, covering "common heritage of mankind" and "commercial utilization" frameworks. Despite the fact that the treaty clearly forbids national appropriation, it remains legally ambiguous concerning the rights of commercial entities to extract and use in-situ resources. This indistinctness has led to different state practices with the

US-led Artemis Accords clearly promoting the right to utilize space resources, while other major space-faring nations, such as Russia and China, show carefulness, possibly interpreting such actions as a de facto form of appropriation [20]. A comprehensive and obligatory international legal framework is essential to avoid a destabilizing "space gold rush" and moderate likely conflict over strategic locations.

19.7.5 Ethical and Environmental Issues

The principle of planetary safety imposes a duty to prevent biological pollution, along with the ethical context for space exploration. While gathering resources like water on Mars, a cautious and highly careful approach is required to safeguard against any kind of contamination. In order to preserve the scientific integrity of astrobiological exploration, imposition of safety measures is of utmost importance [23].

19.7.6 Equitable Benefits

The Outer Space Treaty enshrines that the usage of outer space resources must be carried out for the use and benefit of all nations. As the commercial space technologies are rapidly increasing, the condition of reasonable distribution brings concerning issues. The lead space-faring nations and commercial space players will derive significant benefits of resource operations due to advanced technological and economic abilities. In order to ensure space exploration jointness rather than increasing global space disparities, joint development models and regulatory regimes are very much necessitated [24].

19.8 CONCLUSION

The association between space-based resources and hydrological structures is a shared necessity for sustainable advancement. The "Space for Water" model affords an extraordinary complement for monitoring, modelling, and management of Earth's hydrosphere with improved precision, directly contributing to the fulfilment of water security (SDG 6), food security (SDG 2), and climate action (SDG 13). On the other hand, the "Water for Space" imperative classifies in-situ water resources as the critical facilitator for a scalable and economically viable human presence beyond Earth, crucial for life support arrangements and interplanetary movement.

19.9 RECOMMENDATIONS

19.9.1 Governments and Space Agencies

Growth and disposition of space technologies for water security encompass both terrestrial and extraterrestrial domains. Core obligation for national governments and space agencies involves targeted outlay in dedicated hydrological satellite missions, integrating next-generation emerging and disruptive technologies such as progressive gravity field monitors (e.g., successors to GRACE) [25].

19.9.2 Private Sector

The private sector is critical to fostering innovation throughout the space for water and other space-exploration industries. Its best strength is in making advanced data analytics tools and an intuitive set of user-friendly software. They are critical when integrating complicated satellite data into actionable, user-friendly knowledge for end users such as farmers, water utility managers, and urban planners, and when using space-based information to bring it to the maximum practical value. Moreover, commercial space organizations are pioneers in the core technologies for In-Situ Resource Utilization (ISRU). They are in a unique position to drive down costs and help expedite the development and demonstration of realistic systems to extract and process resources, including water resources on the Moon and Mars, the building block of sustainable, long-duration space missions.

19.9.3 Academic and Research Institutions

In space technology-water resources, academia and research organizations are the backbone. Their importance is primarily their pioneering research that crosses disciplines, such as hydrology, remote sensing, space systems engineering, and international policy and law. The breaking down of academic fragmentation to yield new methodologies and procedures is required to discourse complex global water challenges for a better understanding of space-based competences. Academic institutions fulfil this role by nurturing interdisciplinary fusion, breaking down traditional disciplinary divisions to produce the innovative procedures and foundational knowledge essential to address complex global water challenges and develop space-based capabilities.

19.9.4 International Bodies

United Nations Office for Outer Space Affairs (UNOOSA) and UN-Water constitute a critical organizing and normative forum in the area of global space and water. Their role is to organize pluralistic multilateral discussions among member countries on mutual challenges. There is a primary emphasis on capacity-building, ensuring these developing countries will have access to space-based data that meets their specific water security needs and access and use such data in the capacity area. They are also crucial to achieving global consensus on the complex and growing world governments governing space resource access and use. They serve as a neutral platform to negotiate these challenges, including how to ensure that the dividends of the space economy are equitably distributed and conducted in an environmentally friendly and peaceful manner.

19.10 A VISION: AN INTEGRATED FUTURE

The challenges that are systemic, embedded, complex problems that cannot be dealt with within the limits of traditional disciplinary or national silos are the problems of the twenty-first century. This evolving water-space connection demonstrates the model through its fundamental connection between planetary stewardship and deliberated space domain application. A sustainable course of human progression is dependent upon a strategy-oriented method for handling Earth's vibrant water resources merged with the ability to exploit interplanetary resources.

REFERENCES

1. UN-Water, *The United Nations World Water Development Report 2023: Partnerships and Cooperation for Water.* Paris, France: UNESCO, 2023.
2. Morgan Stanley Research, *Space: Investing in the Final Frontier.* Morgan Stanley, 2021.
3. G. F. Sowers and C. B. Dreyer, "The value of lunar water," *New Space*, vol. 7, no. 4, pp. 205–214, 2019.
4. D. Entekhabi et al., "The Soil Moisture Active Passive (SMAP) mission," *Proceedings of the IEEE*, vol. 98, no. 5, pp. 704–716, May 2010.
5. J.-F. Pekel, A. Cottam, N. Gorelick, and A. S. Belward, "High-resolution mapping of global surface water and its long-term changes," *Nature*, vol. 540, pp. 418–422, 2016.
6. J. S. Famiglietti, "The global groundwater crisis," *Nature Climate Change*, vol. 4, pp. 945–948, 2014.
7. M. Rodell, I. Velicogna, and J. S. Famiglietti, "Satellite-based estimates of groundwater depletion in India," *Nature*, vol. 460, pp. 999–1002, 2009.
8. N. Zhang, M. Wang, and N. Wang, "Precision agriculture: A worldwide overview," *Computers and Electronics in Agriculture*, vol. 200, p. 107123, 2020.
9. J. Alsalam et al., "The role of satellite IoT in sustainable development," *IEEE Internet Things Magazine.*, vol. 4, no. 4, pp. 46–51, 2021.
10. K. E. Joyce, S. E. Belliss, S. V. Samsonov, S. J. McNeill, and P. J. Glassey, "A review of the status of satellite remote sensing and image processing techniques for mapping natural hazards and disasters," *Progress in Physical Geography and Environment*, vol. 33, no. 2, pp. 183–207, 2009.
11. K. G. Wheeler et al., "Cooperative filling approaches for the Grand Ethiopian Renaissance Dam," *Water International*, vol. 43, no. 3, pp. 331–351, 2018.
12. M. Sultan et al., "Monitoring the filling of the Grand Ethiopian Renaissance Dam using satellite data," *Science of the Total Environment*, vol. 697, 134145, 2019.
13. L. C. Jensen et al., "Status of the regenerative ECLSS water recovery system," 45th International Conference on Environmental Systems, 2015.
14. P. Eckart, *Spaceflight Life Support and Biospherics.* Torrance, CA, USA: Microcosm Press, 1996.
15. A. Colaprete et al., "Detection of water in the LCROSS ejecta plume," *Science*, vol. 330, no. 6003, pp. 463–468, 2010.
16. D. Kornuta et al., "Commercial lunar propellant architecture: A collaborative study of lunar propellant production," *Reach*, vol. 13, 100026, pp. 2352–3093, 2019.
17. M. Anand et al., "A brief review of chemical and mineralogical resources on the Moon and likely initial in situ resource utilization (ISRU) applications," *Planetary and Space Science*, vol. 74, no. 1, pp. 42–48, 2012.
18. I. A. Crawford, "Lunar resources: A review," *Progress in Physical Geography*, vol. 39, no. 2, pp. 137–167, 2015.
19. L. Schlüter and A. Cowley, "Review of techniques for in-situ resource utilization (ISRU) on the moon," *Planetary and Space Science*, vol. 182, 104753, pp. 0032–0633, 2020.
20. Din, A. Ud. The Artemis Accords: The End of Multilateralism in the Management of Outer Space? Astropolitics, vol. 20 (2-3), pp. 135–150, 2022.
21. UNCOPUOS, *Guidelines for the Long-term Sustainability of Outer Space Activities.* Committee on the Peaceful Uses of Outer Space, 2019.
22. B. Weeden and V. Samson, *Global Space Situational Awareness: A Review of Current Capabilities and Future Directions.* Secure World Foundation, 2022.
23. M. S. Race et al., "Planetary protection and the search for life: A review of NASA's mission responsibilities," *Astrobiology*, vol. 12, no. 7, pp. 603–610, 2012.
24. P. Martinez, "The sustainable development of outer space: A policy framework," *Space Policy*, vol. 44–45, pp. 1–3, 2018.
25. OECD, *The Space Economy in Figures: How Space Contributes to the Global Economy.* Paris, France: OECD Publishing, 2019.

Index

E

F

G

H

I

K

L

M

V

W

For Product Safety Concerns and Information please contact our EU
representative GPSR@taylorandfrancis.com
Taylor & Francis Verlag GmbH, Kaufingerstraße 24, 80331 München, Germany

www.ingramcontent.com/pod-product-compliance
Ingram Content Group UK Ltd.
Pitfield, Milton Keynes, MK11 3LW, UK
UKHW052051230926
475702UK00025B/1093
9781041245100